Current Trends in Biomedical Engineering

Christiane Bertachini Lombello •
Patricia Aparecida da Ana
Editors

Current Trends in Biomedical Engineering

Springer

Editors
Christiane Bertachini Lombello
Federal University of ABC
São Bernardo do Campo, SP, Brazil

Patricia Aparecida da Ana
Federal University of ABC
São Bernardo do Campo, SP, Brazil

ISBN 978-3-031-38742-5 ISBN 978-3-031-38743-2 (eBook)
https://doi.org/10.1007/978-3-031-38743-2

This Springer imprint is published by the registered company Springer Nature Switzerland AG
The registered company address is: Gewerbestrasse 11, 6330 Cham, Switzerland

Paper in this product is recyclable.

Foreword

This book includes work in key areas of Biomedical Engineering. Relevant aspects in Biomechanics, Rehabilitation Engineering, Digital Health, Telemedicine, Instrumentation, Signal Processing, Tissue Engineering, 3D Printing and Regenerative Medicine, among others, are covered by the authors with exceptional competence.

Technological innovation characterizes the clinical-surgical integration present in public and private systems in the twentieth and twenty-first centuries, resulting in better patient care. In addition, technological development definitely creates wealth for society.

The book *Current Trends in Biomedical Engineering* is rich in cutting-edge content (state of the art) and will certainly contribute to the training of professionals (undergraduate, postgraduate "*stricto and lato sensu*") in the transdisciplinarity inherent in the topics covered.

The publication by Springer ensures national and international coverage of the book's content. I therefore congratulate the editors for the feat.

University of Strathclyde, Glasgow, UK

Alberto Cliquet Jr., PhD

Full Professor, Department of Orthopedics,
Rheumatology and Traumatology,
Faculty of Medical Sciences,
UNICAMP, Campinas, SP, Brazil

Full Professor, Department of Electrical and
Computer Engineering,
School of Engineering of São Carlos,
USP, São Paulo, SP, Brazil

CNPq Researcher #1A (Biomedical Engineering),
Brasília, Brazil

(Overseas) Fellow, The Royal Society of Medicine,
London, UK
e-mail: cliquet@unicamp.br; cliquet@usp.br

Preface

Although Biomedical Engineering is a well-known area in the USA and in some European countries, this is still a subject under development in Latin American, especially in Brazil. Scientific advances, combined with the need for technological transformations in areas related to people's health and well-being, lead to the emergence of research at an accelerated pace, as well as the need for professionals who are increasingly better prepared to work in different sectors of society. In this sense, we are motivated to organize this book, which aims to show, in a simple and easy-to-read way, the main scientific and technological advances that are developed in Brazil, within the scope of a *stricto sensu* graduate program. Such advances are the result of master's dissertations by qualified professionals, that are divulgated herein as a well-founded and up-to-date Biomedical Engineering book. Our goal, therefore, is to inspire these and many other professionals to go on, investing in research and development, as this is the only way we will bring real benefit to people's lives. The multidisciplinary and comprehensive aspect of Biomedical Engineering must also add collaborative activity with other professionals and, even more, talk clearly with health professionals so that such advances are seen and felt in practice by our main customers, that is, the patients. In this way, this book is aimed at a very broad group of professionals and students both in biomedical engineering and related areas, such as electronic, medicine, dentistry, electrical engineering, mechanical engineering, materials engineering, physiotherapy, biology, chemistry, biochemistry, genetics, proteomics, and pharmacy, among others, seeking to contextualize and show the latest scientific advances. Each subfield of Biomedical Engineering is comprised, including modern topics such as neuroengineering, regenerative medicine, additive manufacturing orthosis, postural analysis of Parkinson's patients, modeling and simulation using biomechanical open data, regenerative medicine, advanced drug delivery systems, bioprinting, biomedical optics, and electrical impedance tomography. To achieve these objectives, the researchers intensively work from the bench to the clinics, on better quality diagnosis, more efficient medical devices, faster rehabilitation protocols, biomimetic biomaterials, bioengineered tissues and organs, addressing safety and efficiency. Along with technological innovations, regulatory affairs are also addressed, resulting in the consolidation of medical devices in industry reliability. The authors present concepts and applied technologies to Biomedical Engineering that represents innovative solutions and to inspire future developments to diagnose and treat different adverse medical conditions, such as traumas, injuries, genetically based disease, disabilities, and others. Once the chapters show the in-progress research and applied technologies in the different areas of Biomedical Engineering, the book also stimulates students and young researchers to stand for this career, enhancing future developments. We hope everyone enjoys reading!

São Bernardo do Campo, SP, Brazil — Christiane Bertachini Lombello
São Bernardo do Campo, SP, Brazil — Patricia Aparecida da Ana

About the Book

The book *Current Trends in Biomedical Engineering* was organized on behalf of the 11th year of the Graduation Program on Biomedical Engineering at Federal University of ABC (PPGEBM UFABC), in Brazil. This graduation Program has a single concentration area, Biomedical Engineering, and three lines of research: Biomechanics, Rehabilitation, and Assistive Technology; Medical Devices; Computational Systems Applied to Health. The book aims to contextualize and disseminate the state of the art of research carried out within the broad area of Biomedical Engineering, focusing on different subjects, based on these three main areas, organized as book sections. Also, technological innovations are presented as a book section, including new technologies from bioprinting and electrical impedance tomography.

The purpose of the book is to highlight high-quality and contemporary scientific content for professionals in biomedical engineering and related areas regarding these subjects, as well as to motivate undergraduate and graduate students to scientific research and update their knowledge within one of the subareas of Biomedical Engineering. Furthermore, we aim to disseminate studies in biomedical engineering that are carried out in Brazil.

Contents

Contributors

Alexandre Davi Santos Dias Graduate Program in Biomedical Engineering, Federal University of ABC (UFABC), São Bernardo do Campo, SP, Brazil

Amanda Castro Juraski Federal University of ABC (UFABC), São Bernardo do Campo, SP, Brazil
University of São Paulo, Brazil

Ana Clara Castro Pimentel Silva Araújo Federal University of ABC (UFABC), São Bernardo do Campo, SP, Brazil

Anand Oliveira Masson Federal University of ABC (UFABC), São Bernardo do Campo, SP, Brazil
University of Calgary, Calgary, AB, Canada

Andrea Cecilia Dorion Rodas Undergraduate Course in Biomedical Engineering, Federal University of ABC (UFABC), São Bernardo do Campo, SP, Brazil

Andressa Francine Martins Graduate Program in Biomedical Engineering, Federal University of ABC (UFABC), São Bernardo do Campo, SP, Brazil

Camila Campos Santos Federal University of ABC (UFABC), São Bernardo do Campo, SP, Brazil

Camila Carvalho da Silva Federal University of ABC (UFABC), São Bernardo do Campo, SP, Brazil

Camila Cliquet Federal University of ABC (UFABC), São Bernardo do Campo, SP, Brazil

Carolina Benetti Undergraduate Course in Biomedical Engineering, Federal University of ABC (UFABC), São Bernardo do Campo, SP, Brazil

Christiane Bertachini Lombello Graduate Program in Biomedical Engineering, Federal University of ABC (UFABC), São Bernardo do Campo, SP, Brazil

Claudia Eunice Neves de Oliveira Neuroscience and Cognition Program, Federal University of ABC (UFABC), São Bernardo do Campo, SP, Brazil

Claudionor Bernardo Graduate Program in Biomedical Engineering, Federal University of ABC (UFABC), São Bernardo do Campo, SP, Brazil

Daniel Boari Coelho Graduate Program in Biomedical Engineering, Federal University of ABC (UFABC), São Bernardo do Campo, SP, Brazil

Débora Carajiliascov Ferraraz Federal University of ABC (UFABC), São Bernardo do Campo, SP, Brazil

Débora da Silva Fragoso de Campos Neuroscience and Cognition Program, Federal University of ABC (UFABC), São Bernardo do Campo, SP, Brazil

Edson Rodrigues Federal University of ABC (UFABC), São Bernardo do Campo, SP, Brazil

Emanuele Los Angeles Neuroscience and Cognition Program, Federal University of ABC (UFABC), São Bernardo do Campo, SP, Brazil

Erick Dario León Bueno de Camargo Graduate Program in Biomedical Engineering, Federal University of ABC (UFABC), São Bernardo do Campo, SP, Brazil

Felipe Nogueira Ambrosio Federal University of ABC (UFABC), São Bernardo do Campo, SP, Brazil

Fellipe Soares de Oliveira Federal University of ABC (UFABC), São Bernardo do Campo, SP, Brazil

Fernanda Nascimento Almeida Graduate Program in Biomedical Engineering, Federal University of ABC (UFABC), São Bernardo do Campo, SP, Brazil

Fernando Silva de Moura Graduate Program in Biomedical Engineering, Federal University of ABC (UFABC), São Bernardo do Campo, SP, Brazil

Harki Tanaka Graduate Program in Biomedical Engineering, Federal University of ABC (UFABC), São Bernardo do Campo, SP, Brazil

Idágene Aparecida Cestari Bioengineering Department, Heart Institute (InCor), Faculty of Medicine, University of São Paulo (USP), São Paulo, SP, Brazil

Ilija Uzelac Georgia Institute of Technology, School of Physics, Atlanta, GA, USA

Ilka Tiemy Kato Prates Graduate Program in Biomedical Engineering, Federal University of ABC (UFABC), São Bernardo do Campo, SP, Brazil

Ítalo Sandoval Ramos de Oliveira Graduate Program in Biomedical Engineering, Federal University of ABC (UFABC), São Bernardo do Campo, SP, Brazil

Jimena Gabriela Siles Paredes Federal University of ABC (UFABC), São Bernardo do Campo, SP, Brazil

João Antonio Marques Costa Graduate Program in Biomedical Engineering, Federal University of ABC (UFABC), São Bernardo do Campo, SP, Brazil

João Lameu Undergraduate Course in Biomedical Engineering, Federal University of ABC (UFABC), São Bernardo do Campo, SP, Brazil

João Loures Salinet Jr. Graduate Program in Biomedical Engineering, Federal University of ABC (UFABC), São Bernardo do Campo, SP, Brazil

John Andrew Sims Undergraduate Course in Biomedical Engineering, Federal University of ABC (UFABC), São Bernardo do Campo, SP, Brazil

Jorge Vicente Lopes da Silva Center for Information Technology Renato Archer, Campinas, SP, Brazil

Juliana Kelmy Macário Barboza Daguano Graduate Program in Biomedical Engineering, Federal University of ABC (UFABC), São Bernardo do Campo, SP, Brazil

Kaline do Nascimento Ferreira Center for Information Technology Renato Archer, Campinas, SP, Brazil

Karina Feliciano Santos Federal University of ABC (UFABC), São Bernardo do Campo, SP, Brazil

Laurent Rodrigues Rezende Graduate Program in Biomedical Engineering, Federal University of ABC (UFABC), São Bernardo do Campo, SP, Brazil

Layla Cupertino Salloum e Silva Neuroscience and Cognition Program, Federal University of ABC (UFABC), São Bernardo do Campo, SP, Brazil

Luana dos Santos de Oliveira Neuroscience and Cognition Program, Federal University of ABC (UFABC), São Bernardo do Campo, SP, Brazil

Luciana Pastena Giorno Federal University of ABC (UFABC), Santo André, SP, Brazil

Marcelo Mazzeto Bioengineering Department, Heart Institute (InCor), Hospital das Clínicas da Faculdade de Medicina, Universidade de São Paulo (USP), São Paulo, SP, Brazil

Marcos Antonio Sabino Gutierrez Simón Bolívar University, Caracas, Venezuela

Marcos Duarte Graduate Program in Biomedical Engineering, Federal University of ABC (UFABC), São Bernardo do Campo, SP, Brazil

Maria Elizete Kunkel Graduate Program in Biomedical Engineering, Federal University of ABC (UFABC), São Bernardo do Campo, SP, Brazil

Matheus Del-Valle Federal University of ABC (UFABC), São Bernardo do Campo, SP, Brazil
University of SãoPaulo (USP), SP, Brazil

Mirela Oliveira Tomazini Graduate Program in Biomedical Engineering, Federal University of ABC (UFABC), São Bernardo do Campo, SP, Brazil

Mohamed Bouri École Polytechnique Fédérale de Lausanne, Lausanne, Switzerland
BioRobotics Institute and Department of Excellence in Robotics and AI, Scuola Superiore Sant'Anna, Pisa, Italy

Monica Helena Monteiro do Nascimento Undergraduate Course in Biomedical Engineering, Federal University of ABC (UFABC), São Bernardo do Campo, SP, Brazil

Olavo Luppi Silva Graduate Program in Biomedical Engineering, Federal University of ABC (UFABC), São Bernardo do Campo, SP, Brazil

Pablo Deoclecia dos Santos Graduate Program in Biomedical Engineering, Federal University of ABC (UFABC), São Bernardo do Campo, SP, Brazil

Patricia Aparecida da Ana Graduate Program in Biomedical Engineering, Federal University of ABC (UFABC), São Bernardo do Campo, SP, Brazil

Pedro Henrique Pires Da Silva Graduate Program in Biomedical Engineering, Federal University of ABC (UFABC), São Bernardo do Campo, SP, Brazil

Priscyla Waleska Simões Graduate Program in Biomedical Engineering, Federal University of ABC (UFABC), São Bernardo do Campo, SP, Brazil

Renato Naville Watanabe Graduate Program in Biomedical Engineering, Federal University of ABC (UFABC), São Bernardo do Campo, SP, Brazil

Solaiman Shokur École Polytechnique Fédérale de Lausanne, Lausanne, Switzerland
BioRobotics Institute and Department of Excellence in Robotics and AI, Scuola Superiore Sant'Anna, Pisa, Italy

Sônia Maria Malmonge Graduate Program in Biomedical Engineering, Federal University of ABC (UFABC), São Bernardo do Campo, SP, Brazil

Tainan Cerqueira Federal University of ABC (UFABC), São Bernardo do Campo, SP, Brazil

Thayna Magalhães Novaes Neuroscience and Cognition Program, Federal University of ABC (UFABC), São Bernardo do Campo, SP, Brazil

Thiago Kenzo Fujioka Shida Federal University of ABC (UFABC), São Bernardo do Campo, SP, Brazil

Vinicius Silva Federal University of ABC (UFABC), São Bernardo do Campo, SP, Brazil

Chapter 1
Biomedical Engineering: History and Areas of Expertise

Patricia Aparecida da Ana and Christiane Bertachini Lombello

Abstract Biomedical engineering can be summarily defined as the modality of engineering that integrates the knowledge of electronics, chemistry, physics, and biology to solve problems in the health area. The profession of biomedical engineer is in constant growth worldwide, and its recognition can be noted by its presence in different sectors, such as hospitals, clinics, medical laboratories, medical equipment, and pharmaceutical industry. The biomedical engineer is, therefore, a versatile and multidisciplinary professional, and this allows their performance to be broad in the job market, being able to work in consulting activities, research, and scientific development, in the academic career, among others. For success in their career, it is essential that the biomedical engineer is always up-to-date with the new techniques and technologies available on the market. In this chapter, we will describe the brief history of this occupation, highlighting the main milestones that boosted biomedical engineering in the world. Also, we will mention the basic concepts and main areas of activity of the biomedical engineer in the market, as well as in scientific research and development. In the remaining chapters of this book, examples of scientific advances in different subareas will be addressed.

Keywords Biomedical engineering · Bioengineering · Medical devices · Rehabilitation · Computational systems · Human health

1.1 Introduction

Biomedical engineering is one of the professions with the highest projected growth in the world. By 2030, an average growth of 6% in jobs for American biomedical engineers is estimated, according to data from the US Bureau of Labor Statistics (2022). In Brazil, it is a profession in constant demand considering more than 6600

P. A. da Ana (✉) · C. B. Lombello
Graduate Program in Biomedical Engineering, Federal University of ABC – UFABC Al. da Universidade, São Bernardo do Campo, SP, Brazil
e-mail: patricia.ana@ufabc.edu.br

C. B. Lombello, P. A. da Ana (eds.), *Current Trends in Biomedical Engineering*,
https://doi.org/10.1007/978-3-031-38743-2_1

hospitals and 23,000 health centers in the Brazilian territory (CNSAUDE 2022). The biomedical engineer's employability is due to the increased development of health-oriented technologies. Also, the need for such professionals has expanded thanks to their multidisciplinary knowledge, as well as their ability to move between different functions. Among these, we can mention consulting activities, hospital management, design and maintenance of medical equipment, team training, project of hospital installations, development of prostheses and other wearable devices, and development of new biomaterials, artificial organs, and tissues. Biomedical engineers can act in regulatory agencies, such as in the Brazilian Health Regulatory Agency, ANVISA (2022).

Biomedical engineering is defined as the branch of engineering that applies classical engineering principles and definitions to solving problems in biology and medicine. Institutions such as the NIBIB (National Institute of Biomedical Imaging and Bioengineering 2023) define the biomedical engineer as a professional capable of not only applying classical engineering knowledge but integrating physical, chemical, mathematical, and computational sciences to study and understand aspects related to biology, medicine, behavior, and health. This broader view makes the profession very attractive to different sectors of society, such as hospitals, clinics, medical equipment companies, research and development sectors, research institutes, universities, among others. It is worth noting, however, that biomedical engineering is a very vast area and that, currently, it is impossible for a single professional to be able to act in the entire range of knowledge of this profession.

The performance of the biomedical engineer, together with health professionals, allows the increasingly earlier diagnosis of diseases, as well as the most effective treatment in a less invasive way and with fewer side effects. Also, the development of new technologies allows the visualization and understanding of the molecular mechanisms involved in the pathophysiology of diseases, which also enables the discovery of new drugs and treatment strategies. Associated with bioinformatics and computing sciences, software and hardware development, the biomedical engineers may present equipment and technical innovations, viewing a better life quality for the society, in general (Munzner 2004; O'Cearbhaill et al. 2014; Bellazzi et al. 2023).

The multidisciplinarity of the biomedical engineer is ensured by the knowledge of disciplines such as bioinstrumentation, biomaterials, biomechanics, cellular, tissue, and genetic engineering, clinical engineering, rehabilitation engineering, systems physiology, biophotonics, regulatory of medical devices, medical imaging systems, telemedicine, bioinformatics, modeling, biosensors processing, medical and health informatics, and others (Wigertz et al. 1989; Lhotska 2007; Montesinos et al. 2023). For this, the interface with other traditional knowledge areas, such as physics, chemistry, and biology, as well as with other engineering modalities, such as mechanical, chemical, materials, and computing engineering, enhances the technological advances that can be obtained, as can be seen in Fig. 1.1.

Internationally, according to the American Institute for Medical and Biological Engineering (AIMBE 2023), biomedical engineering involves 19 subareas: bioinstrumentation, bioinformatics, biomechanics, biomaterials, biomechatronics,

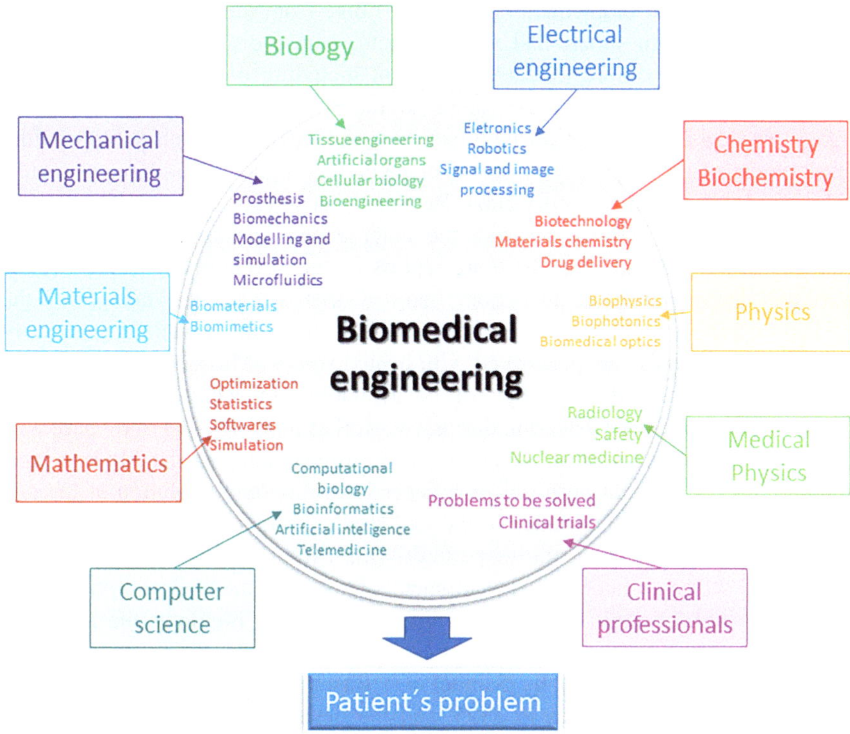

Fig. 1.1 Collaboration of biomedical engineering with other areas. (Modified from https://bioe.uw.edu/academic-programs/about-bioengineering/)

biomimetics, bionanotechnology, biotechnology, clinical engineering, bioprocess engineering, medical engineering, neuroengineering, rehabilitation engineering, pharmaceutical engineering, synthetic biology, systems biology, systems physiology, telehealth, and tissue engineering and regenerative medicine. On the other hand, in Brazil, five subareas of biomedical engineering are seen in practice: clinical engineering, rehabilitation engineering, medical instrumentation, bioengineering, and bioinformatics.

1.2 A Little Bit of History

The second half of the twentieth century was a great milestone for technological development applied to the health areas. Subsequently, with the addition of electronic and instrumentation methods, the emergence of the first centers of biomedical engineering in the world can be observed in some countries of Europe and North America (Lee et al. 2017; Pecchia et al. 2019; Aljamali and Almuhana 2021).

The achievements of biomedical engineering range from the earliest devices such as crutches, platform shoes, and wooden teeth, whose discoveries date back to prehistoric times, to more modern equipment including pacemakers, heart–lung machines, dialysis machines, diagnostic equipment, imaging technologies, artificial organs, medical implants, advanced prostheses, and techniques such as bioprinting, development of drug-delivery systems, telemedicine, organs-on-a-chip, organoids, and microfluidics devices (Leder 2004; Requena–Carrion and Leder 2009).

Although many authors describe the emergence of biomedical engineering from the Second World War, there are reports of the development of foot, lower limb, and dental prostheses in periods before Christ, which were made with the available simple materials such as wood, bronze, iron, and leather. In prehistory times, the hemostasis was performed with heated stones (Massarweh et al. 2006). Although such advanced technology was not available, human beings always needed instruments for curing or detecting disease, as well as replacing lost body parts.

Despite the great scientific development occurred mainly in the Modern Age, technologies for medical purposes experienced an important evolution at the end of the 1700s, with the occurrence of the first studies of bioelectricity and with the development of the first galvanoscope built with a frog limb by Luigi Galvani in 1971. Only 60 years later, the development of the first medical equipment was observed, which boosted the area. At this time, we can highlight the invention of the ophthalmoscope in 1851 by Hermann von Helmholtz, the invention of the sphygmomanometer in 1881 by Samuel von Basch, the accidental discovery of X-rays by Wilhelm Roentgen in 1895, and the development of the first electrocardiograph in history in 1903 by Willem Einthoven, as a galvanometer equipment. The discovery of X-rays raised medical imaging, with an important contribution to diagnosis in orthopedics, cardiology, among other medical fields. Still in 1917, the German mathematician Johan Radon developed the concept of back projection, which would be useful for the future application of the concept of three-dimensional reconstruction, so applied in medical equipment today (Nebeker 2002; Bronzino 2005).

The need to develop other medical devices and to improve existing equipment led to the first formal training in some biomedical engineering disciplines, carried out at the Oswalt Institute for Physics in Medicine, Frankfurt, Germany, in 1921, known today as the Max Planck Institute for Biophysics (Schwan 1984). A few years later, new technological developments were reported, with a focus on the construction of the first steel lungs (Drinker respirators, developed by the American engineer Philip Drinker) in 1927, which were essential during the polio epidemics in the 1950s (Drinker and McKhann 1929). Later, another important technological milestone was noted with the development of the first electroencephalography equipment by Hans Berger in 1929, as well as the evidence that X-rays could be used to visualize internal organs with the use of radiopaque materials. Still in 1929, the first commercial equipment for cutting and coagulating biological tissues was developed by William T. Bovie, which can be considered the first electric scalpel. The creation of the first refrigerators, also in 1930, enabled the emergence of the first blood banks, as well as the conservation of drugs. The discovery of penicillin and other antibiotics in the mid-1930s also revolutionized medicine, drastically reducing the number of deaths from cross-infection in hospitals.

About 10 years later, during the Second World War, the need for new technologies drove the formation of the first society related to the area, the International Biophysical Society, in 1943. The term "biomedical engineering" was only created in 1945, in the postwar period. The term biomedical engineering has the broadest meaning, considering that the bioengineer is more related to basic sciences, such as research and development, biotechnology, genetic engineering, tissue engineering, among other activities. The first conference on biomedical engineering was held in the USA only in 1948. Before that, the initial method of cardiac catheterization had already been developed in 1940. Also in the 1940s, the preceeding studies reported the application of intravenous barium salt contrast, which enabled the development of the hemodynamic technique, that is, the observation of blood vessels with X-rays. In this same decade, the emergence of renal dialysis was observed, with studies conducted by Willem Kolff.

In the 1950s, it was observed the development of nuclear medicine, but it was only in the 1980s that a major expansion of scintigraphy (gamma camera), positron emission tomography (PET), and single-photon emission computed tomography (SPECT) equipment was observed. Concomitantly, in 1953, the development of the cardiopulmonary bypass (heart–lung machine) was carried out by John H. Gibbon Jr. and, in 1956, the first surgery was performed using a total cardiopulmonary bypass machine. Still in 1952, the first artificial heart valve was implanted, developed by Charles A. Hufnagel, and in 1958, the primary implantation of an external cardiac pacemaker was performed.

The electronics revolution, which took place mainly in the 1960s, enabled the replacement of thermionic valves by bipolar transistors, which allowed the construction of lower-cost equipment, with higher energy efficiency and lesser heat dissipation. With the development of the first computer chip by the IBM company in 1964, the incorporation of computers in medical equipment began. The early computerized tomography benefited from this technology and was developed in 1972 by electrical engineer Godfrey Newbold Hounsfield and physicist Allan Cormack. Later, in 1977, the former nuclear magnetic resonance equipment was developed by the Americans Raymond Damadian, Larry Minkoff, and Michael Goldsmith. In the mid-1990s, the first endovascular coronary stent was developed.

The developments that occurred in the 1960s motivated the emergence of undergraduate courses in biomedical engineering around the world. In the USA, the first undergraduate course in biomedical engineering appeared in 1965 at the University of Illinois, followed by courses at the University of Virginia, Case Western Reserve University, Johns Hopkins University, and Duke University in the late 1960s. Currently, there are hundreds of biomedical engineering courses in the USA, and in this country, there was a great diffusion of this profession. Even today, the USA has the largest number of societies linked to the field of biomedical engineering.

The entrance of a biomedical engineer in a hospital environment only occurred at the end of the 1960s, due to the concern of electrical accidents of patients while using different medical equipment. As a result, different organizations were created to standardize and ensure the safety of medical equipment within hospitals. With this, there was the emergence of the discipline of clinical engineering (Bronzino 2005), whose greatest expansion occurred during the 1970s and 1980s. During the 1990s, the American College of Clinical Engineering (ACCE) and the Clinical Engineering Division within the International Federation of Medical and Biological Engineering (IFMBE) were created.

In Latin America, the first biomedical engineering course began in 1975 at Universities Autónoma Metropolitana Iztapalapa and Universidad Iberoamericana, both in Mexico. In 1978, undergraduation courses were registered at the National University of Tucumán, in Argentina. Seven years later (1985), in Argentina, the course was created at the National University of Entre Rios and, in 1992, at the National University of San Juan, University of Favaloro, and National University of Córdoba.

In Brazil, the first studies were directed to the bioengineering area in the 1950s. In the 1970s, the first graduate programs related to the area were observed. In 1975, the Brazilian Society of Biomedical Engineering was founded, and the *Brazilian Journal of Biomedical Engineering* was established. It was only in the early 1980s that biomedical engineering was introduced in the public health and in the hospital environment, giving rise to the subareas of Health Systems Engineering and Clinical Engineering (SBEB. Sociedade Brasileirade Engenharia Biomédica 2023).

Undergraduate courses were created later, in 2001, with the first biomedical engineers graduating in 2005. The biomedical engineer profession was regulated in the country only in 2007, by the Federal Council of Engineering and Agronomy (CONFEA). There are currently 18 undergraduate courses in biomedical engineering available in Brazil, and there are 16 graduate programs officially regulated in biomedical engineering or bioengineering.

The Federal University of ABC (UFABC) initiated the biomedical engineering undergraduate course in 2006, and the graduation activities in 2012. UFABC is in a highly industrialized region of the state of São Paulo with an estimated population of 46.6 million people; of these, 12 million are in the city of São Paulo (IBGE 2021). Although traditionally developed, the region is still noticeably lacking in infrastructure for accessing and maintaining healthcare for the entire population. In this context, the presence of academic programs in biomedical engineering facilitates the generation and application of techno-scientific knowledge and human resources viewing the need for scientific and technological innovation and the improvement of hospital management to solve different problems of access and maintenance of health, including in the public hospital network. According to Brazilian Association of the Medical Device Industry (ABIMO) there has been a significant increase in the market demand for instruments and materials for medical and dental use and optical items, as well as electromedical and electrotherapeutic devices and irradiation equipment considering the last 5 years, with a direct impact on the market for work (ABIMO 2021). The biomedical engineering courses at UFABC has also the potential to contribute to the regional public hospital network, allowing partnerships with other academic institutions, industry, hospitals, and,

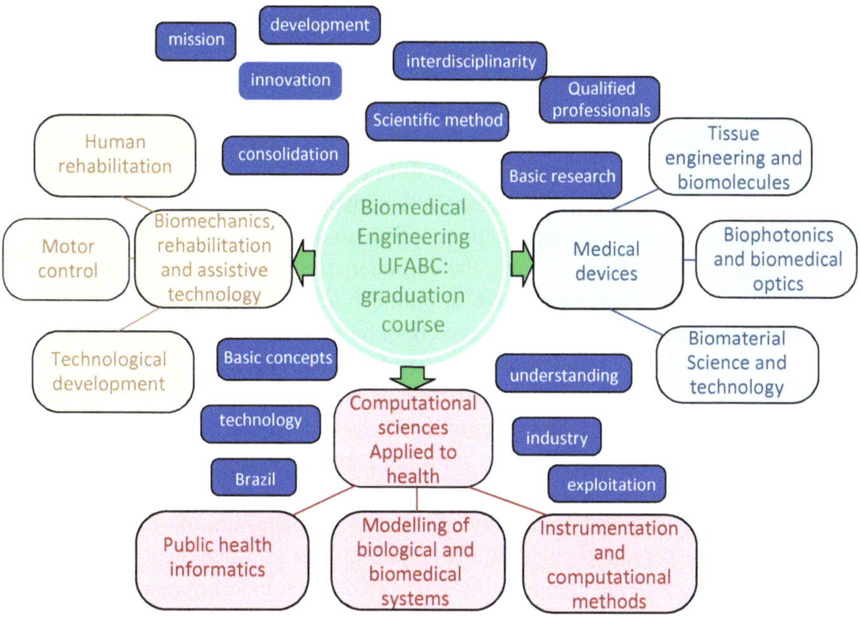

Fig. 1.2 Biomedical engineering main research areas at UFABC graduate course

through advances in basic and applied research, it is also possible to observe contributions to the Unified Health System (SUS – Sistema Único de Saúde).

Although biomedical engineering aims to help medicine in improving the quality of life of patients, through early diagnosis, clinical treatments, device development, and innovative techniques, it is substantially established on basic research. As such, universities and research institutes play a central role in this science consolidation, along with clinics and hospitals, in clinical engineering. The biomedical engineering graduate program at UFABC is based on three main research areas: biomechanics, rehabilitation, and assistive technology; medical devices; information systems applied to healthcare (Fig. 1.2). In the next topics, we will summarize the main subareas of biomedical engineering adopted as the research topics at the UFABC.

1.3 Biomechanics, Rehabilitation, and Assistive Technology

It is reported that the rehabilitation engineering originated in the Second World War, as a need to assist the wounded ones. However, from records of prostheses and other rehabilitation devices, it is known that the rehabilitation procedures date back to at least 300 BC, when it was described the use of a staff to aid an individual's mobility (Childress 2002).

By definition, a rehabilitation engineer is a professional able to use the principles and methods of engineering to improve the quality of life of people, mainly those with disabilities. For that, the professional must have qualified knowledge about musculoskeletal, neuromuscular, and sensory disorders. According to NIBIB (National Institute of Biomedical Imaging and Bioengineering 2023), the research in rehabilitation engineering can involve rehabilitation robotics, virtual rehabilitation, physical prosthetics, advanced kinematics, sensory prosthetics, brain computer interfaces, modulation of organ function, and secondary-disorder treatment.

To create a rehabilitation device, the engineer must know the different types of injury to be treated. Thus, knowledge of biomechanics is associated, which applies the principles of mechanics to understand both human movement and the biological processes involved in these movements. Therefore, the association of biomechanical knowledge with rehabilitation engineering allows the design of prostheses and orthoses adapted to certain clinical conditions, as well as can predict the durability of these devices by correlating them with the forces developed during use. Often, simulation and modeling techniques of biological systems are employed. Studies in biomechanics are also widely used in sports medicine through the development of methods and devices that prevent injuries to athletes or improve their performance.

The concepts of biomechanics and rehabilitation engineering can be applied in assistive technology. This technology is described as a device or system that assists an individual with disabilities, and it can be medical equipment or not. According to the World Health Organization (WHO 2023), assistive technology is a very broad term, but it intends to create devices not only for people with disabilities but includes the elderly, people with diseases such as diabetes and stroke, people with neurological disorders such as autism, as well as individuals with gradual functional decline.

For this reason, studies focusing on biomechanics, rehabilitation, and assistive technology must always be interconnected, and be carried out collaboratively with other sciences, to identify a group of patients to be treated, understand the disability and the causative pathology of such a deficiency, as well as always being up to date with scientific advances that occur in other areas, such as robotics, bioprinting, artificial intelligence, etc. At the UFABC, the research line is focused on the study of human movement, integrating specific concepts and principles for the investigation and identification of useful tools for human rehabilitation, as well as the adaptation mechanisms generated by the locomotor system in response to the mechanical demands generated by human movement. The research group studies technologies for the assistance and rehabilitation of individuals with disabilities, allowing for an improvement in the quality of life or the reintegration of these individuals into society.

1.4 Computational Systems Applied to Health

Its objective is the development and application of mathematical and computational methods in various health sectors, including modeling of biological and biomedical systems, bioinformatics and biological signals, medical images, and health informatics. This area includes topics such as computational medicine, biomedical signal processing, mobile health and telemedicine, and bioinformatics (Lynn et al. 2003; Nadon and Sladek 2003; Chang 2005; Zeng et al. 2021; Deng et al. 2022). Bioinformatics may lead to the identification of nucleotides, genes, and proteins that act based on different diseases, allowing a better understanding of its genetic basis. The Human Genome Project is an important example of this science. Knowing individual gene sequences may lead to the prediction of disease susceptibility, such as some types of cancer or other genetic-related diseases. The screening of biomarkers for breast cancer using bioinformatics analysis is a clinical and commercial reality nowadays. In the same way, computational modeling can direct the understanding of cellular, tissue, and organ response to a certain stimulus, predicting the response of a specific treatment or proposing a direction to function restoration, for example. This area of research results in more reliable protocols, diminishing costs and time for an applied device or technique establishment. The development of drugs based on bioinformatics is another current field of research, which predicts and proposes molecules effective for a specific clinical challenge.

Mobile health devices and telemedicine are widely used for monitoring, collecting data, and driving treatments. Even those designed for individual use, such as smartwatches that indicate health parameters as blood pressure, heart rate, arrhythmias or glycemia, and the clinically assessed, and usually more precise, devices may help to control critical physiological parameters. The frequent monitoring of these parameters results in a more precise understanding of patient clinical conditions (Crico et al. 2018; Sim 2019), as well as it reduces the risks of this patient.

The computational area contributes also with data for public health and epidemiology through biobanks, clinical and genomic integrated data bases, information analyses, and big data. The analysis of information and management of the large amount of healthcare data is an important tool to address healthcare (Maojo and Martin-Sanchez 2004; Batko and Ślęzak 2022; Rehman et al. 2022). Another important contribution is the development of artificial intelligence techniques, which drives computer-aided diagnosis. In this way, it is possible to improve medical analysis and other electronic healthcare systems, as well as to integrate these technologies with telemedicine, providing access to diagnosis and monitoring even in remote locations.

The development and application of mathematical and computational methods in various health sectors, including modeling of biological and biomedical systems, bioinformatics and biological signals, medical images and health informatics are areas of research developed at the UFABC.

1.5 Medical Devices

The purpose is the research, development, application, and evaluation of new medical devices for diagnosis, monitoring, and therapies in various areas of health, as well as the development of applied instrumentation. It includes research topics focused on biomedical devices for diagnosis and therapy, also including implantable ones, biocompatible materials, tissue engineering, and regenerative medicine. Therefore, it seeks integration with the areas of biomechanics and computational systems for the adoption of biofunctional parameters and requirements, as well as modeling in the evaluation of the performance of devices and methods.

The more traditional research areas include biomaterial research, development, characterization, and biological evaluation, which aims at device enhancement or development (Lysaght 1995; Ni and Pu 2009; Aljamali and Almuhana 2021). Equipment used for diagnosis and treatment, such as ultrasound, multiparametric monitor, or ventilators, largely adopted during COVID pandemic, are included in this field of research, early from the project, design, raw materials choice and characterization, electrical functioning, hardware and software, safety, and efficacy. The pulmonary ventilators used during the pandemic is a good example of practical application of biomedical engineering knowledge to adapt medical devices. The fast response of different institutions, from academic ones to industry, enabled the attendance of the population affected by the disease. It was necessary to achieve appropriate mechanical ventilation settings to optimize gas exchange, also avoid hypoventilation and prevent further injury (Cronin et al. 2022).

Another perspective of medical device research areas includes a concept that usually is named as *bioengineering*, once it has the objective to restore not only tissue and organ functions, but also its original constitution, based on the use of biotechnology, biomaterials, and biomolecules (Langer and Vacanti 1993; Pittenger et al. 2019; Dai et al. 2021; Gaiseanu 2021). Cell therapy, tissue engineering, innovative biomaterials, 3D and 4D printing, phototherapy, microfluidic circuits, and other developments and techniques are clinically applied to rebuild the original parts of the human body, assisting traditional medicine. A long way has been made since the proposal of the first tissue substitutes that directed wound closure or bone restoration, until the first tissue engineering commercial product, Integra (approved by the FDA at 1996), and furthermore, to the development of in vitro tissue constructs (Fernandez de Grado et al. 2018; Ashammakhi et al. 2022; Pan et al. 2022). Safety and efficacy are also crucial stages of these developments; biocompatibility, along with biofunctionality, is an area of fundamental importance to achieve clinical feasibility (Helmus et al. 2008; Müller 2008; Williams 2008; Reeve and Baldrick 2017). Once again, the main goal is to improve the life quality of the population, translating basic research to clinical practice.

The purpose of the research area at UFABC is based on the development, application, and evaluation of new medical devices for diagnosis, monitoring, and therapies in various areas of health, as well as the development of applied instrumentation. It also includes research topics focused on biomedical devices for diagnosis and therapy, implantable devices, biocompatible materials, tissue engineering, and regenerative medicine.

1.6 Conclusion

We can note the immense evolution of biomedical engineering, especially in the last 60 years, as well as its consolidation as a science and a professional area, which ensured significant improvements in scientific and technological knowledge related to human health and, thus, improved people's quality of life. However, there is still much to be done, and that the integration of knowledge of distinct areas, the collaborative activity among different professionals, combined with the multidisciplinary nature of the biomedical engineer is certainly very promising to further improve access to treatments and technologies that promote the health and well-being of individuals.

Funding Declaration/Acknowledgments The authors would like to thank to FAPESP (2017/21887-4), CAPES (PROCAD-CAPES 88881.068505/2014-01), and CNPq (INCT # 465763/2014-6 – Photonics National Institute for Science and Technology – INFO – and 440228/2021-2 National System of Photonics Laboratories – Sisfoton-MCTI).

References

ABIMO. Associação Brasileira da Indústria de Dispositivos Médicos (2021) Dados Econômicos. Available at: https://abimo.org.br/dados-do-setor/dados-economicos/#. Accessed: 30/05/2023.

AIMBE. American Institute for Medical and Biological Engineering (2023). Available at: https://aimbe.org/. Accessed: 30/05/2023.

ANVISA. Agência Nacional De Vigilância Sanitária. Institutional (2022) Available at: https://www.gov.br/anvisa/pt-br/english. Accessed: 05/04/2023

Ashammakhi N, GhavamiNejad A, Tutar R, Fricker A, Roy I, Chatzistavrou X, Apu EH, Nguyen K, Ahsan K, Pountos I, Caterson EJ (2022) Highlights on advancing frontiers in tissue engineering. Tissue Eng Part B: Reviews, 633–664

Aljamali NM, Almuhana WHY (2021) Review on biomedical engineering and engineering technology in bio-medical devices. J Advanc Electric Dev 6(2): 18–24

Batko K, Ślęzak A (2022). The use of Big Data Analytics in healthcare. J Big Data 9(1):3

Bellazzi R, Cecconi M, Costantino ML, Veltri P (2023) Bioengineering and medical informatics education in MD programs: perspectives from three Italian experiences. Int J Med Inform 172:105002

Bronzino JD (2005) Biomedical engineering: a historical perspective. In: Enderle JD, Blanchard SM, Bronzino JD (eds) Introduction to Biomedical Engineering (2nd ed), Academic Press, p 1–29

Bureau of Labor Statistics, U.S. (2022) Department of Labor, *Occupational Outlook Handbook*, Bioengineers and Biomedical Engineers. Available at: https://www.bls.gov/ooh/architecture-and-engineering/biomedical-engineers.htm. Accessed: 30/05/2023

Chang PL (2005) Clinical Bioinformatics. Chang Gung Med J 28:201–11

Childress DS (2002). Development of rehabilitation engineering over the years: as I see it. J Rehabil Res Dev 39(6 Suppl):1–10
CNSAUDE (2022) Cenário dos Hospitais no Brasil 2021-2022. Available at: http://cnsaude.org.br/wp-content/uploads/2022/07/CNSAUDE-FBH-CENARIOS-2022.pdf. Accessed: 30/05/2023
Crico C, Renzi C, Graf N, Buyx A, Kondylakis H, Koumakis L, Pravettoni G (2018) mHealth and telemedicine apps: in search of a common regulation. Ecancermedicalscience 11;12:853
Cronin JN, Camporota L, Formenti F (2022) Mechanical ventilation in COVID-19: A physiological perspective. Exp Physiol 107(7):683–693
Dai G, Feinberg AW, Wan LQ (2021) Recent Advances in Cellular and Molecular Bioengineering for Building and Translation of Biological Systems. Cel Mol Bioeng 14 293–308
Deng J, Yang Z, Ojima I, Samaras D, Wang F (2022) Artificial intelligence in drug discovery: applications and techniques. Brief Bioinformatics 23(1)
Drinker P, McKhann CF, (1929) The Use of a New Apparatus for the Prolonged Administration of Artificial Respiration: I. A Fatal Case of Poliomyelitis. JAMA 92.20: 1658–1660
Fernandez de Grado G, Keller L, Idoux-Gillet Y, Wagner Q, Musset AM, Benkirane-Jessel N, Bornert F, Offner D (2018) Bone substitutes: a review of their characteristics, clinical use, and perspectives for large bone defects management. J Tissue Eng 9:2041731418776819.
Gaiseanu F (2021) New perspectives in biomedical engineering and biotechnology: information in human and biological structures. Arch Biomed Eng Biotechnol 6. 1–3
Helmus MN, Gibbons DF, Cebon D (2008) Biocompatibility: meeting a key functional requirement of next-generation medical devices. Toxicol Pathol 36(1):70–80
IBGE. Instituto Brasileiro de Geografia e Estatística (2021) Available at: https://cidades.ibge.gov.br/brasil/sp/sao-paulo/panorama. Accessed: 30/05/2023
Langer R, Vacanti JP (1993) Tissue engineering. Science 260(5110):920–926
Leder R (2004) The importance of biomedical engineering history. IEEE Eng Med Biol Mag 23(6):18
Lee EJ, Huh BK, Kim SN, Lee JY, Park CG, Mikos AG, Choy YB (2017) Application of Materials as Medical Devices with Localized Drug Delivery Capabilities for Enhanced Wound Repair. Prog Mater Sci 89:392–410
Lhotska L (2007) Biomedical engineering as a healthcare profession. Annu Int Conf IEEE Eng Med Biol Soc 2007:5937–40
Lynn DJ, Lloyd AT, O Farrelly C (2003) Bioinformatics: implications for medical research and clinical practice. Clin Investigat Med 26(2):70–74
Lysaght, MJ (1995) Product development in tissue engineering. Tissue Eng 2:221–228
Maojo V, Martin-Sanchez F (2004) Bioinformatics: towards new directions for public health. Meth Inf Med 43(3):208–14
Massarweh NN, Cosgriff N, Slakey DP (2006). Electrosurgery: history, principles, and current and future uses. J Am Col Surg 202:520–30
Montesinos L, Salinas-Navarro DE, Santos-Diaz A (2023) Transdisciplinary experiential learning in biomedical engineering education for healthcare systems improvement. BMC Med Educ 23(1):207
Müller U (2008) In vitro biocompatibility testing of biomaterials and medical devices. Med Dev Technol 19(2):32–34
Munzner RF. Industry careers for the biomedical engineer (2004) Conf Proc IEEE Eng Med Biol Soc 2004:5138
Nadon R, Sladek R (2003) Bioinformatics in research and clinical practice. Clinic Investigat Medic 26(2):75–77
Nebeker F (2002). Golden accomplishments in biomedical engineering. IEEE Engineering in Medicine and Biology Magazine, 21(3), 17–47.
NIBIB. National Institute of Biomedical Imaging and Bioengineering (2023). Available at: https://www.nibib.nih.gov/. Accessed: 30/05/2023
Ni Q, Pu Y (2009) Career development of biomedical engineers in the medical device industry. Annu Int Conf IEEE Eng Med Biol Soc 2009:149–50

O'Cearbhaill ED, Ng KS, Karp JM (2014) Emerging medical devices for minimally invasive cell therapy. Mayo Clin Proc 89(2):259–73
Pan RL, Martyniak K, Karimzadeh M, Gelikman DG, DeVries J, Sutter K, Coathup M, Razavi M, Sawh-Martinez R, Kean TJ (2022) Systematic review on the application of 3D-bioprinting technology in orthoregeneration: current achievements and open challenges. J Exp Ortop 9:95
Pecchia L, Pallikarakis N, Magjarevic R, Iadanza E (2019) Health Technology Assessment and Biomedical Engineering: Global trends, gaps and opportunities. Med Eng Phys 72:19–26
Pittenger MF, Discher DE, Péault BM, Phinney DG, Hare JM, Caplan AI (2019) Mesenchymal stem cell perspective: Cell biology to clinical progress. NPJ Regen Med 4:22
Rehman A, Naz S, Razzak I (2022) Leveraging big data analytics in healthcare enhancement: trends, challenges, and opportunities. Multimed Syst 28:1339–1371
Requena-Carrion J, Leder RS (2009) The natural history of the Engineering in Medicine and Biology Society from a modern perspective. Annu Int Conf IEEE Eng Med Biol Soc 2009:1086–1088
Reeve L, Baldrick P (2017) Biocompatibility assessments for medical devices – evolving regulatory considerations. Expert Rev Med Devices 14(2):161–167
Schwan HP (1984) The development of biomedical engineering: historical comments and personal observations. IEEE Trans Biomed Eng 31(12):730–736
Sim I (2019) Mobile Devices and Health. N Engl J Med 381(10):956–968
SBEB. Sociedade Brasileira de Engenharia Biomédica (2023). Available at https://sbeb.org.br/graduacao/. Accessed: 30/05/2023
Wigertz O, Persson J, Ahlfeldt H (1989) Teaching medical informatics to biomedical engineering students: experiences over 15 years. Methods Inf Med 28(4):309–312
Williams DF (2008) On the mechanisms of biocompatibility. Biomaterials 29(20):2941–53
WHO. World Health Organization (2023). Assistive technology. Available at: https://www.who.int/news-room/fact-sheets/detail/assistive-technology. Accessed: 30/05/2023
Zeng X, Shi G, He Q, Zhu P (2021) Screening and predicted value of potential biomarkers for breast cancer using bioinformatics analysis. Sci Rep 11, 20799

Part I
Biomechanics, Rehabilitation and Assistive Technology

Chapter 2
Postural Control in Humans: Theories, Modeling, and Quantification

Marcos Duarte and Renato Naville Watanabe

Abstract We are still far from a complete understanding of how exactly humans control their posture, and many questions are yet to be satisfactorily answered. The goal of this chapter is to briefly review in a narrative form some of the knowledge of how humans control their posture during standing, with a particular focus on the main theories and modeling of the postural control system. Some aspects of the main technique for quantification of the postural sway, the posturography, will also be addressed. By reviewing a few of the debates in the study of postural control, we also hope to evince how active and intense the study of postural control is and to suggest that much is yet to be done.

Keywords Posturography · Stabilography · Posture · Balance · Equilibrium

2.1 Introduction

The near infinity of movements we can perform, from standing as still as possible to moving vigorously and putting ourselves in challenging positions, while retaining the ability to control our posture, gives us a glimpse of the difficulty and complexity of the control of posture in humans. We have faced this challenge every single day since we adopted the bipedal upright posture at least 4.4 million years ago (Lovejoy et al. 2009). Not that our "choice" of a bipedal standing posture has turned out to be a mistake, but we have to bear the serious consequences of that choice every single day. Deficits in posture control, with serious consequences in most of the cases, are a common problem in aging and in certain diseases including vestibular dysfunction, neuropathies, Parkinson's disease, and musculoskeletal disorders (Horak et al. 1989; Voermans et al. 2007; Mancini et al. 2012). For example, with respect to the standing posture and the difficulty in maintaining it, it is known that about one in every three

M. Duarte (✉) · R. N. Watanabe
Graduate Program in Biomedical Engineering, Federal University of ABC – UFABC, São Bernardo do Campo, SP, Brazil
e-mail: marcos.duarte@ufabc.edu.br

C. B. Lombello, P. A. da Ana (eds.), *Current Trends in Biomedical Engineering*,
https://doi.org/10.1007/978-3-031-38743-2_2

people 65 years and older fall, and falls are the first cause of accidental deaths for this population (Burns and Kakara 2018; James et al. 2020).

Our view of how humans maintain a standing posture has changed considerably over the years. If, in the past, we barely recognized that in fact we sway while standing, or viewed it simply as a phenomenon entirely regulated by automatic reflex mechanisms, today we recognize postural control as a complex skill that is yet to be fully understood. Although we often refer to the postural control system as the entity responsible for the control of equilibrium during a certain body posture, neither is this system a single neuroanatomic structure in our body nor is postural control a single task we perform. Postural control is dependent on a rich and fine integration of many sensorimotor processes in our body and several factors related to the goal we are trying to accomplish in the surrounding environment. Such complexity complicates our understanding of this phenomenon. Nevertheless, the goal of this chapter is to briefly review in a narrative form some of the knowledge on how humans control their posture with a particular focus on the main theories and modeling of the postural control system. Some aspects of the main technique for quantification of the postural sway, posturography, will also be addressed.

2.2 Theories

The exact nature of the control mechanisms that allow humans to stand in an upright posture is still unknown, or at least, it is a matter of controversy. The fundamental question for the understanding of human postural control is: what is the goal of the postural control during the upright posture? The adoption of a standing posture implies at least two explicit mechanical requirements: first, that the body should be somewhat aligned with the vertical axis defined by the gravitational force (the definition of upright); and second, that the vertical projection of the body center of gravity must lie inside the area of the base of support defined by the area that circumscribes both feet. Other important requisites of the posture control mechanism are that whatever the solution adopted it must be robust with regard to resistance to perturbations, flexible to comply with our different goals (e.g., to lean, squat, or initiate a movement), and efficient to minimize the energetic cost of this task.

The concept of a center of gravity (COG) employed above is useful for the mechanical description of the body as a whole, and it is defined as a specific point in a system of particles (segments for the human body) that behaves as if the weight of all particles were concentrated at that point. From simple mechanics, if the vertical projection (at the horizontal plane) of the center of gravity (COGv) steps out of the base of support, the body will be unable to apply restoring forces to maintain the upright posture. In this sense, the contour of the base of support can be viewed as the stability boundary for controlling our posture. In fact, the area in which the COGv must lie inside is lower than the area that contours both feet. This is because the subject may not have enough muscle force to return to the initial position once close to the boundaries of the base of support (Kuo and Zajac 1993), or because the subject may have substantial outward velocity of the COGv so that he or she will not be able to avoid the COGv moving outside the base of support (Patton et al. 1999). Another physical concept related to COG is the center of pressure (COP). The COP expresses the position of the resultant ground reaction force applied to the body at the ground surface. Different from the COG, which is a measure of a three-dimensional position independent of the acceleration of the body and its segments; the COP is a two-dimensional position dependent on the acceleration of the body and its segments.

However, the maintenance of the COGv inside the base of support is a necessary but insufficient requirement for postural control during standing. The upright posture requires that the joint angles present specific values. Even though the joint angles are related to the body's center of gravity, this relation is not unique. Nevertheless, the large majority of theories and modeling concerning posture control in the literature is concerned with the stabilization of COGv and not explicitly with the control of the many body joints during standing.

2.2.1 *The Need of an Active Control of the Upright Posture*

Due to the action of the force of gravity, during upright standing our body is a mechanically unstable system in the sense that if a perturbation is applied to it, the body does not return to its initial position unless an intentional correction is performed. If nothing is done after a perturbation, the body does not return to the upright position and it will fall. Our bones and joints do not have enough contact area nor the necessary shape to equilibrate one segment on top of the other during standing, as in a pile of boxes. In addition, our soft tissues (skin, ligaments, tendons, muscles, etc.) do not have the necessary stiffness to passively support the segments during standing, as in a pole supported by a rope or spring against gravity. It is worth mentioning here that Winter et al. (2001) proposed that in reality our soft tissues do present the necessary stiffness to passively control our posture during standing, but it seems that the majority of the scientific community does not agree. Consequently, muscle contractions are necessary to maintain our posture during standing. However, when a muscle is active, it does not generate a constant force; its force fluctuates due to neural noise (Jones et al. 2002; Wu et al. 2014). In addition, there are other varying forces acting on our body (blood flow, breathing, inaccurate estimation of the necessary muscle force, etc.). As a result, the position of our mechanically unstable segments during standing would vary over time, and then we would move and fall. Consequently, an active dynamic regulation of the muscle contractions is necessary to maintain our posture during standing over time.

Note, however, that the considerations above do not imply that our sway is exclusively due to constant physiological perturbations to our body, neither that we regulate our posture around the vertical axis or around a fixed reference position, nor that we simply sense and react to perturbation in a pure feedback control mechanism. The description above is only a tentative enumeration of what our body is subjected to during upright standing. What exactly we do to achieve our goal, i.e., to stand upright, has been intensely debated over recent years and some of these hypotheses will be discussed next.

2.2.2 The Minimization of the Deviation from Vertical

Probably the most straightforward hypothesis about how we control our posture is that we do it in such a manner as to keep our body as close as possible to a vertical axis or a reference point. So, to comply with the mechanical requirements described above, we would constantly regulate our posture around a fixed reference representing the vertical axis of our body. In fact, during normal upright standing, the axis passing through our ankles and the body's center of gravity is not vertical at the sagittal plane but leans forward a few degrees. A simple model to test this hypothesis is the inverted pendulum model, which will be explored in detail later.

2.2.3 The Sway Around a Moving Reference

Lestienne and Gurfinkel (1988) suggested that different subsystems are used to control the equilibrium during standing. First, a reference position for the equilibrium would be specified by a subsystem called conservative (this equilibrium position could be understood as a set of joint angles). Second, the equilibrium around this position would be maintained by another subsystem, referred to as operative. Some evidence would suggest that these two subsystems could be manipulated to operate at different temporal scales (Lestienne and Gurfinkel 1988; Gurfinkel et al. 1995).

The idea of two subsystems participating in posture control suggests that COP and COGv are in fact composed of two components. Motivated by this idea, Zatsiorsky and Duarte (1999, 2000) suggested a method of decomposing the COP displacement into two components. The reference-point migration was called *rambling,* and the COP migration around the reference was coined as *trembling*. Although the conceptual idea of the rambling–trembling decomposition, the sway around a moving reference point, sounds interesting, the technical implementation to find these two components seems to be inadequate. The rambling trajectory does not seem to capture the migration of the reference position around which the center of gravity would sway. The rambling component seems to be, in fact, equivalent to the trajectory of the COGv itself.

Motivated by similar ideas, researchers have proposed several other methods to decompose the COP and COGv displacements into two components (Dijkstra 2000; Bottaro et al. 2005; Milton et al. 2009). However, whether any of these decompositions indeed represent two different subsystems in the postural control system remains yet to be proven. These decompositions are different in nature from the well-known COP-COGv decomposition (Winter 1995; Morasso et al.

1999; Rougier 2008) where the COP displacement is decomposed as a sum of the COGv and the difference between COP and this displacement (COP-COGv). The COP-COGv decomposition is directly derived from the field of mechanics, and the existence of these components does not seem to be in dispute. However, the COP-COGv decomposition does not imply that our postural control system would explicitly control these components.

In a different context, this hypothesis of two subsystems in the postural control system has been applied to interpret the COP displacement during unconstrained standing, where people can stand naturally, without the constraint of standing as still as possible (Duarte and Zatsiorsky 1999, 2000, 2001; Duarte and Freitas 2005). In this task, due to the postural changes, people usually produce the sway of a reference position around which the individuals oscillate.

2.2.4 The Exploratory Behavior

Yet another hypothesis for postural control is that at least part of the sway during upright posture is, in fact, an intentional sway (Riccio et al. 1992; Stoffregen et al. 2005; Bonnet et al. 2009). This hypothesis advocates that a certain amount of sway, exploratory in nature, is necessary for the perception of the state of the body. An implication of this hypothesis is that in persons with a deteriorated sensorial system or in a challenging condition, the posture control system would purposefully increase the body sway in order to facilitate the perception of its current state (Riley et al. 1997; Bonnet et al. 2009).

2.2.5 The Maximization of Virtual Time to Contact

It is also possible that we do not explicitly care to regulate our body around a vertical axis but that we adopt a strategy not to fall. For example, some authors (Slobounov et al. 1997; DiDomenico et al. 2015) have proposed that we regulate our upright posture by maximizing the time the body's center of pressure would take to contact the stability boundaries at any instant, given the instantaneous position, velocity, and acceleration of the center of pressure on that instant (this time was termed the virtual time to contact). By maximizing the virtual time to contact, we would avoid falling.

2.3 Modeling

The purpose of the following section is to introduce in very simple terms the most common physical-mathematical modeling of postural control during standing. By no means has this text intended to review the state of the art on the modeling of postural control. The sole idea is to introduce some of the basic concepts that will be useful to review some ongoing debates in our area.

2.3.1 The Inverted Pendulum Model

The human musculoskeletal system is a multi-segmented system with about 206 bones, 640 muscles, and 244 degrees of freedom. The exact modeling of the dynamics of such a complex system, if possible, would be too intricate and impractical. Despite this enormous complexity, the main characteristics of the mechanical behavior of the human body during the standing posture can be elegantly captured by a physical-mathematical model of an inverted pendulum with rigid segments articulated by joints. In this simplified approach, muscles are lumped as single-joint torque actuators. The model can be developed to have two or more segments, to move in one or more directions, and the equations of motion are typically expressed in the Newton–Euler or Lagrangian formalism. In the most common version of this mechanical modeling of the human body during the standing posture, the human body in the sagittal plane is reduced to a single inverted pendulum articulated by only one joint (representing the foot, with the rest of the body articulated by the ankle joint).

Several studies have used the inverted pendulum model to understand human postural control (Gurfinkel 1973; Morasso and Schieppati 1999; van der Kooij et al. 1999; Loram and Lakie 2002; Maurer and Peterka 2005; Asai et al. 2009; Elias et al. 2014; Morasso et al. 2019). The attractiveness of the physical-mathematical inverted pendulum model is its physical simplicity, the ability to represent the COP and COG physical quantities and their relation, and the straightforwardness to implement computational simulations to control such a system. For a better understanding of the concept of an inverted pendulum model, a simplification of such a model will be presented next.

Let us consider for now only the movement in the anterior–posterior direction (on the sagittal plane) of a person standing still. Let's represent the human body as composed by two rigid segments articulated by a single hinge joint (feet, rest of the body, and ankle joint). Given this simplification, all the mechanical quantities important to understand the motion of the body are represented in Fig. 2.1.

If we apply the second Newton–Euler equation of motion to the foot segment (which does not move) in this two-dimensional problem, ignore the foot weight, and consider the torques around the ankle joint due to the horizontal (F_x) and vertical (F_y) components of the ground reaction force vector, the following equation of motion can be deducted:

$$T_a + F_a h + F_y \text{COP} \approx 0 \tag{2.1}$$

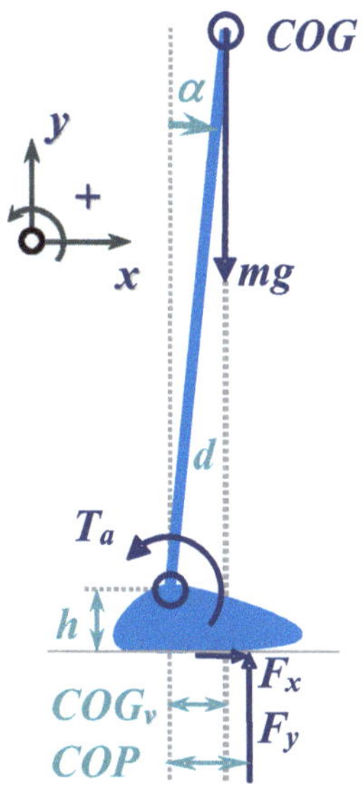

Fig. 2.1 Single inverted pendulum model for the representation of human standing. *COG* center of gravity; *COGv* COG vertical projection (at the horizontal plane) in relation to the ankle joint; *COP* center of pressure in relation to the ankle joint; *m* body mass, *g* acceleration of gravity; *Fx, Fy* horizontal and vertical components of the resultant ground reaction force; *Ta* torque at the ankle joint; *d* distance between *COG* and ankle joint; *h* height of the ankle joint to the ground; α angle of the body

With the exception of the parameter h (the height of the ankle joint), all the variables in Eq. (2.1) (see Fig. 2.1) in principle. vary in time; however, for simplicity, the explicit temporal dependence will be omitted in the equations. During standing still, the horizontal component of the ground reaction force (F_x, which is proportional to the horizontal acceleration of the human body) can be neglected because this component is much smaller than the vertical component (F_y, which its magnitude is approximately equal to the body weight, *mg*, where *g* is the gravity acceleration, ~10 m/s^2), then:

$$T_a \approx -mg\mathrm{COP} \tag{2.2}$$

That is, the amplitude of the torque at the ankle acting on the foot is approximately equal to the body weight times the *COP* position in the anterior–posterior direction (relative to the ankle joint position).

If now we also apply the second Newton–Euler equation of motion to the body segment, then:

$$T_a + mgd\sin(\alpha) = I\frac{d^2\alpha}{dt^2} \tag{2.3}$$

where *I* is a parameter expressing the moment of inertia of the whole body around the ankle joint (for the human body represented as a rigid bar its moment of inertia is 1.33 md^2, and *d* is the height of the center of gravity in relation to the ankle joint). Note that in Eq. (2.3), all terms were originally negative: due to the action-reaction law, the torque at the rest of the body is equal in magnitude but opposite in sign to the torque at the ankle and due to the axes convention (see Fig. 2.1), the torque due to the gravitational force and the body acceleration are negative.

For standing still, the body sways very little and for small values of α, $\sin\alpha \approx \alpha$, and α can be approximated to $\mathrm{COG_v}/d$ (because $\sin\alpha = \mathrm{COG_v}/d$). As a result of these approximations, the former equation can be rewritten as:

$$T_a + \mathrm{mgCOG}_v \approx \frac{I}{d} \frac{d^2\mathrm{COG}_v}{dt^2} \tag{2.4}$$

What this equation tells us is that during standing still (for small body accelerations), the torque at the ankle joint is approximately equal to the torque due to our weight (with opposite signs). Equation (2.4) is similar to Eq. (2.2), but we cannot neglect the acceleration of the body in the horizontal direction this time as we did for the foot segment. In fact, this acceleration is the source of the postural sway we are interested in. If we take the expression for the ankle torque found in Eq. (2.1) and substitute it in Eq. (2.4), we can deduct a relation between *COGv* and *COP*:

$$\mathrm{COG}_v - \mathrm{COP} \approx \frac{I}{\mathrm{mgd}} \frac{d^2\mathrm{COG}_v}{dt^2} \tag{2.5}$$

Finally, considering that the horizontal acceleration of the body is given by the horizontal force acting on the body (*Fx*) divided by its mass (*m*), substituting that in Eq. (2.5), and substituting the constant term *I*/(*m2gd*) by a constant k, then:

$$\mathrm{COG}_v - \mathrm{COP} \approx kF_x \tag{2.6}$$

This equation expresses a very simple relation between COGv, COP, and the horizontal force (or the body acceleration). From Eqs. (2.5) and (2.6), it is possible to predict some important relations among the variables in these equations which have been observed for postural sway during standing:

1. For a continuously regulated inverted pendulum (like the standing-still task), the common frequencies of *COGv* and *COP* signals are in phase (Morasso and Schieppati 1999);
2. *COGv* behaves as a low-pass filtered version of the *COP* signal (Caron et al. 1997);
3. *COGv–COP* is positively correlated with the horizontal ground reaction force in the anterior–posterior direction (Winter et al. 1998; Zatsiorsky and Duarte 2000).
4. When the horizontal force is zero, *COGv* and *COP* coincide. This has been used in a method to derive *COGv* from the *COP* signal (King and Zatsiorsky 1997).

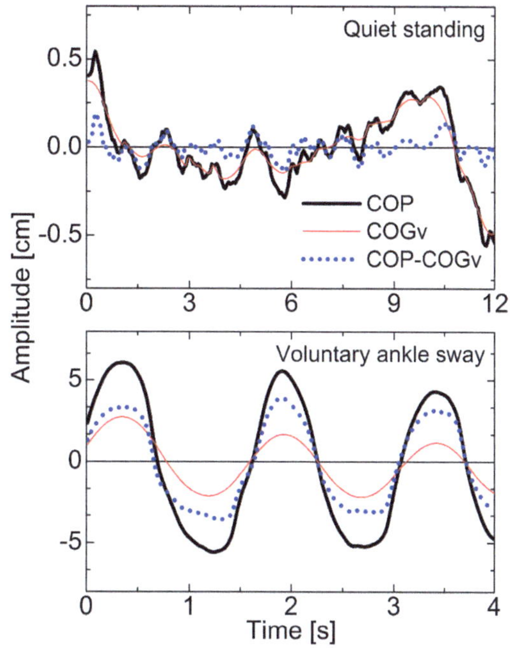

Fig. 2.2 Exemplary time series of COP, COGv, and COP-COGv during quiet standing (top) and forced voluntary sway around the ankles during standing (bottom). The COGv trajectory was estimated from the COP trajectory (Zatsiorsky and Duarte 2000)

Figure 2.2 shows examples of time series of COP, COGv, and COP-COGv during two different tasks: quiet standing and voluntary sway around the ankles while standing by a young adult. The differences between COP and COGv are quite small during quiet standing (because the horizontal body accelerations are very small) but quite large differences between COP and COGv can be observed during voluntary ankle sway.

2.3.2 *A Model for Control of the Inverted Pendulum*

The inverted pendulum model of the previous section is only a model of the mechanics of the body during standing still. This model does not tell us anything about how such an inverted pendulum would be controlled (the control of the torque at the ankle (T_a) necessary to maintain the upright position against the torques due to gravity (T_g) and perturbations (T_{noise})). Taking into account these terms, we will have the following equation of motion: $T_a + T_g + T_{noise} = I\alpha$ (where α is the angular acceleration of the inverted pendulum). T_a can be divided into two components: the passive component due to the viscoelastic properties of the muscle-tendon units and of other soft tissues, and the active component due to the muscle activation: $T_{a,\,passive} + T_{a,\,active} + T_g + T_{noise} = I\alpha$. In other words, the question would be: given $T_g + T_{noise}$, what should be T_a to stabilize the inverted pendulum in the upright position?

Note that in the question itself, we have already implied a goal for the controller system to maintain the inverted pendulum in a fixed upright position. As discussed in the previous section, some authors would not even agree with the way this question was formulated. Nevertheless, if we accept that, the next debate will be around the relative contribution of the passive and active components to the ankle torque. The main contribution to the passive component is the intrinsic stiffness of the muscle-tendon unit which would behave like a spring characterized by its stiffness, but we can also consider the passive damping of the muscle-tendon unit. Estimates of the muscle-tendon stiffness vary considerably across the literature. A critical value for the muscle-tendon stiffness is the minimum necessary to support the inverted pendulum against the gravitational torque. For our model, this critical value is mgd (in units of Nm/rad). In fact, a stiffness-control mechanism to mimic the characteristics of the postural sway observed during standing would require even greater values (Morasso and Schieppati 1999). Winter et al. (2001) seem to be the only ones to argue that the muscle-tendon stiffness of the ankle muscles is indeed greater than this critical value and consequently our posture during standing would be controlled only passively, with no need for an active control. Several other authors believe the muscle-tendon stiffness is subcritical and, consequently, an active control is needed for posture stabilization (Koozekanani et al. 1980; Morasso and Schieppati 1999; Loram and Lakie 2002; Maurer and Peterka 2005; Kiemel et al. 2006; Asai et al. 2009; Elias et al. 2014).

Considering the need for an active controller, and using the theories of control in engineering, the type of controller most employed is one that perceives the angle deviation (error) from the reference position and sends a signal directly proportional to the error ($K_p\dot{\theta}$) and to the error derivative ($K_d\dot{\theta}$) to control muscle activation. The correction is based on the perception of the error, and this type of control is called the *PD* feedback controller. There is, however, an ongoing debate whether a pure, continuous *PD* feedback controller would be adequate to control our posture during standing, or rather a feedback controller with an intermittent actuation with on and off periods (Bottaro et al. 2005; Asai et al. 2009), or whether a feedforward controller would also be necessary (Gatev et al. 1999; Loram and Lakie 2002). A feedforward controller for the inverted pendulum would imply a torque at the ankle independent of the current angle deviation (the error), possibly to anticipate perturbations or as exploratory behavior.

2.3.3 *On the Open-Closed Loop Control*

Collins and De Luca (1993) interpreted the COP displacements as a stochastic process and modeled the COP as a fractional Brownian motion. As a result, they were able to generate plots of the mean square displacement of the COP as a function of the time interval. These plots, called stabilogram diffusion analysis, showed two different behaviors of the COP diffusion in temporal scale (a time interval of about one second marked the transition between these two different behaviors). Collins and De Luca (1993) interpreted these two temporal regimes in the stabilogram diffusion analysis as evidence for two different control mechanisms acting on different time scales during standing, namely, open loop (on the short-time scale) and closed loop (on the long-time scale). However, alternative explanations have been presented that discard such interpretations. For example, Newell et al. (1997) proposed a model composed of a random walk with only a single stiffness component (i.e., a single process), which produced the same results as the two-process, random-walk model of Collins and De Luca (1993). With a different approach, Peterka (2000) showed that a simple closed-loop control model of upright stance, taking into account the physical properties of the human body, typical time delays representing conduction, processing, and muscle activation, and with a neural controller solely based on feedback, is also able to generate the same results when analyzed with the stabilogram diffusion analysis.

The work of Collins and De Luca (1993) suggested a very interesting technique for looking at the temporal structure of the COP displacements and it has been commonly used since. Evidently, the results produced by the stabilogram diffusion analysis are not being questioned. The change in the temporal behavior of the COP displacement is really there and the same behavior can be observed if one looks at the autocorrelation function of the COP displacement. What is under discussion is the interpretation of this behavior, and this case serves as an illustration of how much can be inferred from the COP displacement about the nature of the postural control mechanism.

2.3.4 On the COP and COGv Relation

Different authors Winter et al. 1998; Zatsiorsky and Duarte 2000) argued that the in-phase relationship between COP and COGv and the strong correlation between the acceleration of the COGv and the COP-COGv would rule out the active/reactive control of balance, because the delays in the sensory feedback would cause the COP to lag behind the COGv. However, such relationships seem to be a consequence of physical laws (the mechanical nature of the COGv and COP displacements) and cannot be used to prove one control theory over another (Morasso and Schieppati 1999; Peterka 2002). In addition, computational simulations solely based on a closed-loop control of the upright posture on the inverted pendulum model with a feedback control also produce simulated COGv and COP displacements that are in phase (Morasso et al. 1999; Peterka 2002).

2.3.5 On the COP, COGv, and Muscle Activity Relationship

Another point of debate about postural control was the finding that COGv and COP displacements are correlated with the gastrocnemius electromyographic (EMG) activity with a time lag of about 250 ms, i.e., the EMG activity preceded the COP and COGv displacements Gatev et al. 1999. The authors' interpretation of this finding was that "the gastrocnemius activity suggests a central program of control of the ankle joint stiffness working to predict the loading pattern" (i.e., feedforward control). However, it might be the case that if COP and COGv are displaced forward due to a perturbation, a gastrocnemius activity might be necessary to correct the body position, which in turn will always displace the COGv backward. So, we still might have a feedback control mechanism of quiet standing which might produce a positive value of the cross-correlation function between COGv and muscle activity due to the nature of the closed-loop system being controlled. This indeed has been observed. Inverted pendulum models with pure feedback control have been simulated and the COP and COGv displacements are lagged in relation to the simulated muscle activity (Kohn 2005; van der Kooij et al. 2005).

Note that we are not advocating that the control of posture during standing is equal to the control of an inverted pendulum with a fixed reference position by a feedback controller. We are only arguing that if a simple, feasible, and realistic model of the posture control during standing can produce the same results, these results alone cannot be used to support a single theory of posture control. Once again, this case illustrates the pitfalls of inferring about the nature of the controller based solely on the output of the system. Unfortunately, it seems that for the postural control system, it is nearly impossible to infer anything about the nature of the controller by looking only at the outcome of the entire system, the controller plus what is being controlled (see for example, van der Kooij et al. 2005).

2.4 Quantification

Posturography is a general term for all techniques concerned with quantifying postural sway. It is possible to quantify a certain property of a phenomenon, like the postural control during upright standing, without any underlying scientific theory on what generated that property and what the implication of that measurement for the understanding of the phenomenon is. Sometimes, this is the only approach to start with. However, as our knowledge on the phenomenon advances, the need for a scientific theory or a hypothesis to interpret the results in a more meaningful way becomes evident. It is relatively easy to perform many measurements of the body sway while upright standing; far more difficult is to interpret those measurements to understand what they mean. The most widespread empirical interpretation of typical measurements of posture sway is that more sway means more instability which is seen as an indication of a deteriorated postural control system. This rationale is based on many experiments on aging and pathological conditions that, indeed, have observed increased sway in those conditions (see for example, the reviews of Horak et al. (1989) and Bonnet et al. (2009))). However, bear in mind that it might not always be the case. For example, patients with Parkinson's disease in some cases present reduced postural sway compared to elderly adults, despite the fact that patients with Parkinson's disease do present severe problems of postural control (Romero and Stelmach 2003).

2.4.1 COP and COGv Quantification

Although more than one hundred years have passed since Romberg developed a method to test postural control (Lanska and Goetz 2000), only in the last decades has the precise and accurate quantification of the body sway been implemented. The technological development of force plates and video cameras and the advancement in signal processing were the major drive to this improvement. Typically in posturography, instead of measuring the sway of each segment, measures of the whole body sway have been used. The displacements of COP and COGv are the most common measures of body sway. While the COP displacement can be easily measured with a force plate, the direct measurement of the COGv is more complicated and typically subject to a larger error. The measurement of the COGv is computed by measuring the position of each body segment and by estimating each segment's mass using an anthropometric model. More commonly, the displacement of the COGv is determined indirectly from the COP displacement and different methods are available for such, which produce similar results (Lafond et al. 2004).

In the last decade, many markerless systems based on video capture records and their processing using modern techniques such as deep learning neural networks appeared (Cronin 2021). This kind of system makes it possible to perform records outside a laboratory since they do not require either expensive or big equipment and provides estimations of the position and orientation of the limbs. However, the accuracy of the limb centers and joint angles using these systems is not sufficient for clinical and research purposes Wade et al. 2022.

Although COP and COGv express different concepts, the difference between these two quantities is approximately proportional to the body's horizontal acceleration; consequently, less the frequency of body sway, the smallest are the differences between COP and COGv (Gurfinkel 1973). Because of that, it is not uncommon to see studies employing different measures, COP or COGv, to describe the same phenomenon, the body sway during the standing posture. This situation is somewhat risky, as inconsistent results can be obtained since COP and COGv are indeed different quantities, and some studies have shown that the differences between COP and COGv, small as they are, may reveal subtle characteristics of the postural control (Rougier 2008).

2.4.2 Measurements for the Quantification of Body Sway

Evidently, selecting which exact measurement will be used to quantify a certain characteristic of the postural control depends on which experiment we design and which task the individual under evaluation is performing. In this sense, posturography has been divided into two experimental paradigms: dynamic and static posturography. In static posturography, the individual is typically trying to stay as quiet as possible while his or her sway is measured, and no perturbation is applied during the task. If a perturbation, defined as a disturbance of the conditions needed to perform the task, is applied, this is typically done before the task initiation, such as removing visual information or an unstable base of support. In dynamic posturography, a momentary perturbation is applied and how the subject responded to that perturbation is measured. The perturbation can be mechanical, e.g., a displacement or rotation of the floor, or sensorial, e.g., a displacement of the visual scene or an electrical stimulus to the vestibular system.

Several different measures have been proposed to quantify postural sway. For a review see for example (Prieto et al. 1996; Zatsiorsky and Duarte 1999; Baratto et al. 2002; Piirtola and Era 2006; Rougier 2008). Although there is no standardization for which measures should be used in posturography, the most commonly employed measures seem to be: quantification of spatial displacement (usually, standard deviation or root mean square (RMS) in the anterior–posterior (AP), and medio-lateral (ML) directions or total area of sway); mean speed or velocity; and a frequency variable (usually, mean or median frequency).

Baratto and collaborators (2002) examined 38 posturographic measures and looked at the reliability and power to discriminate among three different groups of subjects: normal individuals, Parkinsonian patients, and osteoporotic patients. They concluded that only four measures were valuable for clinical practice: total sway path (similar to the mean speed variable), frequency band, and two measures from the method of COP and decomposition, called sway-density plots. Piirtola and Era (2006) reviewed the literature to determine which posturographic measures have been used as predictors of falls among elderly populations. They concluded that the mean COP speed, range, and RMS at the ML direction were the measures that showed significant associations with future falls in the elderly population. Yamamoto et al. (2015) analyzed 73 posturographic measures. They concluded that the slopes of the low-frequency spectra of the AP and ML directions, the angle of the major direction of the COP motion, the median frequencies of the COP in the AP and ML directions, and the ratio between the short and long axes of the ellipse formed by the COP motion are the characteristics that most differentiate between different subjects.

Clearly, standardization for posturography is needed to define standards for tasks, data collection, and analysis, particularly to allow data comparison across laboratories.

2.4.3 Open Datasets

In the last decade, different research groups published datasets containing kinetic and kinematic data collected from subjects during quiet standing. These datasets have the common characteristics to be available in public repositories like *Figshare* (Santos and Duarte 2016; Dos Santos et al. 2017; de Oliveira et al. 2022; Palermo et al. 2022) and have data both from health subjects and patients from motor diseases. Most likely, this movement toward opening up data and algorithms for analysis will contribute to speeding up the search for answers to the still unanswered questions raised in this chapter.

2.5 Concluding Remarks

Despite what is the exact nature of postural control during standing, the understanding of the mechanical properties relevant to describe such a task is fundamental to avoid some pitfalls in the study of postural control. This is not to say that the physical-mathematical modeling of human postural control is the only valid strategy. But at least the physical-mathematical modeling should be used to test and validate new ideas about human postural control as much as possible.

We have witnessed the proposal of many different methods in the quest for the right set of measures to describe the characteristics of the COP displacement in order to detect changes in the human postural control related to disease, aging, sensorimotor properties, or characteristics of the task to be achieved. Judging by the number of new studies on this topic every year, it seems that this quest is not finished yet. In fact, perhaps the COP signal alone during solely the quiet standing task might not entail sufficient information to capture all these changes. It might be necessary to combine the COP signal with other neurophysiological signals and to study other tasks for a greater understanding of the postural control system.

References

Asai, Y. *et al.* (2009) 'A model of postural control in quiet standing: robust compensation of delay-induced instability using intermittent activation of feedback control', *PloS one*, 4(7), p. e6169.

Baratto, L. *et al.* (2002) 'A new look at posturographic analysis in the clinical context: sway-density versus other parameterization techniques', *Motor control*, 6(3), pp. 246–270.

Bonnet, C., Carello, C. and Turvey, M.T. (2009) 'Diabetes and postural stability: review and hypotheses', *Journal of motor behavior*, 41(2), pp. 172–190.

Bottaro, A. *et al.* (2005) 'Body sway during quiet standing: is it the residual chattering of an intermittent stabilization process?', *Human movement science*, 24(4), pp. 588–615.

Burns, E. and Kakara, R. (2018) 'Deaths from Falls Among Persons Aged ≥65 Years - United States, 2007-2016', *MMWR. Morbidity and mortality weekly report*, 67(18), pp. 509–514.

Caron, O., Faure, B. and Brenière, Y. (1997) 'Estimating the centre of gravity of the body on the basis of the centre of pressure in standing posture', *Journal of biomechanics*, 30(11-12), pp. 1169–1171.

Collins, J.J. and De Luca, C.J. (1993) 'Open-loop and closed-loop control of posture: A random-walk analysis of center-of-pressure trajectories', *Experimental Brain Research*, pp. 308–318. Available at: https://doi.org/10.1007/bf00229788.

Cronin, N.J. (2021) 'Using deep neural networks for kinematic analysis: Challenges and opportunities', *Journal of biomechanics*, 123, p. 110460.
DiDomenico, A., McGorry, R.W. and Banks, J.J. (2015) 'Factors affecting time-to-contact calculations during quiet standing', *Motor control*, 19(1), pp. 1–9.
Dijkstra, T.M.H. (2000) 'A gentle introduction to the dynamic set-point model of human postural control during perturbed stance', *Human Movement Science*, pp. 567–595. Available at: https://doi.org/10.1016/s0167-9457(00)00027-0.
Dos Santos, D.A. *et al.* (2017) 'A data set with kinematic and ground reaction forces of human balance', *PeerJ*, 5, p. e3626.
Duarte, M. and Freitas, S.M.S.F. (2005) 'Speed-accuracy trade-off in voluntary postural movements', *Motor control*, 9(2), pp. 180–196.
Duarte, M. and Zatsiorsky, V.M. (1999) 'Patterns of Center of Pressure Migration during Prolonged Unconstrained Standing', *Motor Control*, pp. 12–27. Available at: https://doi.org/10.1123/mcj.3.1.12.
Duarte, M. and Zatsiorsky, V.M. (2000) 'On the fractal properties of natural human standing', *Neuroscience letters*, 283(3), pp. 173–176.
Duarte, M. and Zatsiorsky, V.M. (2001) 'Long-range correlations in human standing', *Physics letters. A*, 283(1–2), pp. 124–128.
Elias, L.A., Watanabe, R.N. and Kohn, A.F. (2014) 'Spinal mechanisms may provide a combination of intermittent and continuous control of human posture: predictions from a biologically based neuromusculoskeletal model', *PLoS computational biology*, 10(11), p. e1003944.
Gatev, P. *et al.* (1999) 'Feedforward ankle strategy of balance during quiet stance in adults', *The Journal of physiology*, 514 (Pt 3)(Pt 3), pp. 915–928.
Gurfinkel, E.V. (1973) 'Physical foundations of stabilography', *Agressologie: revue internationale de physio-biologie et de pharmacologie appliquees aux effets de l'agression*, 14(Spec No C), pp. 9–13.
Gurfinkel, V.S. *et al.* (1995) 'Kinesthetic reference for human orthograde posture', *Neuroscience*, pp. 229–243. Available at: https://doi.org/10.1016/0306-4522(95)00136-7.
Horak, F.B., Shupert, C.L. and Mirka, A. (1989) 'Components of postural dyscontrol in the elderly: a review', *Neurobiology of aging*, 10(6), pp. 727–738.
James, S.L. *et al.* (2020) 'The global burden of falls: global, regional and national estimates of morbidity and mortality from the Global Burden of Disease Study 2017', *Injury prevention: journal of the International Society for Child and Adolescent Injury Prevention*, 26(Supp 1), pp. i3–i11.
Jones, K.E., de Hamilton, C. A.F. and Wolpert, D.M. (2002) 'Sources of Signal-Dependent Noise During Isometric Force Production', *Journal of Neurophysiology*, pp. 1533–1544. Available at: https://doi.org/10.1152/jn.2002.88.3.1533.
Kiemel, T., Oie, K.S. and Jeka, J.J. (2006) 'Slow dynamics of postural sway are in the feedback loop', *Journal of neurophysiology*, 95(3), pp. 1410–1418.
King, D.L. and Zatsiorsky, V.M. (1997) 'Extracting gravity line displacement from stabilographic recordings', *Gait & Posture*, pp. 27–38. Available at: https://doi.org/10.1016/s0966-6362(96)01101-0.
Kohn, A.F. (2005) 'Cross-correlation between EMG and center of gravity during quiet stance: theory and simulations', *Biological cybernetics*, 93(5), pp. 382–388.
van der Kooij, H. *et al.* (1999) 'A multisensory integration model of human stance control', *Biological cybernetics*, 80(5), pp. 299–308.
van der Kooij, H., van Asseldonk, E. and van der Helm, F.C.T. (2005) 'Comparison of different methods to identify and quantify balance control', *Journal of neuroscience methods*, 145(1-2), pp. 175–203.
Koozekanani, S.H. *et al.* (1980) 'On the role of dynamic models in quantitative posturography', *IEEE transactions on bio-medical engineering*, 27(10), pp. 605–609.
Kuo, A.D. and Zajac, F.E. (1993) 'A biomechanical analysis of muscle strength as a limiting factor in standing posture', *Journal of biomechanics*, 26 Suppl 1, pp. 137–150.

Lafond, D., Duarte, M. and Prince, F. (2004) 'Comparison of three methods to estimate the center of mass during balance assessment', *Journal of biomechanics*, 37(9), pp. 1421–1426.
Lanska, D.J. and Goetz, C.G. (2000) 'Romberg's sign: development, adoption, and adaptation in the 19th century', *Neurology*, 55(8), pp. 1201–1206.
Lestienne, F.G. and Gurfinkel, V.S. (1988) 'Postural control in weightlessness: a dual process underlying adaptation to an unusual environment', *Trends in neurosciences*, 11(8), pp. 359–363.
Loram, I.D. and Lakie, M. (2002) 'Direct measurement of human ankle stiffness during quiet standing: the intrinsic mechanical stiffness is insufficient for stability', *The Journal of physiology*, 545(3), pp. 1041–1053.
Lovejoy, C.O. *et al.* (2009) 'The pelvis and femur of Ardipithecus ramidus: the emergence of upright walking', *Science*, 326(5949), pp. 71e1–6.
Mancini, M. *et al.* (2012) 'Postural sway as a marker of progression in Parkinson's disease: a pilot longitudinal study', *Gait & posture*, 36(3), pp. 471–476.
Maurer, C. and Peterka, R.J. (2005) 'A new interpretation of spontaneous sway measures based on a simple model of human postural control', *Journal of neurophysiology*, 93(1), pp. 189–200.
Milton, J. *et al.* (2009) 'The time-delayed inverted pendulum: implications for human balance control', *Chaos* , 19(2), p. 026110.
Morasso, P., Cherif, A. and Zenzeri, J. (2019) 'Quiet standing: The Single Inverted Pendulum model is not so bad after all', *PloS one*, 14(3), p. e0213870.
Morasso, P.G. *et al.* (1999) 'Internal models in the control of posture', *Neural networks: the official journal of the International Neural Network Society*, 12(7–8), pp. 1173–1180.
Morasso, P.G. and Schieppati, M. (1999) 'Can muscle stiffness alone stabilize upright standing?', *Journal of neurophysiology*, 82(3), pp. 1622–1626.
Newell, K.M. *et al.* (1997) 'Stochastic processes in postural center-of-pressure profiles', *Experimental Brain Research*, pp. 158–164. Available at: https://doi.org/10.1007/bf02454152.
de Oliveira, C.E.N. *et al.* (2022) 'A Public Data Set With Ground Reaction Forces of Human Balance in Individuals With Parkinson's Disease', *Frontiers in neuroscience*, 16, p. 865882.
Palermo, M. *et al.* (2022) 'A multi-camera and multimodal dataset for posture and gait analysis', *Scientific data*, 9(1), p. 603.
Patton, J.L., Pai, Y. and Lee, W.A. (1999) 'Evaluation of a model that determines the stability limits of dynamic balance', *Gait & posture*, 9(1), pp. 38–49.
Peterka, R.J. (2000) 'Postural control model interpretation of stabilogram diffusion analysis', *Biological cybernetics*, 82(4), pp. 335–343.
Peterka, R.J. (2002) 'Sensorimotor Integration in Human Postural Control', *Journal of Neurophysiology*, pp. 1097–1118. Available at: https://doi.org/10.1152/jn.2002.88.3.1097.
Piirtola, M. and Era, P. (2006) 'Force platform measurements as predictors of falls among older people - a review', *Gerontology*, 52(1), pp. 1–16.
Prieto, T.E. *et al.* (1996) 'Measures of postural steadiness: differences between healthy young and elderly adults', *IEEE transactions on bio-medical engineering*, 43(9), pp. 956–966.
Riccio, G.E., Martin, E.J. and Stoffregen, T.A. (1992) 'The role of balance dynamics in the active perception of orientation', *Journal of experimental psychology. Human perception and performance*, 18(3), pp. 624–644.
Riley, M.A. *et al.* (1997) 'Influences of Body Lean and Vision on Unperturbed Postural Sway', *Motor Control*, pp. 229–246. Available at: https://doi.org/10.1123/mcj.1.3.229.
Romero, D. H., & Stelmach, G. E. (2003). Changes in postural control with aging and Parkinson's disease. IEEE engineering in medicine and biology magazine: the quarterly magazine of the Engineering in Medicine & Biology Society, 22(2), 27–31. https:/doi.org/10.1109/memb.2003.1195692
Rougier, P.-R. (2008) 'What insights can be gained when analysing the resultant centre of pressure trajectory?', *Neurophysiologie clinique = Clinical neurophysiology*, 38(6), pp. 363–373.
Santos, D.A. and Duarte, M. (2016) 'A public data set of human balance evaluations', *PeerJ*, 4, p. e2648.

Slobounov, S.M., Slobounova, E.S. and Newell, K.M. (1997) 'Virtual Time-to-Collision and Human Postural Control', *Journal of motor behavior*, 29(3), pp. 263–281.
Stoffregen, T.A., Yang, C.-M. and Bardy, B.G. (2005) 'Affordance Judgments and Nonlocomotor Body Movement', *Ecological Psychology*, pp. 75–104. Available at: https://doi.org/10.1207/s15326969eco1702_2.
Voermans, N.C. *et al.* (2007) 'Why old people fall (and how to stop them)', *Practical neurology*, 7(3), pp. 158–171.
Wade, L. *et al.* (2022) 'Applications and limitations of current markerless motion capture methods for clinical gait biomechanics', *PeerJ*, 10, p. e12995.
Winter, D.A. (1995) 'Human balance and posture control during standing and walking', *Gait & Posture*, pp. 193–214. Available at: https://doi.org/10.1016/0966-6362(96)82849-9.
Winter, D.A. *et al.* (1998) 'Stiffness Control of Balance in Quiet Standing', *Journal of Neurophysiology*, pp. 1211–1221. Available at: https://doi.org/10.1152/jn.1998.80.3.1211.
Winter, D.A. *et al.* (2001) 'Ankle muscle stiffness in the control of balance during quiet standing', *Journal of neurophysiology*, 85(6), pp. 2630–2633.
Wu, H.G. *et al.* (2014) 'Temporal structure of motor variability is dynamically regulated and predicts motor learning ability', *Nature neuroscience*, 17(2), pp. 312–321.
Yamamoto, T. *et al.* (2015) 'Universal and individual characteristics of postural sway during quiet standing in healthy young adults', *Physiological reports*, 3(3). Available at: https://doi.org/10.14814/phy2.12329.
Zatsiorsky, V.M. and Duarte, M. (1999) 'Instant equilibrium point and its migration in standing tasks: rambling and trembling components of the stabilogram', *Motor control*, 3(1), pp. 28–38.
Zatsiorsky, V.M. and Duarte, M. (2000) 'Rambling and trembling in quiet standing', *Motor control*, 4(2), pp. 185–200.

Chapter 3
Postural Control in Parkinson's Disease

Daniel Boari Coelho, Thiago Kenzo Fujioka Shida, João Antonio Marques Costa, Layla Cupertino Salloum e Silva, Luciana Pastena Giorno, Débora da Silva Fragoso de Campos, Claudia Eunice Neves de Oliveira, Emanuele Los Angeles, Claudionor Bernardo, Luana dos Santos de Oliveira, Thayna Magalhães Novaes, Solaiman Shokur, and Mohamed Bouri

Abstract Balance impairment is a common symptom of Parkinson's disease (PD), suggesting the reclassification of the PD to a tetrad: rest tremor, rigidity, bradykinesia, and balance impairment. Falling is the most evident symptom of inadequate balance. Falls are expected to be twice as likely in individuals with PD compared to age-matched controls, and PD has a fourfold increased risk of hip fractures compared to age-matched controls. Balance is crucial for walking and, as a result, influences the performance of various daily activities around the

D. B. Coelho (✉)
Biomedical Engineering, Federal University of ABC, São Bernardo do Campo, SP, Brazil

Center for Mathematics, Computation, and Cognition, Federal University of ABC, São Bernardo do Campo, Brazil

Centre for Engineering, Modeling and Applied Social Sciences (CECS), Federal University of ABC (UFABC), São Bernardo do Campo, SP, Brazil

T. K. F. Shida · J. A. M. Costa · C. Bernardo
Biomedical Engineering, Federal University of ABC, São Bernardo do Campo, SP, Brazil

L. C. S. e Silva · D. da Silva Fragoso de Campos · C. E. N. de Oliveira · E. Los Angeles · L. dos Santos de Oliveira · T. M. Novaes
Center for Mathematics, Computation, and Cognition, Federal University of ABC, São Bernardo do Campo, Brazil
e-mail: daniel.boari@ufabc.edu.br

L. P. Giorno
Center of Natural and Human Sciences, Federal University of ABC, São Bernardo do Campo, SP, Brazil

S. Shokur
École Polytechnique Fédérale de Lausanne, Lausanne, Switzerland

The BioRobotics Institute and Department of Excellence in Robotics and AI, Scuola Superiore Sant'Anna, Pisa, Italy

M. Bouri
École Polytechnique Fédérale de Lausanne, Lausanne, Switzerland

C. B. Lombello, P. A. da Ana (eds.), *Current Trends in Biomedical Engineering*,
https://doi.org/10.1007/978-3-031-38743-2_3

home and community. This chapter is designed around the four postural domains to give researchers a framework for assessing balance control in Parkinson's disease patients: (1) postural transitions; (2) reactive response; (3) quiet and prolonged standing; and (4) dynamic balance during walking.

Keywords Freezing of gait · Anticipatory postural adjustments · Reactive response · Standing · Walking

3.1 Introduction

Parkinson's disease (PD) is a neurodegenerative and progressive disorder of dopaminergic neurons in the substantia nigra, and the presence of intracellular neural inclusions in the dorsal motor nucleus of the vagus nerve, olfactory bulb, and locus coeruleus, which result in different symptoms, both motor and non-motor (Hawkes et al. 2010; Jellinger 2003). PD is characterized as a prolonged and irreversible pathological process, tending to appear in middle age, and can be divided into six stages according to disease progression (for a review, see Del Tredici and Braak 2016). In the initial phase, PD first affects the peripheral nervous system, the autonomic nervous system, and/or the central nervous system (dorsal motor nucleus of the vagus nerve); in stage 2, it spreads to raphe (serotoninergic) and locus coeruleus (noradrenergic) neurons; in stage 3, it affects the substantia nigra (dopaminergic) and the pedunculopontine nuclei (cholinergic, gabaergic, and glutamatergic), also affecting the spinal cord and amygdala; in stage 4, the thalamus and temporal lobe may be affected; in stage 5, sensory association areas and prefrontal regions are affected; and in the final stage, the primary motor and sensory areas are also affected (Mancini et al. 2019). Another form of classification used in clinical practice divides PD into five stages (V) according to the degree of motor impairment: (I) unilateral disease, usually with minimal or no functional impairment; (II) bilateral disease, without balance impairment; (III) mild-to-moderate bilateral disease, some postural instability, physically independent; (IV) severe disability, still able to stand upright unaided; (V) needs a wheelchair or bed, physically dependent (Hoehn and Yahr 1967).

The clinical diagnosis of PD is made from motor symptoms associated with bradykinesia and non-motor symptoms such as mood disorders, cognitive impairment, olfactory impairment, autonomic function disorders, and/or sleep disorders (Kehagia et al. 2013). The common motor symptoms of PD include bradykinesia and/or akinesia, joint stiffness, resting tremors, postural instability, and gait disorders (Jankovic 2008). In the early stages, motor symptoms tend to be asymmetrical, predominantly affecting one side of the body (Kalia and Lang 2015). However, with the progression of the disease, this impairment becomes bilateral, negatively interfering with the individuals' daily activities, quality of life, and motor independence (Hoehn and Yahr 1967). Figure 3.1 shows the balance dysfunction in individuals with Parkinson's disease than in neurologically healthy individuals.

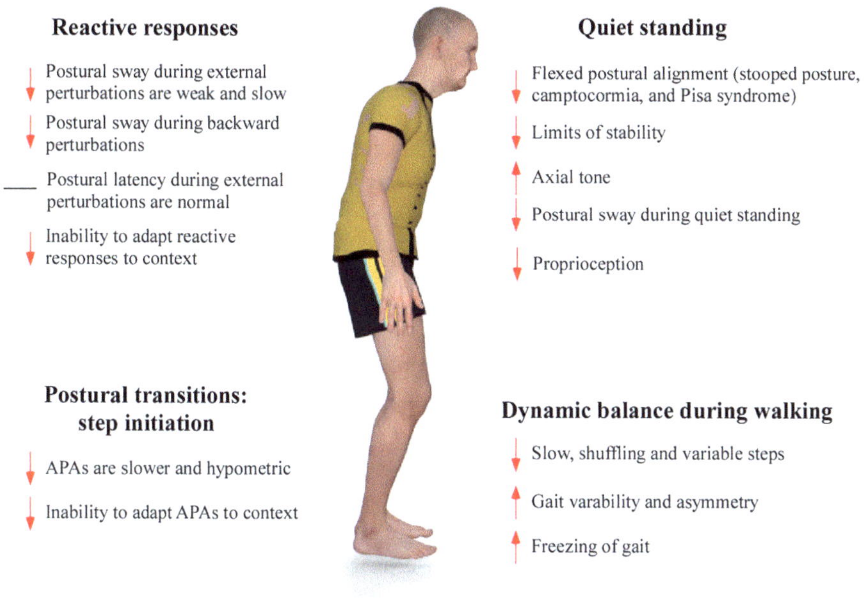

Fig. 3.1 Balance dysfunction in individuals with Parkinson's disease than in neurologically healthy individuals. The red color indicates worse indices for PD. The up arrow indicates an increase, while the down arrow indicates a decrease. The horizontal black stroke indicates that there is no difference. *APA* anticipatory postural adjustments

Freezing of gait (FoG) is a common symptom in PD that further affects gait, increasing the risk of falling and worsening the quality of life and functional independence of individuals with this symptom. Often, individuals report the sensation of their feet being stuck to the ground, despite the intention to take the step and continue walking. Usually, FoG is a transient and short-lived episode. It can be triggered in different situations, such as at the beginning of the movement, approaching the final destination of the gait, during the turn, and when passing through narrow places or obstacles. Still, it can also occur for no reason. The FoG can be classified according to its behavioral manifestation associated with the movement of the legs in FoG with short and fast steps; FoG with alternating leg tremor; and akinetic FoG when there is no movement in the legs (Schaafsma et al. 2003; Thompson and Marsden 1995). In addition, individuals with FoG have greater postural instability (Bekkers et al. 2017), more severe motor and cognitive symptoms, longer duration of the disease, and use a higher dose of antiparkinsonian medication (Lord et al. 2020). When analyzing prospective studies that followed individuals with PD from the early stages, a meta-analysis by Gao et al. (2020) showed that gait disorders could predict the development of FoG. In addition, this same study showed that non-motor factors, such as depression and anxiety, cognitive impairment, and lower educational level, can also be considered risk factors for the development of BC, but with limited evidence in the literature.

One of the main drug treatments for PD is the dopaminergic replacement from the administration of levodopa, the immediate dopamine precursor capable of crossing the blood-brain barrier and entering the brain. Once in the brain, levodopa quickly converts to dopamine through simple enzymatic reactions. Dopaminergic replacement therapy is composed of two main components. The first is characterized by short-lived effects (about a few hours), related to the concentration of circulating dopamine. The second is characterized by long-lasting effects (about days to weeks), related to neural plasticity induced by dopaminergic signaling (Albin and Leventhal 2017). However, no evidence demonstrates a decrease in disease progression with conventionally used drug treatment (Tarazi et al. 2014).

Increased levodopa concentration in the body produces acute motor responses, such as reduced bradykinesia, rigidity, resting tremor, and greater movement vigor (Albin and Leventhal 2017). This fluctuating motor response dependent on medication administration is called the ON–OFF phenomenon, characterized by an improvement in the motor pattern in the presence of the medication (ON state) and motor worsening with the decrease in the blood concentration of the medication (OFF state). The motor response to drug treatment occurs due to short-term and long-term effects. However, as the disease progresses, the response to long-term effects decreases, and short-term effects become more important in the treatment (Nutt et al. 1997, 2002). In addition, studies show that continued treatment can induce side effects leading to the development of motor fluctuations, involuntary, dyskinetic movements, and acceleration of dementia (Bastide et al. 2015; Tarazi et al. 2014).

The postural control system must be able to regulate balance in unstable situations and, on the other hand, must be versatile enough to allow rapid movement initiation. Perhaps the postural control system performs the most obvious task of maintaining the bipedal upright posture. Still, this system also acts during walking, for example. As a result, individuals with PD frequently experience balance issues. Because of this, some medical professionals have proposed that the PD triad of rest tremor, stiffness, and bradykinesia be changed to a tetrad, adding balance impairment as a fourth symptom. This observation suggests that decreased balance is a key component of Parkinson's disease. Nearly all PD individuals will experience balance issues at some point during the disease, and balance control will deteriorate as the disease advances, the doctor may anticipate with confidence.

3.2 Freezing of Gait

Freezing of gait (FoG) is a unique and disabling clinical phenomenon characterized by "brief, episodic absence or marked reduction of forward progression of the feet despite the intention to walk" (Nutt et al. 2011). This definition includes the three subtypes of FoG: a patient who suddenly becomes unable to start walking or to move forward, with a start hesitation; a complete absence of movement (akinesia); or a shuffling gait with small steps (Bloem et al. 2004). Some possible hypotheses on the pathogenesis of FoG are the abnormal coupling of posture with gait (Jacobs et al. 2009a, b), impaired gait rhythmicity and gait cycle coordination (Plotnik and Hausdorff 2008), or impairment in movement automaticity (Hallett 2008; Vandenbossche et al. 2012). FoG is triggered by postural transitions, such as gait initiation, turning, narrow passages, obstacle crossing, or approaching a destination (Nutt et al. 2011). Freezing episodes are often described as "the feeling that the feet are being glued to the floor" while the center of mass continues to move forward, resulting that FoG being one the most common reasons for falls in individuals with PD (Michalowska et al. 2005). The physiological mechanisms that trigger FoG have not yet been fully elucidated, and there are several hypotheses in the literature based on the different FoG phenotypes (Gao et al. 2020; Nieuwboer and Giladi 2013). Two models are worth mentioning: (a) threshold model (Plotnik et al. 2012) and (b) interference model (Lewis and Barker 2009). The first model is based on the gait disturbances presented by individuals with FoG (i.e., decreased stride amplitude, impaired gait coordination, and increased stride time variability). This model assumes that when these gait disturbances accumulate to a certain threshold, a point of motor breakdown triggers the FoG. Thus, according to the authors of the model, by improving the gait of individuals with PD, it is possible to reduce the propensity to FoG (Plotnik et al. 2012). The second model is based on the competitive and complementary relationship between the motor, cognitive, and limbic circuits through the basal ganglia, which is responsible for integrating information from different sensory inputs and coordinating an efficient functional response (Lewis and Barker 2009). The authors of this model suggest that the simultaneous processing of cognitive and/or limbic information during motor tasks would overload a system already affected by the dopaminergic deficit. This overload would result in the inactivation of the pedunculopontine nucleus (PPN), responsible for regulating the functional motor response, thus serving as trigger-to-trigger FoG episodes. Furthermore, the authors add that this overload can be reversed by inducing the focus of attention of the individual with PD to a single external sensory cue (a line on the floor or an obstacle to be overcome), thus reducing the number of sensory inputs and stimulating the PPN again. It is worth mentioning that the models proposed in the literature are not exclusive and the understanding of the possible physiological mechanisms that trigger FoG can help in treatment strategies, pharmacological or not.

Co-existing postural impairments may affect the occurrence and severity of FoG. FoG and postural instability are interconnected (Coelho et al. 2021; Peterson et al. 2020), can affect one another behaviorally, and may share a neural basis (for a review, see Bekkers et al. 2018a). The most prevalent FoG-related postural deficits comprised weight-shifting impairments and insufficient scaling and timing of postural responses, which were especially evident when impending postural alterations when time limitations were present. A negative cycle of coupled and more severe postural impairments will likely exacerbate postural instability in individuals with FoG. Consequently, the deficits in large-scale brain networks associated with FoG may concurrently impact postural stability.

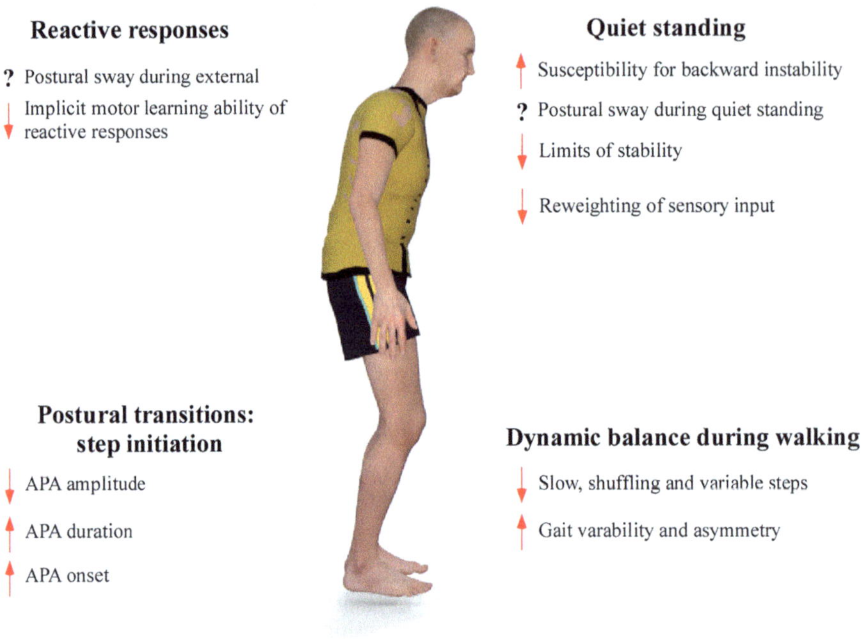

Fig. 3.2 Balance dysfunction in individuals with Parkinson's disease with freezing of gait (freezers) than without freezing of gait (nonfreezers). The red color indicates worse indices for the freezers. The up arrow indicates an increase, while the down arrow indicates a decrease. The question mark indicates that the studies are controversial. *APA* anticipatory postural adjustments

Individuals with FoG (freezers) have worse performance on clinical balance tests (Bekkers et al. 2017), more frequent episodes of falls and near-falls (Gazibara et al. 2017), slower and longer duration of anticipatory postural adjustments (APA) (Schlenstedt et al. 2018), smaller step length in automatic postural responses (APR) (Smulders et al. 2014), poorer flexed postural alignment, smaller voluntary limits of stability that could reflect poorer proprioception (de Souza Fortaleza et al. 2017), and worst variability and coordination of stepping during walking (Weiss et al. 2015) when compared to individuals without FoG (nonfreezers) (Fig. 3.2). A recent review showed that freezers have worse performance on clinical scales that assess domains of balance and gait, such as stability limits, APR, APA, and sensory orientation (Bekkers et al. 2017). Specifically, during quiet stance, Pelykh et al. (2015) showed reduced time-to-boundary and more regular postural sway, as observed by lower sample entropy and lower adaptability of the postural sway in freezers. As a certain amount of dynamic irregularity is considered a highly automatized functioning of postural control systems (Donker et al. 2007), it can be assumed that postural control in quiet standing in freezers needs more attention since it may explain the risk of falling in these patients. When analyzing postural responses after a sudden perturbation, the reactive postural control between freezers and nonfreezers is similar (Bekkers et al. 2018b). Given the similarity of postural responses, the authors hypothesize that reactive postural control seems to affect different mechanisms than those governing FoG. However, this hypothesis remains speculative and has not been directly assessed. Although freezers had smaller mediolateral APA during self-initiated gait without start hesitation, trials with freezing showed larger mediolateral APA when compared to nonfreezers (Schlenstedt et al. 2018). Schlenstedt et al. found a positive correlation between the size of mediolateral APA and the New Freezing of Gait Questionnaire (NFoG-Q), and a start hesitation in FoG is not caused by an inability to weight shifting when preparing gait. Therefore, postural instability is behaviorally exacerbated by FoG and vice versa (Bekkers et al. 2018a).

Freezers were shown to have impairments in circuitries involving the frontal cortex (for a review, see Fonoff et al. 2019). Studies have shown that freezers have altered supraspinal locomotor neural networks, including the frontal motor regions (supplementary motor area, presupplementary motor area, and primary motor cortices) and subcortical areas (basal ganglia, pontomedullary reticular formation, and mesencephalic and cerebellar locomotor region) (Butler et al. 2017; Ehgoetz Martens et al. 2018; Fling et al. 2013; Gilat et al. 2015; Peterson et al. 2014; Shine et al. 2013b). Resting-state functional magnetic resonance findings pointed to a decreased activity in the premotor (PM) and orbitofrontal cortex (Gallardo et al. 2018; Matsui et al. 2005) and increased functional connectivity between supplementary motor area (SMA) and areas known to be involved in gait initiation such as cerebellar (CLR) and mesencephalic (MLR) locomotor regions (Fling et al. 2014), which was positively correlated to FoG severity. Such an increase in the SMA and locomotor centers connectivity might indicate a greater influence of cognitive processing on gait initiation. The activity of the frontal cortex, including SMA, becomes higher during motor arrests, compared to PD without FoG (Maidan et al. 2016; Shine et al. 2013a; Vercruysse et al. 2014). Freezers showed decreased structural and functional connectivity between SMA and subthalamic nucleus (STN), the hyperdirect pathway that modulates response inhibition (Fling et al. 2013, 2014). Accordingly, FoG has been associated with cognitive function impairments, especially inhibitory control (Cohen et al. 2014). Decreased connectivity between the dorsolateral prefrontal cortex, a key region involved in the process of response inhibition, with basal ganglia (Shine et al. 2013c) and its correlation with the frequency of FoG supports the notion that there is a decoupling between inhibition control with movement. Furthermore, findings in a resting state functional magnetic resonance imaging demonstrate higher functional connectivity between areas that process sensory information (middle temporal gyrus) and emotion (amygdala) with MLR and basal ganglia, respectively (Gilat et al. 2018; Wang et al. 2016), and increased insula activation during FoG (Shine et al. 2013a). All this evidence shows the influence of cognition, sensory, and emotional processing in the pathophysiology of FoG.

3.3 Anticipatory Postural Adjustments During Step Initiation

During the step initiation (SI), the nervous system organizes a series of anticipatory postural adjustments (APAs), preparing the body for the disturbances caused by voluntary movement. Postural instability rapidly rises during SI as the base of support shifts and the center of mass (CoM) travels forward and outside the latter. Before foot-off, the initial stage of the SI process can be thought of as a forward fall that needs to be stopped but is essential for further forward body movement. These APAs are important because they release the load on the swing leg, establishing the conditions necessary for the step. Healthy subjects always behave in a rather stereotypical manner before SI. The CoM is properly shifted to the support leg to start a step, releasing the movement leg. Initially, there is a short shift from the center of mass to the moving leg, then the force is directed to the support leg (Breniere et al. 1987). As the first displacement of the CoM is toward the movement leg, inhibitory control mechanisms are required to direct the force–displacement toward the supporting leg. For example, if the objective is to initiate the step with the right leg, initially, the CoM is shifted to the right leg. This movement is inhibited and redirected to the left leg, which will serve as a support, allowing the release of the right leg. This change in body weight causes the center of pressure (CoP) to move backward and toward the swing leg. The CoP is then seen to move first toward the stance leg and subsequently forward. Toe-off of the swing leg happens right before the forward CoP displacement, while the heel-off happens at the beginning of the second phase of the CoP displacement (i.e., with a lateral shift toward the stance leg) (Breniere and Do 1991; Delval et al. 2005). The CoM lies outside the base of support during the single support phase of SI (when the subject lifts the swing leg off the ground), which causes the body to be imbalanced. The CoM fall must be stopped for movement to continue and prevent the body from falling to the ground. Before the swing limb leg makes contact with the ground, healthy adults' ankle plantar flexors are activated to produce this braking action (Welter et al. 2007).

In individuals with PD (Fig. 3.3), different APA disturbances are observed, causing an abnormal coupling between posture and the motor program of APAs (Cohen et al. 2017; Hass et al. 2005; Mancini et al. 2009; Schlenstedt et al. 2018): bradykinetic and hypokinetic APAs, with the mediolateral and anteroposterior CoP changes are longer and weaker; pure akinesia, extended delays between the commencement of the APA and the onset of the step; and multiple APAs. The onset of these anomalies in PD can be quite early. Untreated early-to-moderate PD (Carpinella et al. 2007; Mancini et al. 2009) has been associated with a lower lateral, but not backward, APA magnitude, suggesting that the pathology of the disease may specifically affect how the legs are loaded and unloaded. Later in the disease, individuals with PD experience bradykinetic lateral and backward APAs. It is still controversial whether APAs are smaller or larger than normally associated with freezing of gait (Schlenstedt et al. 2018). Four out of eight studies reported smaller

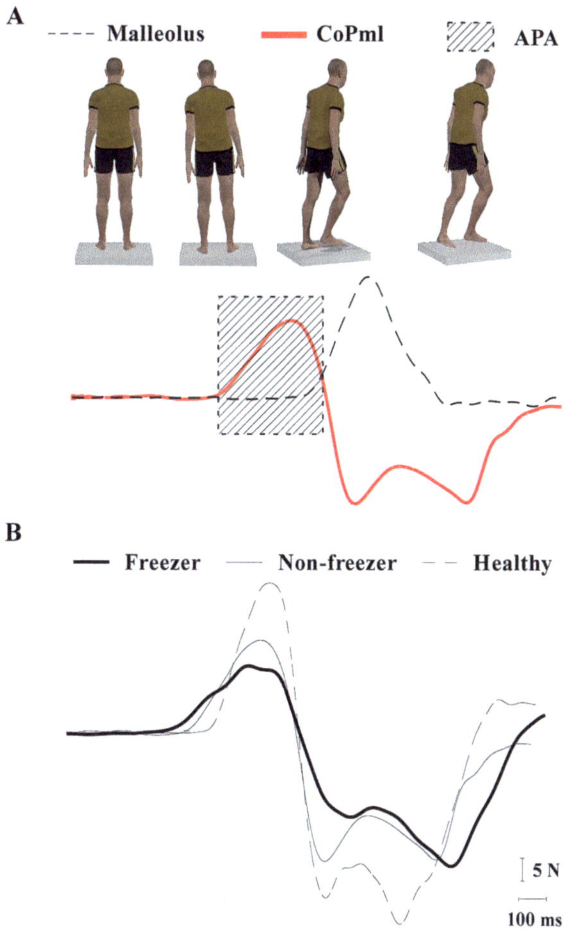

Fig. 3.3 Anticipatory postural adjustment (APA) before step initiation. (**a**) Three phases of step initiation: quiet stance, weight transfer to the contralateral leg, and step. The red line shows mediolateral force amplitude during the APA, and the dashed line shows the forward displacement of the reflective marker attached to the ankle during the step. The shaded area indicates the APA. (**b**) Result of the average lateral force of the freezers, nonfreezers, and healthy

APA amplitudes and three with a longer duration in freezers than nonfreezers (for a review, see Bekkers et al. 2018a).

The motor program for SI must change when the motor task's parameters change. APA is generated according to the context and inhibited if the support is directed to an inappropriate place (Cohen et al. 2011, 2017). Patients with PD seem to have difficulty releasing the step after it has been previously inhibited, which is further evidence of changes in the inhibitory control of step initiation present in patients with PD. Recent research has shown that SI of the healthy (Stins and Beek 2011) and PD (Lagravinese et al. 2018; Naugle et al. 2012) individuals are influenced by the valence of emotional inputs, suggesting that the limbic system may be involved in SI.

APA ranges from the participation of cortical and subcortical structures (Jacobs and Horak 2007; Yiou et al. 2017) to afferent signals that also emerge in the spinal cord, where they are modulated. Some important regions of the brain have been described as essential to initiate gait, such as mesencephalic, subthalamic, and cerebellar locomotor regions (Takakusaki 2017) in the subcortical level and premotor (PM) and supplementary motor area (SMA) (de Lima-Pardini et al. 2020; Jacobs et al. 2009a, b; Takakusaki 2017) in the cortical level. These cortical areas act in a feedforward control, projecting to the corticospinal tract (CST) (He et al. 1995) and having substantial connections with subcortical locomotor regions through cortico-reticular tracts. PM and SMA modulate step initiation by sending information on the required muscle tone to prepare the body for stepping according to the context via cortico-reticular and reticular spinal tracts (Keizer and Kuypers 1989). In parallel, SMA and PM contribute to the output from CST that elicit voluntary motor commands. Therefore, there are two important systems involved in gait initiation: one, comprising cortical and subcortical levels, that prepares the body by adjusting the muscle tone properly, and the other, that cortically modulates the voluntary movement of the leg. Given the challenge of shifting from an upright stance with both feet on the ground to a more unstable condition of moving the body forward, these two systems must be coupled to elicit a correct muscle tone background during the forward movement of the leg. When these systems work properly, gait initiation is characterized by an initial backward movement toward the moving leg, then to the support leg. The numerous connections between the basal ganglia and the supplementary motor area, and the premotor area, both of which are implicated in movement preparation (Massion 1992; Nakano et al. 2000; Schlenstedt et al. 2017), or with the peduncular pontine nucleus in the brainstem, which is implicated in locomotion initiation, may explain why APAs are impaired in PD (Pahapill and Lozano 2000). Changes in the basal ganglia, which slow down the sequential execution of the preparation and stepping task's subcomponents, may also cause gait beginning issues in PD patients. Additionally, spinal control mechanisms are involved in SI and are important in executing motor commands (Fling et al. 2013; Honeine et al. 2016). Lira et al. (2020) show that loss of presynaptic inhibition (i.e., facilitation) during deficient APAs in the freezers may be due to FoG. Their results support the notion that lack of central inhibition of stance posture to allow stepping to commence is reflected in a loss of presynaptic inhibition in freezers, which is associated with more FoG severity. During the APA, only freezers showed facilitation rather than inhibition. This conclusion may be explained by three different aberrant changes in the processes underpinning the FoG in Parkinson's disease: abnormal proprioceptive inputs, abnormal supraspinal motor commands, or a mix of both abnormalities (i.e., deficits in sensorimotor integration).

3.4 Reactive Response

Smaller than normal strength postural responses are the most evident abnormality of postural responses to external perturbations in PD. PD is linked to less stability in response to perturbations in all directions, but the difference is greatest in the backward direction (Carpenter et al. 2004). Postural response latencies are normal in individuals with PD despite the slowness to produce force and lower peak forces of postural responses, especially as the disease advances with increased bradykinesia and more falls. These findings also show that bradykinesia and akinesia significantly impact the rate of development and maximal muscular torque exerted by postural reactions (Horak et al. 1996).

Patients with Parkinson's disease also exhibit abnormal muscle activation patterns in response to postural responses (Horak et al. 1992). For example, when using a feet-in-place ankle strategy, control subjects and subjects with Parkinson's disease activate a distal to proximal pattern of muscle activation. Contrary to control subjects, individuals with PD frequently add short bursts of muscular activation in muscles that are antagonistic to one another, causing coactivation that would stiffen joints. However, antagonist muscle activation stops when people with Parkinson's disease take levodopa medication. As a result, postural response muscle activation temporal patterns resemble normal, and joints are less rigid, even though levodopa does not increase the magnitude of postural responses to normal levels.

When responding to large forward perturbations, the body uses its hip, trunk, and knee joints and its ankle joint motion to restore equilibrium quickly. Individuals with PD keep their joints rigid, whereas control patients flex the hips/trunk in response to sideways perturbations and the knees in reaction to backward perturbations (Horak et al. 2005). Due to this rigidity, the same disturbances result in higher body displacements.

Lack of protective arm reaching reflexes to postural perturbations may also be caused by rigidity of joint mobility and decreased kinesthesia. When their postural stability is disturbed, healthy participants respond quickly and automatically, flexing or extending and abducting their shoulders. Hence, their arms lead their trunks and are likely to fall to the ground if their postural adjustments are ineffective. However, despite the PD's much earlier beginning of deltoid muscle responses, this did not provide any practical protection since the PD's arm motions were irregularly oriented, slower moving, and had smaller peak displacements (Carpenter et al. 2004).

3.5 Quiet Standing

For balance, the postural control system needs information about the relative positions of body segments and the magnitude of forces acting on the body. Simply put, the task of the postural control system is to maintain the horizontal projection of the individual's CoM within the base of support defined by the area of the base of the feet during static upright posture. Stability is achieved by generating moments of force on the joints of the body to counteract the effect of gravity or any other disturbance in a continuous and dynamic process during the permanence of a certain posture. For balance regulation, the system needs information about the relative positions of body segments and the magnitude of forces acting on the body. For this, the body can use three sensors: somatosensory, visual, and vestibular. These sensors act in a complex, integrated, redundant, and differentiated way for each disturbance on the human body. The passive properties of the musculoskeletal system, especially the rigidity of biological structures, also play an important role in maintaining balance. The control of postural balance in a person is highly affected by the nature of the task, the environmental conditions, and the sensory information available. Body sway during upright posture is usually investigated using a force plate, a measuring instrument on which subjects remain standing during the experiments. The most common variable to analyze this oscillation is the position of the center of pressure (CoP), the point of application of the resultant forces acting on the support surface. The displacement of the CoP represents a sum of the actions of the postural control system and the force of gravity.

Sway is also increased in many neurological diseases impacting sensory and/or motor systems. Postural control during quiet standing is controversial in individuals with PD. While some studies show that postural sway can be abnormal before levodopa administration (Mancini et al. 2011) and even in the prodromal period (Maetzler and Hausdorff 2012), other studies show that it is similar to normal (Bronte-Stewart et al. 2002), at least at the earlier stages (Frenklach et al. 2009). The increased mediolateral sway in PD has been a more persistent result (Mitchell et al. 1995), associated with a history of falls (Visser et al. 2008). It is still unclear what the underlying mechanisms involved in freezing of gait (FoG) and postural control are. Quiet standing is largely similar between freezers and nonfreezeres, but the susceptibility for backward postural instability in people with FoG increases (for a review, see Bekkers et al. 2018a).

Individuals with severe PD frequently suffer from standing on unstable surfaces (Frenklach et al. 2009), which suggests reduced proprioception (Bronstein et al. 1990). In addition, there is growing evidence that the basal ganglia are related to reduced tactile discrimination (Tagliabue et al. 2009), suggesting that supraspinal circuits could be responsible for abnormal sensory integration for postural control in PD.

Patients with PD, whether in the ON or OFF state, exhibit an initially rigid strategy independent of the conditions of their surface support, changing gradually over time and with several repeats (Horak et al. 1992). This rigidity also explains why PD patients frequently experience less improvement than anticipated with manual assistance and find it challenging to benefit from canes and walkers without extensive training.

3.6 Prolonged Standing

Daily activities like waiting in line or talking to someone usually involve standing for a long time (more than a few minutes). In working conditions as well as in activities of daily living, some people remain standing for a long time, mostly confined to a small area. In the natural standing posture, people usually adopt asymmetrical postures and change their body position periodically while maintaining a relatively fixed body posture. The continuous, slow, low-amplitude oscillation is commonly interrupted by postural changes characterized by rapid and wide-ranging movements (Duarte et al. 2000). These postural changes are thought to be performed to alleviate the discomfort caused by psychological (tension, mental stress, and decreased motivation and concentration) and physiological factors (increase of venous pooling in the lower extremities, occlusion of blood flow, vertigo, muscular fatigue, and increased joint pressure) (Cavanagh et al. 1987; Edwards 1988; Kraemer et al. 2000). In this sense, during prolonged standing, postural shifts and an increase in body sway are considered effective responses of the postural control system to carry out the activity with the least amount of effort. Less postural changes and sway areas are conservative strategies (de Freitas et al. 2009). Aging reduces the adaptability of postural control, resulting in fewer postural shifts and, as a result, less body sway (Duarte and Sternad 2008). Even though standing still for a long period is a common task in people's daily lives, it can present a high degree of difficulty for those with balance deficits, such as individuals with PD. Moretto et al. (2021) compared the postural control during the prolonged posture of individuals with PD to healthy individuals. Their results show that individuals with PD reduced body sway (smaller sway area), faster sway, and postural control complexity compared with the control group during prolonged standing. These results indicate impaired adaptation and a lack of postural control to produce a sufficient response in individuals with PD during prolonged standing. Faster sway during prolonged standing in Parkinson's disease is an aging effect, implying a decline in neural processing and delayed muscle activation, which affects feedback and feedforward control and reduces sway.

3.7 Dynamic Balance During Walking

The gait pattern altered by PD is one of the motor signs that most affect the quality of life of individuals with PD, with more than 50% of falls in these individuals occurring during gait (Lord et al. 2017). The gait of individuals with PD tends to be slower, characterized by narrow and short steps, flexed trunk, little or no arm swing, and slow and spasmodic turning (Mancini et al. 2019). When analyzing the spatiotemporal parameters of gait in individuals with PD compared to healthy individuals (Fig. 3.4), studies have shown decreased speed (Cheng et al. 2014; Mondal et al. 2019; Morris et al. 1999), increased number of steps (Mondal et al. 2019), decrease in step length (Cheng et al. 2014; Mondal et al. 2019), and stride length (Mondal et al. 2019); shorter duration of the swing phase and unipedal stance phase (Mondal et al. 2019); and longer duration of the double support phase (Mondal et al. 2019; Sofuwa et al. 2005). Galna et al. (2015) evaluated 16 gait parameters of individuals with PD in the early stages and identified that 12 were different between older adults with PD and neurologically healthy older adults. These differences are in the measures of step speed, step length, variability of swing phase duration, stance phase duration, stride duration, stride length, and stride width,

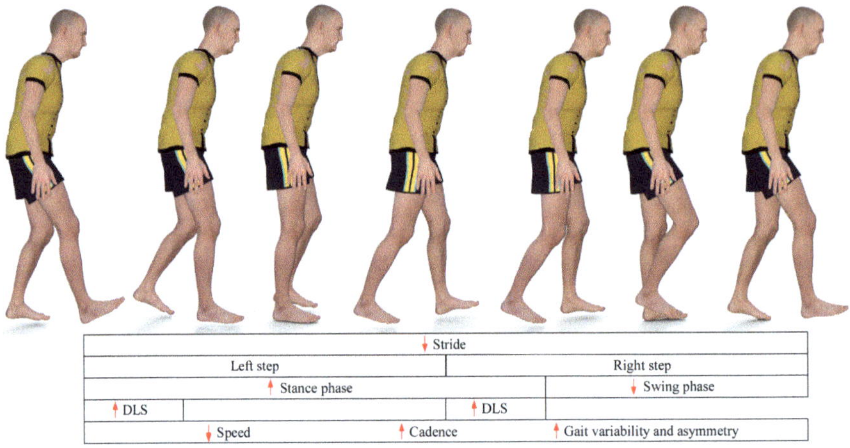

Fig. 3.4 Subdivision of the gait cycle illustrating the differences in the spatiotemporal parameters of the gait of an individual with Parkinson's disease to neurologically healthy individuals. The red color indicates worse indices for PD. The up arrow indicates an increase, while the down arrow indicates a decrease

in stride duration time and stride phase duration, stance and asymmetry of step duration, and swing phase and stance phase.

Regarding kinematic parameters of gait in PD, studies indicate a lower range of motion of the hips in the coronal plane and of the pelvic obliquity, lower range of flexion-extension of the knees with a high degree of flexion in the initial contact, and the stance phase; and greater ankle dorsiflexion during the stance phase, compared to healthy individuals (Pistacchi et al. 2017; Sale et al. 2013; Zanardi et al. 2021). Our group analyzed the joint angles of the lower limb of individuals with PD compared to healthy individuals (Fig. 3.5). Our results show greater differences in the distal joints to the proximal ones. Albani et al. (2014) found that individuals in the early stage of PD have lower ankle power during terminal stance compared to healthy individuals and a lower maximum dorsiflexion moment during the stance phase. Morris et al. (2005) found a significant reduction in range of motion in the sagittal plane of the hip, knee, and ankle joints, in addition to lower pelvic obliquity and rotation, and reduced hip abduction in individuals with PD compared to healthy older adults.

Individuals with PD have deficits in gait at a constant speed (Hausdorff 2009), and in gait planning and modulation in challenging tasks (Rochester et al. 2014; Vitorio et al. 2010, 2013, 2014). This indicates that PD affects gait automation, defined as the ability of the nervous system to stably control gait with minimal use of attentional resources (Clark et al. 2014; Schneider and Shiffrin 1977). Furthermore, it is suggested that this deficiency in gait automation is responsible for increased variability in spatiotemporal gait parameters and greater activation of prefrontal cortical regions during locomotion of individuals with PD compared to

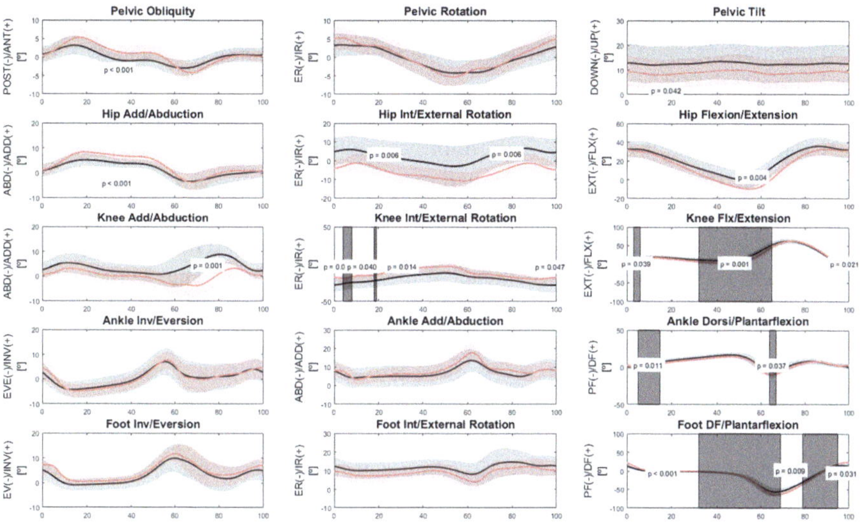

Fig. 3.5 Angular kinematics during the gait of the Parkinson's disease group in the ON medication (grey) and the age-matched healthy elderly control group (red). Mean and standard deviation of the angular displacement of the pelvis, hip, knee, and foot joints in the three joint axes of the most affected limb. The hatched areas indicate the difference between the groups

their neurologically healthy peers (Hausdorff 2009; Maidan et al. 2016; Peterson and Horak 2016).

Regarding the effect of the medication on the gait of individuals with PD, in general, during the ON state of the medication, the self-selected gait speed increases (Curtze et al. 2015; Mondal et al. 2019). However, during gait in the OFF state, individuals with PD show greater cortical activation when compared to healthy individuals (Stuart et al. 2019), which could be a possible explanation for the change in gait speed as an effect of medication. In a recent study, Orcioli-Silva et al. (2021) identified lower activation of the prefrontal cortex in subjects with PD in the OFF state during obstacle clearance compared to activation during unobstructed gait, an effect that was reversed after drug ingestion. On the other hand, the authors identified that individuals with PD in the ON state showed an increase in prefrontal cortex activation during gait with overcoming obstacles when compared with activation during unobstructed gait. Furthermore, it was identified that medication-related changes (i.e., ON–OFF) in posterior parietal cortex activation were associated with medication-related changes in step length for both walking conditions. These results suggest that dopaminergic medication may facilitate the recruitment of attentional-executive resources from the prefrontal cortex in challenging situations (overcoming obstacles), in addition to increasing sensorimotor integration (increased activation of the posterior parietal cortex) during gait. Furthermore, Gilat et al. (2017) showed that individuals in the ON state of the medication show improvement in gait automaticity, with greater cerebellar activation at times of greater gait variability.

Furthermore, their results showed a high correlation between attentional cortical participation and step time variability in the OFF state, without the same correlation occurring in the ON state.

On the other hand, some authors argue that the increase in gait speed in the ON state may result not only from improved automatism but also increased task engagement from dopaminergic signaling in the perception of effort (Albin and Leventhal 2017). For example, Mazzoni et al. (2007) showed that individuals with PD, compared to healthy individuals, are more likely to move more slowly (lower motor vigor) when the energy demands of movement increase. That is, they present a possible alteration in the mechanism of perception of effort and reward to healthy individuals. On the other hand, Chong et al. (2015) showed that individuals with PD in the ON state were willing to invest more effort in a given task than individuals in the OFF state. The authors then concluded that medication motivated the behavior of individuals who chose to exert greater effort while performing the task. Thus, regardless of which dopaminergic signaling pathway acts (motor or motivational), there is a consensus in the literature that drug treatment can increase gait speed in individuals with PD.

Despite the consensus in the literature about the improvement in gait speed in people with PD treated with levodopa, when specifically analyzing spatiotemporal parameters such as stride length, cadence, step initiation, step time, gait phases, among others, the results are not always consistent. For example, Curtze et al. (2015) showed that individuals increased stride speed and length in the ON state without influencing cadence, step initiation, double support time, and swing time. Mondal et al. (2019) showed that in the ON state, there was a decrease in the number of steps, an increase in step length, an increase in stride length, and a decrease in double support time. In the same study, the medication did not affect cadence, single-leg stance time, step time, cycle time, swing time, and base of support width. This variety of results makes it difficult to conclude anything about the effect of dopaminergic medication on spatiotemporal gait parameters. While Curtze et al. (2015) argued that levodopa improves gait without changing parameters related to its dynamic stability, Mondal et al. (2019) discussed that parameters related to gait rhythm are resistant to levodopa and parameters that require caloric expenditure (i.e., stride length) are sensitive to medication.

The pathological substrates of gait impairments in PD are still not fully understood. However, it is suggested that the degeneration of dopaminergic neurons in the substantia nigra with projection to the basal nuclei affects gait, as these brain structures are responsible, among other functions, for context-specific adaptation, motor coordination, body posture, and gait automation (Mancini et al. 2019). Furthermore, alterations in the cerebral cortex in the generation and propagation of neural rhythms have also been identified as involved in the locomotor deficits of individuals with PD (Arbuthnott and Garcia-Munoz 2009; Galna et al. 2015; la Fougere et al. 2010).

3.8 Conclusion

Poor balance control and postural instability are among the most incapacitating aspects of PD. Balance issues in PD go beyond simply being a symptom of motor failure. Balance issues in PD result from dysfunctions in the cognitive, emotional, sensory, and autonomic systems. The regulation of balance involves a wide variety of brain networks. The basal ganglia play four key functions: posture-movement coupling, energization/scaling (vigor), automatization, and context-dependent adaptability. Multiple components of balance regulation, such as standing balance, reactive postural responses, anticipatory postural changes before gait initiation, and dynamic balance during walking and turning, are all affected by PD:

(a) FoG and falls go together frequently. More balance impairments are present in freezers than in nonfreezers, especially during dynamic balance during walking and anticipatory postural adjustments.
(b) PD have delayed, hypometric, and rigid anticipatory postural adjustments before step initiations.
(c) Although PD has normal onset latency, postural responses to external perturbations are weak and slow.
(d) Limits of stability are lowered, particularly in the forward motion.
(e) The axial tone is higher than usual, and increased neck tone is associated with decreased functional movement.
(f) Proprioception may be difficult for PD to integrate for standing balance swiftly.
(g) Parkinsonian gait shows slow, shuffling, and variable steps.

Conflict of Interest Statement None.

References

Albani, G., Cimolin, V., Fasano, A., Trotti, C., Galli, M., & Mauro, A. (2014, Apr–Jun). "Masters and servants" in parkinsonian gait: a three-dimensional analysis of biomechanical changes sensitive to disease progression. *Functional Neurology, 29*(2), 99–105. https://www.ncbi.nlm.nih.gov/pubmed/25306119

Albin, R. L., & Leventhal, D. K. (2017, Jul). The missing, the short, and the long: Levodopa responses and dopamine actions. *Annals of neurology, 82*(1), 4–19. doi: https://doi.org/10.1002/ana.24961

Arbuthnott, G., & Garcia-Munoz, M. (2009, Dec). Dealing with the devil in the detail – some thoughts about the next model of the basal ganglia. *Parkinsonism & related disorders, 15 Suppl 3*, S139–142. doi: https://doi.org/10.1016/S1353-8020(09)70801-6

Bastide, M. F., Meissner, W. G., Picconi, B., Fasano, S., Fernagut, P. O., Feyder, M., Francardo, V., Alcacer, C., Ding, Y., Brambilla, R., Fisone, G., Jon Stoessl, A., Bourdenx, M., Engeln, M., Navailles, S., De Deurwaerdere, P., Ko, W. K., Simola, N., Morelli, M., Groc, L., Rodriguez, M. C., Gurevich, E. V., Quik, M., Morari, M., Mellone, M., Gardoni, F., Tronci, E., Guehl, D., Tison, F., Crossman, A. R., Kang, U. J., Steece-Collier, K., Fox, S., Carta, M., Angela Cenci, M., & Bezard, E. (2015, Sep). Pathophysiology of L-dopa-induced motor and non-motor complications in Parkinson's disease. *Progress in neurobiology, 132*, 96–168. doi: https://doi.org/10.1016/j.pneurobio.2015.07.002

Bekkers, E. M. J., Dijkstra, B. W., Dockx, K., Heremans, E., Verschueren, S. M. P., & Nieuwboer, A. (2017, Jul). Clinical balance scales indicate worse postural control in people with Parkinson's disease who exhibit freezing of gait compared to those who do not: A meta-analysis. *Gait & Posture, 56*, 134–140. doi: https://doi.org/10.1016/j.gaitpost.2017.05.009

Bekkers, E. M. J., Dijkstra, B. W., Heremans, E., Verschueren, S. M. P., Bloem, B. R., & Nieuwboer, A. (2018a, Nov). Balancing between the two: Are freezing of gait and postural instability in Parkinson's disease connected? *Neuroscience and biobehavioral reviews, 94*, 113–125. doi: https://doi.org/10.1016/j.neubiorev.2018.08.008

Bekkers, E. M. J., Van Rossom, S., Heremans, E., Dockx, K., Devan, S., Verschueren, S. M. P., & Nieuwboer, A. (2018b). Adaptations to Postural Perturbations in Patients With Freezing of Gait. *Frontiers in Neurology, 9*, 540. doi: https://doi.org/10.3389/fneur.2018.00540

Bloem, B. R., Hausdorff, J. M., Visser, J. E., & Giladi, N. (2004, Aug). Falls and freezing of gait in Parkinson's disease: a review of two interconnected, episodic phenomena. *Movement Disorders, 19*(8), 871–884. doi: https://doi.org/10.1002/mds.20115

Breniere, Y., Cuong Do, M., & Bouisset, S. (1987, Mar). Are dynamic phenomena prior to stepping essential to walking? *Journal of motor behavior, 19*(1), 62–76. doi: https://doi.org/10.1080/00222895.1987.10735400

Breniere, Y., & Do, M. C. (1991, Dec). Control of gait initiation. *Journal of motor behavior, 23*(4), 235–240. doi: https://doi.org/10.1080/00222895.1991.9942034

Bronstein, A. M., Hood, J. D., Gresty, M. A., & Panagi, C. (1990, Jun). Visual control of balance in cerebellar and parkinsonian syndromes. *113 (Pt 3)*, 767–779. doi: https://doi.org/10.1093/brain/113.3.767

Bronte-Stewart, H. M., Minn, A. Y., Rodrigues, K., Buckley, E. L., & Nashner, L. M. (2002, Sep). Postural instability in idiopathic Parkinson's disease: the role of medication and unilateral pallidotomy. *125*(Pt 9), 2100–2114. http://www.ncbi.nlm.nih.gov/pubmed/12183355

Butler, J. S., Fearon, C., Killane, I., Waechter, S. M., Reilly, R. B., & Lynch, T. (2017, Mar). Motor preparation rather than decision-making differentiates Parkinson's disease patients with and without freezing of gait. *128*(3), 463–471. doi: https://doi.org/10.1016/j.clinph.2016.12.019

Carpenter, M. G., Allum, J. H., Honegger, F., Adkin, A. L., & Bloem, B. R. (2004, Sep). Postural abnormalities to multidirectional stance perturbations in Parkinson's disease. *75*(9), 1245–1254. doi: https://doi.org/10.1136/jnnp.2003.021147 75/9/1245 [pii]

Carpinella, I., Crenna, P., Calabrese, E., Rabuffetti, M., Mazzoleni, P., Nemni, R., & Ferrarin, M. (2007, Dec). Locomotor function in the early stage of Parkinson's disease. *IEEE transactions on neural systems and rehabilitation engineering: a publication of the IEEE Engineering in Medicine and Biology Society, 15*(4), 543–551. doi: https://doi.org/10.1109/TNSRE.2007.908933

Cavanagh, P. R., Rodgers, M. M., & Iiboshi, A. (1987, Apr). Pressure distribution under symptom-free feet during barefoot standing. *7*(5), 262–276. doi: https://doi.org/10.1177/107110078700700502

Cheng, K. Y., Lin, W. C., Chang, W. N., Lin, T. K., Tsai, N. W., Huang, C. C., Wang, H. C., Huang, Y. C., Chang, H. W., Lin, Y. J., Lee, L. H., Cheng, B. C., Kung, C. T., Chang, Y. T., Su, C. M., Chiang, Y. F., Su, Y. J., & Lu, C. H. (2014, Jan). Factors associated with fall-related fractures in Parkinson's disease. *Parkinsonism & related disorders, 20*(1), 88–92. doi: https://doi.org/10.1016/j.parkreldis.2013.09.024

Chong, T. T., Bonnelle, V., Manohar, S., Veromann, K. R., Muhammed, K., Tofaris, G. K., Hu, M., & Husain, M. (2015, Aug). Dopamine enhances willingness to exert effort for reward in Parkinson's disease. *Cortex; a journal devoted to the study of the nervous system and behavior, 69*, 40–46. doi: https://doi.org/10.1016/j.cortex.2015.04.003

Clark, D. J., Christou, E. A., Ring, S. A., Williamson, J. B., & Doty, L. (2014, Nov). Enhanced somatosensory feedback reduces prefrontal cortical activity during walking in older adults. *The journals of gerontology. Series A, Biological sciences and medical sciences, 69*(11), 1422–1428. doi: https://doi.org/10.1093/gerona/glu125

Coelho, D. B., Ribeiro de Souza, C., de Lima-Pardini, A. C., Treza, R. C., Shida, T. K. F., Silva-Batista, C., & Teixeira, L. A. (2021, Feb). Is freezing of gait correlated with postural control in patients with moderate-to-severe Parkinson's disease? *European Journal of Neuroscience, 53*(4), 1189–1196. doi: https://doi.org/10.1111/ejn.15010

Cohen, R. G., Klein, K. A., Nomura, M., Fleming, M., Mancini, M., Giladi, N., Nutt, J. G., & Horak, F. B. (2014). Inhibition, executive function, and freezing of gait. *Journal of Parkinson's Disease, 4*(1), 111–122. doi: https://doi.org/10.3233/JPD-130221

Cohen, R. G., Nutt, J. G., & Horak, F. B. (2011, Jun). Errors in postural preparation lead to increased choice reaction times for step initiation in older adults. *The Journals of Gerontology Series A Biological Sciences and Medical Sciences, 66*(6), 705–713. doi: https://doi.org/10.1093/gerona/glr054

Cohen, R. G., Nutt, J. G., & Horak, F. B. (2017). Recovery from Multiple APAs Delays Gait Initiation in Parkinson's Disease. *Frontiers in Human Neuroscience, 11*, 60. doi: https://doi.org/10.3389/fnhum.2017.00060

Curtze, C., Nutt, J. G., Carlson-Kuhta, P., Mancini, M., & Horak, F. B. (2015, Sep). Levodopa Is a Double-Edged Sword for Balance and Gait in People With Parkinson's Disease. *Movement Disorders, 30*(10), 1361–1370. doi: https://doi.org/10.1002/mds.26269

de Freitas, P. B., Freitas, S. M., Duarte, M., Latash, M. L., & Zatsiorsky, V. M. (2009, Aug). Effects of joint immobilization on standing balance. *28*(4), 515–528. doi: https://doi.org/10.1016/j.humov.2009.02.001

de Lima-Pardini, A. C., Coelho, D. B., Nucci, M. P., Boffino, C. C., Batista, A. X., de Azevedo Neto, R. M., Silva-Batista, C., Barbosa, E. R., Cohen, R. G., Horak, F. B., Teixeira, L. A., & Amaro, E., Jr. (2020). Brain networks associated with anticipatory postural adjustments in Parkinson's disease patients with freezing of gait. *NeuroImage. Clinical, 28*, 102461. doi: https://doi.org/10.1016/j.nicl.2020.102461

de Souza Fortaleza, A. C., Mancini, M., Carlson-Kuhta, P., King, L. A., Nutt, J. G., Chagas, E. F., Freitas, I. F. J., & Horak, F. B. (2017, Jul). Dual task interference on postural sway, postural transitions and gait in people with Parkinson's disease and freezing of gait. *Gait & Posture, 56*, 76–81. doi: https://doi.org/10.1016/j.gaitpost.2017.05.006

Del Tredici, K., & Braak, H. (2016, Feb). Review: Sporadic Parkinson's disease: development and distribution of alpha-synuclein pathology. *Neuropathology and applied neurobiology, 42*(1), 33–50. doi: https://doi.org/10.1111/nan.12298

Delval, A., Krystkowiak, P., Blatt, J. L., Labyt, E., Destee, A., Derambure, P., & Defebvre, L. (2005, Nov-Dec). [Differences in anticipatory postural adjustments between self-generated and triggered gait initiation in 20 healthy subjects]. *Neurophysiologie clinique = Clinical neurophysiology, 35*(5–6), 180–190. doi: https://doi.org/10.1016/j.neucli.2006.01.002 (Caracterisation des ajustements posturaux lors d'une initiation de la marche declenchee par un stimulus sonore et autocommandee chez 20 sujets sains.)

Donker, S. F., Roerdink, M., Greven, A. J., & Beek, P. J. (2007, Jul). Regularity of center-of-pressure trajectories depends on the amount of attention invested in postural control. *Experimental Brain Research, 181*(1), 1–11. doi: https://doi.org/10.1007/s00221-007-0905-4

Duarte, M., Harvey, W., & ZatsiorskyVm. (2000, Nov). Stabilographic analysis of unconstrained standing. *43*(11), 1824–1839. doi: https://doi.org/10.1080/00140130050174491

Duarte, M., & Sternad, D. (2008, Nov). Complexity of human postural control in young and older adults during prolonged standing. *191*(3), 265–276. doi: https://doi.org/10.1007/s00221-008-1521-7

Edwards, R. H. (1988). Hypotheses of peripheral and central mechanisms underlying occupational muscle pain and injury. *57*(3), 275–281. doi: https://doi.org/10.1007/BF00635985

Ehgoetz Martens, K. A., Hall, J. M., Georgiades, M. J., Gilat, M., Walton, C. C., Matar, E., Lewis, S. J. G., & Shine, J. M. (2018, Apr 1). The functional network signature of heterogeneity in freezing of gait. *Brain, 141*(4), 1145–1160. doi: https://doi.org/10.1093/brain/awy019

Fling, B. W., Cohen, R. G., Mancini, M., Carpenter, S. D., Fair, D. A., Nutt, J. G., & Horak, F. B. (2014). Functional reorganization of the locomotor network in Parkinson patients with freezing of gait. *9*(6), e100291. doi: https://doi.org/10.1371/journal.pone.0100291

Fling, B. W., Cohen, R. G., Mancini, M., Nutt, J. G., Fair, D. A., & Horak, F. B. (2013, Aug). Asymmetric pedunculopontine network connectivity in parkinsonian patients with freezing of gait. *Brain: a journal of neurology, 136*(Pt 8), 2405–2418. doi: https://doi.org/10.1093/brain/awt172

Fonoff, E. T., de Lima-Pardini, A. C., Coelho, D. B., Monaco, B. A., Machado, B., Pinto de Souza, C., Dos Santos Ghilardi, M. G., & Hamani, C. (2019). Spinal Cord Stimulation for Freezing of Gait: From Bench to Bedside. *Frontiers in Neurology, 10*, 905. doi: https://doi.org/10.3389/fneur.2019.00905

Frenklach, A., Louie, S., Koop, M. M., & Bronte-Stewart, H. (2009, Feb 15). Excessive postural sway and the risk of falls at different stages of Parkinson's disease. *24*(3), 377–385. doi: 10.1002/mds.22358

Gallardo, M. J., Cabello, J. P., Corrales, M. J., Torres-Donaire, J., Bravo, J. J., Talavera, M. P., Leon, A., & Vaamonde-Gamo, J. (2018, Oct). Freezing of gait in Parkinson's disease: functional neuroimaging studies of the frontal lobe. *Neurological Research, 40*(10), 900–905. doi: https://doi.org/10.1080/01616412.2018.1484985

Galna, B., Lord, S., Burn, D. J., & Rochester, L. (2015, Mar). Progression of gait dysfunction in incident Parkinson's disease: impact of medication and phenotype. *Movement disorders: official journal of the Movement Disorder Society, 30*(3), 359–367. doi: https://doi.org/10.1002/mds.26110

Gao, C., Liu, J., Tan, Y., & Chen, S. (2020). Freezing of gait in Parkinson's disease: pathophysiology, risk factors and treatments. *Translational neurodegeneration, 9*, 12. doi: https://doi.org/10.1186/s40035-020-00191-5

Gazibara, T., Kisic Tepavcevic, D., Svetel, M., Tomic, A., Stankovic, I., Kostic, V. S., & Pekmezovic, T. (2017, Oct). Near-falls in people with Parkinson's disease: Circumstances, contributing factors and association with falling. *Clinical Neurology and Neurosurgery, 161*, 51–55. doi: https://doi.org/10.1016/j.clineuro.2017.08.008

Gilat, M., Bell, P. T., Ehgoetz Martens, K. A., Georgiades, M. J., Hall, J. M., Walton, C. C., Lewis, S. J. G., & Shine, J. M. (2017, May 15). Dopamine depletion impairs gait automaticity by altering cortico-striatal and cerebellar processing in Parkinson's disease. *NeuroImage, 152*, 207–220. doi: https://doi.org/10.1016/j.neuroimage.2017.02.073

Gilat, M., Ehgoetz Martens, K. A., Miranda-Dominguez, O., Arpan, I., Shine, J. M., Mancini, M., Fair, D. A., Lewis, S. J. G., & Horak, F. B. (2018, Mar 15). Dysfunctional Limbic Circuitry Underlying Freezing of Gait in Parkinson's Disease. *Neuroscience, 374*, 119–132. doi: https://doi.org/10.1016/j.neuroscience.2018.01.044

Gilat, M., Shine, J. M., Walton, C. C., O'Callaghan, C., Hall, J. M., & Lewis, S. J. G. (2015). Brain activation underlying turning in Parkinson's disease patients with and without freezing of gait: a virtual reality fMRI study. *NPJ Parkinsons Disease, 1*, 15020. doi: https://doi.org/10.1038/npjparkd.2015.20

Hallett, M. (2008). The intrinsic and extrinsic aspects of freezing of gait. *Movement Disorders, 23 Suppl 2*, S439–443. doi: https://doi.org/10.1002/mds.21836

Hass, C. J., Waddell, D. E., Fleming, R. P., Juncos, J. L., & Gregor, R. J. (2005, Nov). Gait initiation and dynamic balance control in Parkinson's disease. *Archives of physical medicine and rehabilitation, 86*(11), 2172–2176. doi: https://doi.org/10.1016/j.apmr.2005.05.013

Hausdorff, J. M. (2009, Jun). Gait dynamics in Parkinson's disease: common and distinct behavior among stride length, gait variability, and fractal-like scaling. *Chaos, 19*(2), 026113. doi: https://doi.org/10.1063/1.3147408

Hawkes, C. H., Del Tredici, K., & Braak, H. (2010, Feb). A timeline for Parkinson's disease. *Parkinsonism & related disorders, 16*(2), 79–84. doi: https://doi.org/10.1016/j.parkreldis.2009.08.007

He, S. Q., Dum, R. P., & Strick, P. L. (1995, May). Topographic organization of corticospinal projections from the frontal lobe: motor areas on the medial surface of the hemisphere. *The Journal of Neuroscience, 15*(5 Pt 1), 3284–3306. https://www.ncbi.nlm.nih.gov/pubmed/7538558

Hoehn, M. M., & Yahr, M. D. (1967, May). Parkinsonism: onset, progression and mortality. *Neurology, 17*(5), 427–442. doi: https://doi.org/10.1212/wnl.17.5.427

Honeine, J. L., Schieppati, M., Crisafulli, O., & Do, M. C. (2016). The Neuro-Mechanical Processes That Underlie Goal-Directed Medio-Lateral APA during Gait Initiation. *Frontiers in human neuroscience, 10*, 445. doi: https://doi.org/10.3389/fnhum.2016.00445

Horak, F. B., Dimitrova, D., & Nutt, J. G. (2005). Direction-specific postural instability in subjects with Parkinson's disease. *Experimental Neurology, 193*(2), 504–521. doi: https://doi.org/10.1016/j.expneurol.2004.12.008

Horak, F. B., Frank, J., & Nutt, J. (1996, Jun). Effects of dopamine on postural control in parkinsonian subjects: scaling, set, and tone. *Journal of Neurophysiology, 75*(6), 2380–2396. http://www.ncbi.nlm.nih.gov/pubmed/8793751 (J. Neurophysiol.)

Horak, F. B., Nutt, J. G., & Nashner, L. M. (1992, Aug). Postural inflexibility in parkinsonian subjects. *Journal of the Neurological Sciences, 111*(1), 46–58. doi: https://doi.org/10.1016/0022-510X(92)90111-W (J. Neurol. Sci.)

Jacobs, J. V., & Horak, F. B. (2007). Cortical control of postural responses. *Journal of Neural Transmission, 114*(10), 1339–1348. doi: https://doi.org/10.1007/s00702-007-0657-0 (J. Neural Transm.)

Jacobs, J. V., Lou, J. S., Kraakevik, J. A., & Horak, F. B. (2009a, Dec 1). The supplementary motor area contributes to the timing of the anticipatory postural adjustment during step initiation in participants with and without Parkinson's disease. *Neuroscience, 164*(2), 877–885. doi: https://doi.org/10.1016/j.neuroscience.2009.08.002 (Neuroscience)

Jacobs, J. V., Nutt, J. G., Carlson-Kuhta, P., Stephens, M., & Horak, F. B. (2009b, Feb). Knee trembling during freezing of gait represents multiple anticipatory postural adjustments. *Experimental Neurology, 215*(2), 334–341. doi: https://doi.org/10.1016/j.expneurol.2008.10.019

Jankovic, J. (2008, Apr). Parkinson's disease: clinical features and diagnosis. *Journal of neurology, neurosurgery, and psychiatry, 79*(4), 368–376. doi: https://doi.org/10.1136/jnnp.2007.131045

Jellinger, K. A. (2003, Sep). Alpha-synuclein pathology in Parkinson's and Alzheimer's disease brain: incidence and topographic distribution – a pilot study. *Acta neuropathologica, 106*(3), 191–201. doi: https://doi.org/10.1007/s00401-003-0725-y

Kalia, L. V., & Lang, A. E. (2015). Parkinson's disease. *386*(9996), 896–912. doi: https://doi.org/10.1016/s0140-6736(14)61393-3

Kehagia, A. A., Barker, R. A., & Robbins, T. W. (2013). Cognitive impairment in Parkinson's disease: the dual syndrome hypothesis. *Neuro-degenerative diseases, 11*(2), 79–92. doi: https://doi.org/10.1159/000341998

Keizer, K., & Kuypers, H. G. (1989). Distribution of corticospinal neurons with collaterals to the lower brain stem reticular formation in monkey (Macaca fascicularis). *Experimental Brain Research, 74*(2), 311–318. doi: https://doi.org/10.1007/BF00248864

Kraemer, W. J., Volek, J. S., Bush, J. A., Gotshalk, L. A., Wagner, P. R., Gomez, A. L., Zatsiorsky, V. M., Duarte, M., Ratamess, N. A., Mazzetti, S. A., & Selle, B. J. (2000, Nov). Influence of compression hosiery on physiological responses to standing fatigue in women. *32*(11), 1849–1858. doi: https://doi.org/10.1097/00005768-200011000-00006

la Fougere, C., Zwergal, A., Rominger, A., Forster, S., Fesl, G., Dieterich, M., Brandt, T., Strupp, M., Bartenstein, P., & Jahn, K. (2010, May 1). Real versus imagined locomotion: a [18F]-FDG PET-fMRI comparison. *50*(4), 1589–1598. doi: https://doi.org/10.1016/j.neuroimage.2009.12.060 S1053-8119(09)01349-4 [pii]

Lagravinese, G., Pelosin, E., Bonassi, G., Carbone, F., Abbruzzese, G., & Avanzino, L. (2018, Apr). Gait initiation is influenced by emotion processing in Parkinson's disease patients with freezing. *Movement disorders: official journal of the Movement Disorder Society, 33*(4), 609–617. doi: https://doi.org/10.1002/mds.27312

Lewis, S. J., & Barker, R. A. (2009, Jun). A pathophysiological model of freezing of gait in Parkinson's disease. *Parkinsonism & related disorders, 15*(5), 333–338. doi: https://doi.org/10.1016/j.parkreldis.2008.08.006

Lira, J. L. O., Ugrinowitsch, C., Coelho, D. B., Teixeira, L. A., de Lima-Pardini, A. C., Magalhaes, F. H., Barbosa, E. R., Horak, F. B., & Silva-Batista, C. (2020, Apr). Loss of presynaptic inhibition for step initiation in parkinsonian individuals with freezing of gait. *The Journal of Physiology, 598*(8), 1611–1624. doi: https://doi.org/10.1113/JP279068

Lord, S., Galna, B., Yarnall, A. J., Morris, R., Coleman, S., Burn, D., & Rochester, L. (2017, Nov). Natural history of falls in an incident cohort of Parkinson's disease: early evolution, risk and protective features. *Journal of neurology, 264*(11), 2268–2276. doi: https://doi.org/10.1007/s00415-017-8620-y

Lord, S. R., Bindels, H., Ketheeswaran, M., Brodie, M. A., Lawrence, A. D., Close, J. C. T., Whone, A. L., Ben-Shlomo, Y., & Henderson, E. J. (2020). Freezing of Gait in People with Parkinson's Disease: Nature, Occurrence, and Risk Factors. *Journal of Parkinson's Disease, 10*(2), 631–640. doi: https://doi.org/10.3233/JPD-191813

Maetzler, W., & Hausdorff, J. M. (2012, Apr 15). Motor signs in the prodromal phase of Parkinson's disease. *Movement disorders: official journal of the Movement Disorder Society, 27*(5), 627–633. doi: https://doi.org/10.1002/mds.24973

Maidan, I., Nieuwhof, F., Bernad-Elazari, H., Reelick, M. F., Bloem, B. R., Giladi, N., Deutsch, J. E., Hausdorff, J. M., Claassen, J. A., & Mirelman, A. (2016, Nov). The role of the frontal lobe in complex walking among patients with Parkinson's disease and healthy older adults: an fNIRS study. *Neurorehabilitation and Neural Repair, 30*(10), 963–971. doi: https://doi.org/10.1177/1545968316650426 (Neurorehabil. Neural Repair.)

Mancini, M., Horak, F. B., Zampieri, C., Carlson-Kuhta, P., Nutt, J. G., & Chiari, L. (2011, Aug). Trunk accelerometry reveals postural instability in untreated Parkinson's disease. *Parkinsonism & Related Disorders, 17*(7), 557–562. doi: https://doi.org/10.1016/j.parkreldis.2011.05.010

Mancini, M., Nutt, J. G., & Horak, F. B. (2019). *Balance dysfunction in Parkinson's disease: basic mechanisms to clinical management*. Cambridge, MA, USA. https://doi.org/10.1016/C2017-0-00054-X

Mancini, M., Zampieri, C., Carlson-Kuhta, P., Chiari, L., & Horak, F. B. (2009, Sep). Anticipatory postural adjustments prior to step initiation are hypometric in untreated Parkinson's disease: an accelerometer-based approach. *16*(9), 1028–1034. doi: https://doi.org/10.1111/j.1468-1331.2009.02641.x

Massion, J. (1992). Movement, posture and equilibrium: interaction and coordination. *Progress in Neurobiology, 38*(1), 35–56. doi: 0301-0082(92)90034-C [pii]

Matsui, H., Udaka, F., Miyoshi, T., Hara, N., Tamaura, A., Oda, M., Kubori, T., Nishinaka, K., & Kameyama, M. (2005, Oct). Three-dimensional stereotactic surface projection study of freezing of gait and brain perfusion image in Parkinson's disease. *Movement Disorders, 20*(10), 1272–1277. doi: https://doi.org/10.1002/mds.20520

Mazzoni, P., Hristova, A., & Krakauer, J. W. (2007, Jul 4). Why don't we move faster? Parkinson's disease, movement vigor, and implicit motivation. *The Journal of neuroscience: the official journal of the Society for Neuroscience, 27*(27), 7105–7116. doi: https://doi.org/10.1523/JNEUROSCI.0264-07.2007

Michalowska, M., Fiszer, U., Krygowska-Wajs, A., & Owczarek, K. (2005, Oct-Dec). Falls in Parkinson's disease. Causes and impact on patients' quality of life. *Functional neurology, 20*(4), 163–168. http://www.ncbi.nlm.nih.gov/pubmed/16483454

Mitchell, S. L., Collins, J. J., De Luca, C. J., Burrows, A., & Lipsitz, L. A. (1995, Sep 8). Open-loop and closed-loop postural control mechanisms in Parkinson's disease: increased mediolateral activity during quiet standing. *197*(2), 133–136. doi: 030439409511924L [pii]

Mondal, B., Choudhury, S., Banerjee, R., Chatterjee, K., Ghosal, S., Anand, S. S., & Kumar, H. (2019). Analysis of gait in Parkinson's disease reflecting the effect of l-DOPA. *Annals of Movement Disorders, 2*(1), 21–27. doi: https://doi.org/10.4103/AOMD.AOMD_19_18

Moretto, G. F., Santinelli, F. B., Penedo, T., Mochizuki, L., Rinaldi, N. M., & Barbieri, F. A. (2021, Jan). Prolonged Standing Task Affects Adaptability of Postural Control in People With Parkinson's Disease. *35*(1), 58–67. doi: https://doi.org/10.1177/1545968320971739

Morris, M., Iansek, R., McGinley, J., Matyas, T., & Huxham, F. (2005, Jan). Three-dimensional gait biomechanics in Parkinson's disease: evidence for a centrally mediated amplitude regulation disorder. *Movement Disorders, 20*(1), 40–50. doi: https://doi.org/10.1002/mds.20278

Morris, M. E., McGinley, J., Huxham, F., Collier, J., & Iansek, R. (1999). Constraints on the kinetic, kinematic and spatiotemporal parameters of gait in Parkinson's disease. *Human Movement Science, 18*(2–3), 461–483. doi: https://doi.org/10.1016/s0167-9457(99)00020-2

Nakano, K., Kayahara, T., Tsutsumi, T., & Ushiro, H. (2000, Sep). Neural circuits and functional organization of the striatum. *Journal of neurology, 247 Suppl 5*, V1-15. doi: https://doi.org/10.1007/pl00007778

Naugle, K. M., Hass, C. J., Bowers, D., & Janelle, C. M. (2012, Mar). Emotional state affects gait initiation in individuals with Parkinson's disease. *Cognitive, affective & behavioral neuroscience, 12*(1), 207–219. doi: https://doi.org/10.3758/s13415-011-0071-9

Nieuwboer, A., & Giladi, N. (2013, Sep 15). Characterizing freezing of gait in Parkinson's disease: models of an episodic phenomenon. *Movement disorders: official journal of the Movement Disorder Society, 28*(11), 1509–1519. doi: https://doi.org/10.1002/mds.25683

Nutt, J. G., Bloem, B. R., Giladi, N., Hallett, M., Horak, F. B., & Nieuwboer, A. (2011, Aug). Freezing of gait: moving forward on a mysterious clinical phenomenon. *The Lancet Neurology, 10*(8), 734–744. doi: https://doi.org/10.1016/S1474-4422(11)70143-0

Nutt, J. G., Carter, J. H., Lea, E. S., & Sexton, G. J. (2002, Jun). Evolution of the response to levodopa during the first 4 years of therapy. *Annals of neurology, 51*(6), 686–693. doi: https://doi.org/10.1002/ana.10189

Nutt, J. G., Carter, J. H., Van Houten, L., & Woodward, W. R. (1997, Sep). Short- and long-duration responses to levodopa during the first year of levodopa therapy. *Annals of neurology, 42*(3), 349–355. doi: https://doi.org/10.1002/ana.410420311

Orcioli-Silva, D., Vitorio, R., Nobrega-Sousa, P., Beretta, V. S., Conceicao, N. R. D., Oliveira, A. S., Pereira, M. P., & Gobbi, L. T. B. (2021, May). Cortical Activity Underlying Gait Improvements Achieved With Dopaminergic Medication During Usual Walking and Obstacle Avoidance in Parkinson Disease. *Neurorehabilitation and neural repair, 35*(5), 406–418. doi: https://doi.org/10.1177/15459683211000736

Pahapill, P. A., & Lozano, A. M. (2000, Sep). The pedunculopontine nucleus and Parkinson's disease. *Brain: a journal of neurology, 123 (Pt 9)*, 1767–1783. doi: https://doi.org/10.1093/brain/123.9.1767

Pelykh, O., Klein, A. M., Botzel, K., Kosutzka, Z., & Ilmberger, J. (2015, Sep). Dynamics of postural control in Parkinson patients with and without symptoms of freezing of gait. *Gait & Posture, 42*(3), 246–250. doi: https://doi.org/10.1016/j.gaitpost.2014.09.021

Peterson, D. S., & Horak, F. B. (2016, Mar). Neural Control of Walking in People with Parkinsonism. *Physiology, 31*(2), 95–107. doi: https://doi.org/10.1152/physiol.00034.2015

Peterson, D. S., Pickett, K. A., Duncan, R., Perlmutter, J., & Earhart, G. M. (2014). Gait-related brain activity in people with Parkinson disease with freezing of gait. *9*(3), e90634. doi: https://doi.org/10.1371/journal.pone.0090634

Peterson, D. S., Van Liew, C., Stuart, S., Carlson-Kuhta, P., Horak, F. B., & Mancini, M. (2020). Relating Parkinson freezing and balance domains: A structural equation modeling approach. *Parkinsonism & Related Disorders*. doi: https://doi.org/10.1016/j.parkreldis.2020.08.027

Pistacchi, M., Gioulis, M., Sanson, F., De Giovannini, E., Filippi, G., Rossetto, F., & Zambito Marsala, S. (2017, Jan/Mar). Gait analysis and clinical correlations in early Parkinson's disease. *Functional Neurology, 32*(1), 28–34. doi: https://doi.org/10.11138/fneur/2017.32.1.028

Plotnik, M., Giladi, N., & Hausdorff, J. M. (2012). Is freezing of gait in Parkinson's disease a result of multiple gait impairments? Implications for treatment. *Parkinson's disease, 2012*, 459321. doi: https://doi.org/10.1155/2012/459321

Plotnik, M., & Hausdorff, J. M. (2008). The role of gait rhythmicity and bilateral coordination of stepping in the pathophysiology of freezing of gait in Parkinson's disease. *Movement Disorders, 23 Suppl 2*, S444–450. doi: https://doi.org/10.1002/mds.21984

Rochester, L., Galna, B., Lord, S., & Burn, D. (2014, Apr 18). The nature of dual-task interference during gait in incident Parkinson's disease. *Neuroscience, 265*, 83–94. doi: https://doi.org/10.1016/j.neuroscience.2014.01.041

Sale, P., De Pandis, M. F., Vimercati, S. L., Sova, I., Foti, C., Tenore, N., Fini, M., Stocchi, F., Albertini, G., Franceschini, M., & Galli, M. (2013, Apr). The relation between Parkinson's disease and ageing. Comparison of the gait patterns of young Parkinson's disease subjects with healthy elderly subjects. *European Journal of Physical and Rehabilitation Medicine, 49*(2), 161–167. https://www.ncbi.nlm.nih.gov/pubmed/22569487

Schaafsma, J. D., Balash, Y., Gurevich, T., Bartels, A. L., Hausdorff, J. M., & Giladi, N. (2003, Jul). Characterization of freezing of gait subtypes and the response of each to levodopa in Parkinson's disease. *European Journal of Neurology, 10*(4), 391–398. doi: https://doi.org/10.1046/j.1468-1331.2003.00611.x

Schlenstedt, C., Mancini, M., Horak, F., & Peterson, D. (2017, Jul). Anticipatory Postural Adjustment During Self-Initiated, Cued, and Compensatory Stepping in Healthy Older Adults and Patients With Parkinson Disease. *Archives of physical medicine and rehabilitation, 98*(7), 1316–1324 e1311. doi: https://doi.org/10.1016/j.apmr.2017.01.023

Schlenstedt, C., Mancini, M., Nutt, J., Hiller, A. P., Maetzler, W., Deuschl, G., & Horak, F. (2018). Are Hypometric Anticipatory Postural Adjustments Contributing to Freezing of Gait in Parkinson's Disease? *Frontiers in Aging Neuroscience, 10*, 36. doi: https://doi.org/10.3389/fnagi.2018.00036

Schneider, W., & Shiffrin, R. M. (1977). Controlled and automatic human information processing: I. Detection, search, and attention. *84*(1), 1–66. doi: https://doi.org/10.1037/0033-295x.84.1.1

Shine, J. M., Matar, E., Ward, P. B., Bolitho, S. J., Gilat, M., Pearson, M., Naismith, S. L., & Lewis, S. J. (2013a, Apr). Exploring the cortical and subcortical functional magnetic resonance imaging changes associated with freezing in Parkinson's disease. *Brain, 136*(Pt 4), 1204–1215. doi: https://doi.org/10.1093/brain/awt049

Shine, J. M., Matar, E., Ward, P. B., Bolitho, S. J., Pearson, M., Naismith, S. L., & Lewis, S. J. (2013b). Differential neural activation patterns in patients with Parkinson's disease and freezing of gait in response to concurrent cognitive and motor load. *Plos One, 8*(1), e52602. doi: https://doi.org/10.1371/journal.pone.0052602

Shine, J. M., Matar, E., Ward, P. B., Frank, M. J., Moustafa, A. A., Pearson, M., Naismith, S. L., & Lewis, S. J. (2013c, Dec). Freezing of gait in Parkinson's disease is associated with functional decoupling between the cognitive control network and the basal ganglia. *Brain: a journal of neurology, 136*(Pt 12), 3671–3681. doi: https://doi.org/10.1093/brain/awt272

Smulders, K., Esselink, R. A., De Swart, B. J., Geurts, A. C., Bloem, B. R., & Weerdesteyn, V. (2014, Feb). Postural inflexibility in PD: does it affect compensatory stepping? *Gait & Posture, 39*(2), 700–706. doi: https://doi.org/10.1016/j.gaitpost.2013.10.003

Sofuwa, O., Nieuwboer, A., Desloovere, K., Willems, A. M., Chavret, F., & Jonkers, I. (2005, May). Quantitative gait analysis in Parkinson's disease: comparison with a healthy control group. *Archives of physical medicine and rehabilitation, 86*(5), 1007–1013. doi: https://doi.org/10.1016/j.apmr.2004.08.012

Stins, J. F., & Beek, P. J. (2011, Jun). Organization of voluntary stepping in response to emotion-inducing pictures. *Gait & posture, 34*(2), 164–168. doi: https://doi.org/10.1016/j.gaitpost.2011.04.002

Stuart, S., Belluscio, V., Quinn, J. F., & Mancini, M. (2019). Pre-frontal Cortical Activity During Walking and Turning Is Reliable and Differentiates Across Young, Older Adults and People With Parkinson's Disease. *Frontiers in neurology, 10*, 536. doi: https://doi.org/10.3389/fneur.2019.00536

Tagliabue, M., Ferrigno, G., & Horak, F. (2009, Feb 18). Effects of Parkinson's disease on proprioceptive control of posture and reaching while standing. *158*(4), 1206–1214. doi: S0306-4522(08)01763-6 [pii] https://doi.org/10.1016/j.neuroscience.2008.12.007

Takakusaki, K. (2017, Jan). Functional Neuroanatomy for Posture and Gait Control. *Journal of Movement Disorders, 10*(1), 1–17. doi: https://doi.org/10.14802/jmd.16062

Tarazi, F. I., Sahli, Z. T., Wolny, M., & Mousa, S. A. (2014, Nov). Emerging therapies for Parkinson's disease: from bench to bedside. *144*(2), 123–133. doi: https://doi.org/10.1016/j.pharmthera.2014.05.010

Thompson, P. D., & Marsden, C. D. (1995). Walking disorders. In W. G. Bradley, R. B. Daroff, & G. M. M. Fenichel, C. D. (Eds.), *Neurology in Clinical Practice – Principles of Diagnosis and Management* (pp. 321–334). Butterworth-Heinemann.

Vandenbossche, J., Deroost, N., Soetens, E., Coomans, D., Spildooren, J., Vercruysse, S., Nieuwboer, A., & Kerckhofs, E. (2012). Freezing of gait in Parkinson's disease: disturbances in automaticity and control. *Frontiers in Human Neuroscience, 6*, 356. doi: https://doi.org/10.3389/fnhum.2012.00356

Vercruysse, S., Spildooren, J., Heremans, E., Wenderoth, N., Swinnen, S. P., Vandenberghe, W., & Nieuwboer, A. (2014, Dec). The neural correlates of upper limb motor blocks in Parkinson's disease and their relation to freezing of gait. *Cerebral Cortex, 24*(12), 3154–3166. doi: https://doi.org/10.1093/cercor/bht170

Visser, J. E., Carpenter, M. G., van der Kooij, H., & Bloem, B. R. (2008, Nov). The clinical utility of posturography. *Clinical neurophysiology: official journal of the International Federation of Clinical Neurophysiology, 119*(11), 2424–2436. doi: https://doi.org/10.1016/j.clinph.2008.07.220

Vitorio, R., Lirani-Silva, E., Barbieri, F. A., Raile, V., Stella, F., & Gobbi, L. T. (2013, Jun). Influence of visual feedback sampling on obstacle crossing behavior in people with Parkinson's disease. *Gait & Posture, 38*(2), 330–334. doi: https://doi.org/10.1016/j.gaitpost.2012.12.019

Vitorio, R., Lirani-Silva, E., Pieruccini-Faria, F., Moraes, R., Gobbi, L. T., & Almeida, Q. J. (2014, Sep 26). Visual cues and gait improvement in Parkinson's disease: which piece of information is really important? *Neuroscience, 277*, 273–280. doi: https://doi.org/10.1016/j.neuroscience.2014.07.024

Vitorio, R., Pieruccini-Faria, F., Stella, F., Gobbi, S., & Gobbi, L. T. (2010, Jan). Effects of obstacle height on obstacle crossing in mild Parkinson's disease. *Gait & posture, 31*(1), 143–146. doi: https://doi.org/10.1016/j.gaitpost.2009.09.011

Wang, M., Jiang, S., Yuan, Y., Zhang, L., Ding, J., Wang, J., Zhang, J., Zhang, K., & Wang, J. (2016, Aug). Alterations of functional and structural connectivity of freezing of gait in Parkinson's disease. *263*(8), 1583–1592. doi: https://doi.org/10.1007/s00415-016-8174-4

Weiss, A., Herman, T., Giladi, N., & Hausdorff, J. M. (2015, Mar). New evidence for gait abnormalities among Parkinson's disease patients who suffer from freezing of gait: insights using a body-fixed sensor worn for 3 days. *Journal of Neural Transmission, 122*(3), 403–410. doi: https://doi.org/10.1007/s00702-014-1279-y

Welter, M. L., Do, M. C., Chastan, N., Torny, F., Bloch, F., du Montcel, S. T., & Agid, Y. (2007, Sep). Control of vertical components of gait during initiation of walking in normal adults and patients with progressive supranuclear palsy. *Gait & posture, 26*(3), 393–399. doi: https://doi.org/10.1016/j.gaitpost.2006.10.005

Yiou, E., Caderby, T., Delafontaine, A., Fourcade, P., & Honeine, J. L. (2017, Nov 18). Balance control during gait initiation: State-of-the-art and research perspectives. *World journal of orthopedics, 8*(11), 815–828. doi: https://doi.org/10.5312/wjo.v8.i11.815

Zanardi, A. P. J., da Silva, E. S., Costa, R. R., Passos-Monteiro, E., Dos Santos, I. O., Kruel, L. F. M., & Peyre-Tartaruga, L. A. (2021, Jan 12). Gait parameters of Parkinson's disease compared with healthy controls: a systematic review and meta-analysis. *Scientific Reports, 11*(1), 752. doi: https://doi.org/10.1038/s41598-020-80768-2

Chapter 4
Narrative Review on the Application of Additive Manufacturing in the Production of Upper Limb Orthoses

Maria Elizete Kunkel and Ana Clara Castro Pimentel Silva Araújo

Abstract The development of upper limb orthoses using additive manufacturing technology, also known as 3D printing, holds significant potential to enhance accessibility, availability, customization, and user acceptance of the devices. This manufacturing method offers advantages over conventional production methods, such as greater design freedom. 3D printing creates complex and customized geometries perfectly tailored to medical needs and individual user preferences. This technology results in orthoses that are more comfortable and aesthetically pleasing. Additionally, customization allows for creating bespoke orthoses for each user, considering specific anatomical characteristics. Moreover, additive manufacturing allows for rapid iteration and modification in orthosis design, facilitating a more agile and enhanced development process. Prototypes can be produced and tested quickly, with reduced manufacturing time and cost. The application of additive manufacturing in developing upper limb orthoses has tremendous potential to improve accessibility, availability, customization, and user acceptance. However, 3D modeling of a custom orthosis requires specific expertise, software, and equipment that limit the production of printed orthoses. Additive manufacturing is not ideal as a large-scale production method for orthoses because the final cost is similar to orthoses made from thermoplastic plates, and the production time is longer.

Keywords Additive manufacturing · Distal radius fracture · Upper limb orthosis · 3D printing

---

M. E. Kunkel (✉)
Graduate Program in Biomedical Engineering, Federal University of ABC – UFABC, São Bernardo do Campo (SP), Brazil
e-mail: Elizete.kunkel@unifesp.br

A. C. C. P. S. Araújo
Graduate Program in Biomedical Engineering, Center for Engineering, Modeling and Applied Social Sciences – CECS, Federal University of ABC – UFABC, São Bernardo do Campo (SP), Brazil

C. B. Lombello, P. A. da Ana (eds.), *Current Trends in Biomedical Engineering*,
https://doi.org/10.1007/978-3-031-38743-2_4

4.1 Introduction

Orthoses are orthopedic devices used to treat various conditions, such as musculoskeletal injuries, deformities, or functional impairments. These devices, composed of materials such as plastic, metal, and fabric, offer support, stability, alignment, or corrective measures for specific body parts (Spaulding et al. 2019). In addition to "orthosis," the term "exoskeleton" is frequently used to describe external devices that aid or enhance motor function. It is essential to note the distinctions between these two concepts. An exoskeleton typically comprises a robotic or mechanical device featuring a rigid structure, mechanical drives, sensors, and user interfaces. Exoskeletons are worn externally on the human body and provide structural and functional support, actively assisting in bodily movements. In contrast, orthoses are orthopedic devices that offer support, stability, or corrective measures for specific body parts without active intervention (Herr 2009; Kalita et al. 2021).

In recent years, thermoplastic polymers have gained significant popularity in orthosis production. This material exhibits malleability when heated and solidifies upon reaching room temperature. The malleable nature of thermoplastics facilitates practical orthoses manufacturing, allowing for an excellent fit to the patient's anatomy and enhanced comfort. Moreover, thermoplastic orthoses have revolutionized the nonsurgical treatment of upper limb fractures, maintaining limb functionality, protecting the fracture from excessive stress, and preventing joint deformities (Freeland et al. 2003; O'Brien and Thurn 2013). Traditionally, occupational therapists or healthcare professionals manually fabricate upper limb orthoses using thermoplastic material. The process involves placing the patient's hand and wrist on a thermoplastic plate and contouring it using a pen. Subsequently, the contoured piece is cut out and heated, accomplished through immersion in hot water or a thermal blower to achieve malleability. Due to the critical and individualized nature of joint positioning, this procedure necessitates execution by qualified professionals (Jacobs and Austin 2003). Thermoplastic materials offer the advantages of lightweight and sanitizable orthoses, and specific variants possess biodegradable characteristics contributing to their increased adoption in recent years (Francisco 2004). However, some drawbacks are associated with this process, including the time required for orthosis creation, the high cost of thermoplastic plates, and material wastage.

In recent years, numerous studies have explored the application of additive manufacturing to fabricate upper limb orthoses using thermoplastic materials. These studies demonstrate that orthoses manufactured through additive manufacturing possess similar qualities to those made from thermoplastic plates, including lightweight, sanitizable, low toxicity risks, and customizability. Furthermore, additive manufacturing allows for a broader range of material options, greater design flexibility, and reduced material waste (Palousek et al. 2014; Chen et al. 2017; Kadioglu et al. 2019; Chen et al. 2020). The poly (lactic acid) polymer (PLA) is the most utilized material in orthosis production, thanks to its biocompatible and biodegradable properties, making it a sustainable alternative to petroleum-based polymers (Han and Huang 2011). Additionally, PLA can be modified by incorporating organic and inorganic components to tailor its physical and chemical properties, resulting in composites better suited for specific applications (Oksman et al. 2003). This chapter provides a narrative review of the existing literature to assess the feasibility of utilizing additive manufacturing to produce customized upper limb orthoses.

4.2 Additive Manufacturing

Additive manufacturing, also known as 3D printing, is a revolutionary manufacturing technique that involves the layer-by-layer addition of materials through various processes (Gibson et al. 2021). These processes are classified based on the machine architecture and energy source employed for material processing. One such process is vat photopolymerization (VPP), which utilizes a liquid photopolymer contained in a vat. By selectively applying energy, specific regions of a part section are cured. Material extrusion (MEX), another prevalent process, involves material deposition through a nozzle, creating cross-sectional patterns that form the desired part. While "Fused Deposition Modeling" (FDM) is commonly used to refer to material extrusion, it is a protected trademark of Stratasys Company's manufacturing process. MEX is extensively employed in the medical field to produce prosthetics and orthoses (Bandyopadhyay et al. 2015). Each additive manufacturing process has distinct characteristics impacting factors such as part precision, material and mechanical properties, post-processing requirements, 3D printer size, and manufacturing costs. Initially, additive manufacturing focused on polymeric materials, waxes, and paper laminates, but it has since expanded to include composites, metals, and ceramics.

In the MEX process, the 3D printer head deposits and extrudes the material onto a table based on coordinates generated by computer-aided design (CAD) software. A thermoplastic polymer filament stored in a cartridge is the preferred material choice. The filament is melted using an electric heating system and then extruded through a nozzle, where it is deposited onto the printing table. Once deposited, the molten strand of material undergoes heat transfer, solidifying onto the previous layer and forming a solidified layer. This layer-by-layer process is repeated until the part is entirely fabricated (Gebhardt 2011; Wong and Hernandez 2012). The fabrication of upper limb orthoses using the material extrusion process necessitates a combination of engineering knowledge (regarding processes, software, and materials) and the involvement of occupational therapists or other healthcare professionals within the team. The following steps outline the process:

(1) Creating the orthosis digital model: This step involves utilizing 3D CAD modeling software. A bio-model of the upper limb structure, which will receive the orthosis, is generated by anthropometric parameters, 3D scanning techniques, or 3D reconstruction of the structure from medical imaging. Subsequently, the orthosis is customized and modeled based on the bio-model, considering the specific medical

requirements for the given case (Palousek et al. 2014). *(2) Conversion of the orthosis file to STL format and transmission to the 3D printer*: Most additive manufacturing processes employ the STL (STereoLithography) file format, derived from the first commercial additive manufacturing technology by 3D Systems. The STL format represents a simplified approach to describing a CAD model, focusing solely on its geometry, and approximating the model surfaces with triangular facets that serve as the foundation for layer calculations. At this stage, file manipulation may be necessary to rectify the part's size, position, and orientation on the printing table or platform. *(3) Configuration of 3D printer parameters and orthosis manufacturing:* The 3D printer must be appropriately configured before manufacturing. This step involves adjusting various parameters, including layer thickness, extruder nozzle temperature, printing speed, and infill type, considering the specific material and power source. Although the 3D printing process is mainly automated, basic monitoring is necessary to ensure the absence of errors such as material or power failures and software crashes**.** *(4) Post-processing of the orthosis:* Upon completion of the 3D printing process, the fabricated part is removed from the printer. If the part includes support structures, these also need to be removed. Additionally, post-processing steps such as curing, priming, painting, or heat treatment may be employed to enhance the texture and surface finish of the orthosis. By following these steps, the fabrication of upper limb orthoses using the material extrusion process can be effectively executed, combining engineering expertise and the specialized knowledge of healthcare professionals (Gibson et al. 2021).

4.3 Upper Limb Orthoses Fabricated Through Advanced Manufacturing

Contemporary orthopedics and rehabilitation practices are increasingly focused on personalized care, which can be effectively implemented through digital technologies such as 3D scanning and additive manufacturing. These advanced techniques enable treatment planning and the on-demand fabrication of orthotic devices tailored to meet the specific needs of individual patients (Chen et al. 2016; Graham et al. 2020). Additive manufacturing has found significant application in the production of upper limb orthoses, particularly in immobilization, spasticity management, fracture treatment, and deformity correction. Standard methods for acquiring 3D models of the upper limb orthosis include 3D scanning, 3D reconstruction of medical images, and digital calipers. Among the various additive manufacturing processes employed, MEX represents the most widely adopted method (90.9%), followed by selective laser sintering (SLS) (13.63%). As for materials, the predominant choice is PLA, followed by acrylonitrile butadiene styrene (ABS), thermoplastic polyurethane (TPU), thermoplastic elastomer (TPE), and polyamide (PA) (Table 4.1).

Table 4.1 Studies on the use of additive manufacturing in the production of upper limb orthosis

References	Application	N	Material	Process	Geometry acquisition	Results
Palousek et al. (2014)	Wrist immobilization	1	ABS	MEX	3D scanner	The technologies enabled the rapid fabrication of a customized orthosis and can be an alternative solution to the standard orthosis design method
Wong et al. (2015)	Phalangeal immobilization	13	ABS	MEX	Digital Pachymeter	The orthoses fit securely without pain or discomfort. They allowed a full active range of motion of the ring finger's proximal interphalangeal joint, extending the distal interphalangeal joint
Baronio et al. (2016)	Hand spasticity treatment	–	ABS	MEX	3D scanner	The orthosis was geometrically satisfactory, with high precision and low cost
Cano (2017)	Wrist immobilization	–	PLA	MEX	–	The orthosis based on the measurements in the image bank had a good size and was reliable, demonstrating the method's feasibility
Chen et al. (2017)	DRF treatment	10	Polypropylene and PA	SLS	CT and MRI	The orthosis met the orthopedic requirements for treating a fracture. The orthosis-maintained alignment of the fractured bone and immobilized the forearm during healing without loss of reduction
Chang et al. (2018)	Neurofacilitation of phalanx extension post-stroke	1	PLA	MEX	3D scanner	After one month of training with the 3D printed orthosis, the patient could open his hand and perform functional training tasks that he could not before
Kim et al. (2018)	Wrist immobilization	20	TPU	MEX	3D scanner	The wrist orthosis scored better than the commercial orthosis. Wrist pain was reduced, and there was no significant difference in the ability to perform tasks. The 3D wrist orthosis was better because of its design, lightness, and comfort
Lee et al. (2019)	Wrist support for patient with hemiparesis	1	TPE	MEX	3D scanner	The wrist dorsiflexion angle maintained significantly decreased the fatigue associated with typing. The patient was delighted and could attach a connector to the device

(continued)

Table 4.1 (continued)

References	Application	N	Material	Process	Geometry acquisition	Results
Nam et al. (2018)	Treatment of deformities in fingers after burns	3	PLA and TPU	MEX	3D scanner	All patients were satisfied with the practical performance and comfort of orthoses. The most significant advantage for post-burn patients is the adjustment of the orthoses to meet the needs of the hand
Portnova et al. (2018)	Assist in hand closure	3	PLA and ABS	MEX	–	The orthosis received high ratings for comfort, aesthetics, ease of fabrication, and customization. User testing highlighted improving performance during daily activities and increasing grip strength
Wang et al. (2018)	Treatment of hand spasticity in a post-stroke patient	13	PLA	MEX	Sand mold and 3D scanner	Most of the patients were satisfied, and none of them experienced extra pain. Grip strength, hand function, and range of motion improved while muscle tension decreased
Dudley et al. (2021)	Hand exoskeleton for post-stroke	1	PLACTIVE™	MEX	–	There was greater extensor activation in the affected arm and less flexor activation during exoskeleton use.
Kadioglu et al. (2019)	Treatment of fractured hands and fingers	25	PLA	MEX	–	All patients recovered fully and returned quickly to their activities of daily living. The orthoses were more comfortable, economical, and had better aesthetics
Sari et al. (2020)	Preventing the evolution of finger deformities	–	ABS	MEX	3D scanner	The orthosis can be easily put on and removed, ensuring the correct positioning of the distal and proximal interphalangeal joints during use. There are advantages in terms of aesthetics, time, and customization
Yan et al. (2019)	DRF treatment	–	PLA	MEX	3D scanner	The orthosis has a topology that uses 40% less material than the conventional design. In load application simulations, the orthosis showed strain/mobility values within the requirements for the treatment of DRF.

Zhang et al. (2019)	Treatment of wrist spasticity in a post-stroke patient	40	Photoactivated resin	MEX	3D scanner	Compared to thermoplastic orthoses, 3D orthoses showed reduced spasticity/swelling of the wrist, improving its motor function and passive extension range in the patients
Chae et al. (2020)	Wrist immobilization of patients with neuropathy	3	TPU	MEX	CT	The patients were satisfied with the fit and aesthetics of the orthosis, and there was an improvement in the function of the limbs
Chen et al. (2020)	DRF treatment	60	PA	SLS	CT and MRI	The orthosis maintained the alignment of the fractured bone and immobilized the forearm without a secondary reduction. The clinical questionnaire showed high scores, as did the functional wrist evaluation
Chu et al. (2020)	Thumb support	1	TPE	MEX	3D scanner	The questionnaire on usability regarding fit, comfort, and efficiency while performing functional tasks revealed the highest score on the average satisfaction level
Portnoy et al. (2020)	Preventing the evolution of finger deformities	–	ABS	MEX	Digital pachymeter	The volunteers reported more satisfaction with the fit, preparation, and aesthetics of the 3D-printed orthosis than the manual preparation method
Popescu et al. (2020)	Hand immobilization	–	PLA	MEX	Digital pachymeter	Based on measurements of the patient's hand, specific parameters of the splint were estimated to automate the design process based on virtual 3D models with significantly reduced time/cost in the manufacturing process
Toth et al. (2020)	Hand spasticity in post-stroke patient	6	TPU and PA	SLSMEX	Digital pachymeter and measuring tape	The orthosis increased the functionality of the users' hands with spasticity, enabling them to perform daily living and self-care tasks

ABS Acrylonitrile butadiene styrene; *CT* Computed tomography; *DRF* Distal radius fracture; *MEX* Material extrusion; *MRI* Magnetic resonance imaging; *PA* Polyamide; *PLA* Poly (lactic acid); *SLS* Selective laser sintering; *TPE* Thermoplastic elastomer; *TPU* Thermoplastic polyurethane

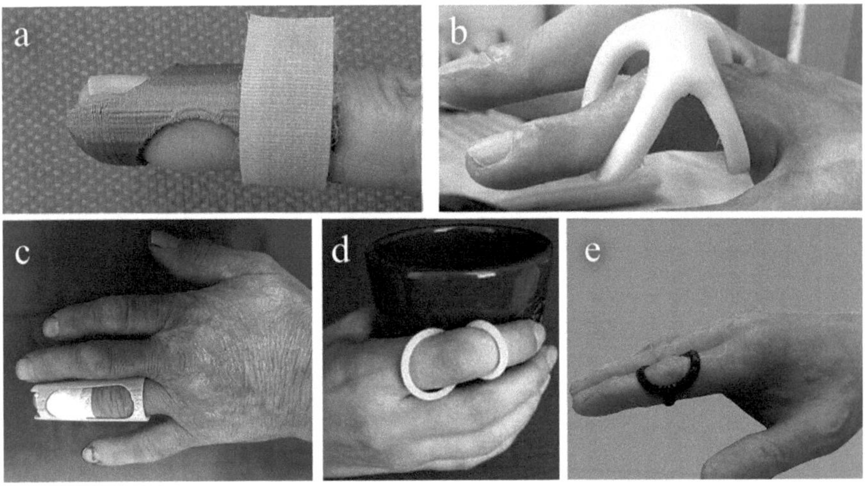

Fig. 4.1 Examples of finger orthoses produced by additive manufacturing. The image credits belong to the respective authors of the following studies: (**a**) Wong (2015); (**b**) Nam et al. (2018); (**c**) Kadioglu et al. (2019); (**d**) Sari et al. (2020); and (**e**) Portnoy et al. (2020)

4.4 Finger Orthoses

Digital manufacturing techniques have facilitated the development of various finger orthoses aimed at immobilizing deformities associated with conditions such as rheumatoid arthritis, fractures, and burns. The acquisition of finger geometry typically involves the utilization of calipers and 3D scanning, while the manufacturing process commonly employed is MEX.

For instance, Wong (2015) designed a custom orthosis utilizing ABS material to immobilize a hammer toe (Fig. 4.1a). Obtaining the patient's anthropometric data using digital calipers achieved a secure and comfortable fit of the orthosis on the finger. The digital manufacturing approach significantly reduced material usage and manufacturing time compared to traditional methods. Nam et al. (2018) produced flexible and rigid orthoses for treating finger deformities in burn victims using TPU and PLA materials, respectively (Fig. 4.1b). Digital manufacturing enabled the production of orthopedic finger devices that are low cost, lightweight, durable, and easily replaceable, surpassing the limitations of conventional production processes.

In another study, Kadioglu et al. (2019) utilized custom PLA orthoses manufactured digitally to treat finger fractures (Fig. 4.1c). In all cases examined, patients recovered and could resume their daily activities. The printed orthoses were more comfortable compared to commercially available alternatives. Sari et al. (2020) developed custom orthotic rings to prevent the progression of finger deformities caused by rheumatoid arthritis (Fig. 4.1d). The orthoses, created through 3D hand scanning, exhibited excellent aesthetic appearance, lightweight design, and perfect fit to the patient's hand. The digital manufacturing process allowed for producing these orthoses quickly and at a lower cost than conventional methods. Furthermore, Portnoy et al. (2020) conducted a study comparing the production of finger orthoses using conventional and digital approaches among 36 occupational therapy students. Participants reported greater satisfaction with the fit and aesthetics of the digitally manufactured orthoses due to their reduced weight compared to those produced using conventional methods (Fig. 4.1e).

4.5 Wrist-Hand Orthoses

Using 3D scanning and additive manufacturing techniques has shown considerable potential in developing customized wrist-hand orthoses, targeting a range of conditions, including musculoskeletal injuries, deformities, and functional impairments. The production of personalized orthoses with enhanced precision and reduced cost represents a promising avenue for wrist and hand joint rehabilitation, ultimately leading to improved therapeutic outcomes and heightened patient satisfaction. Moreover, the integration of reverse engineering techniques, such as 3D scanning, enables the fabrication of orthoses tailored to specific therapeutic requirements, thereby enhancing the effectiveness of rehabilitation therapy. Baronio et al. (2016) employed a reverse engineering process to produce a personalized hand orthosis using ABS material to promote the rehabilitation of post-stroke patients' limbs (Fig. 4.2a). A low-cost 3D scanner was used to capture the anatomy to achieve high precision in orthosis fabrication. The results were considered geometrically satisfactory, with low cost and high precision. Accurate and detailed image acquisition is crucial for high comfort and tolerability, enabling physicians to identify pressure zones and create orthoses responsive to therapeutic needs.

In 2018, Wang et al. produced a personalized hand orthosis using a sand mold and a 3D scanner for finger shape acquisition (Fig. 4.2b). The orthosis was 3D printed with PLA filament and used to prevent and treat finger spasms in 13 patients. The participants evaluated the orthoses as comfortable, with no pain or skin allergies aggravation. Furthermore, the orthosis increased grip strength, improved hand function and range of motion, and reduced muscle tension.

Chang et al. (2018) developed a neurofacilitation-oriented orthosis using PLA filament to promote the extension of the distal phalanges in a post-stroke patient (Fig. 4.2c). Over one month, the 33-year-old volunteer underwent a functional training program, acquiring the ability to perform previously deemed impossible manual tasks, such as stacking cones and grasping/releasing blocks. The orthosis in question assisted in hand opening by utilizing elastic cords that leverage elbow flexion and extension to enable hand opening and closing. Notably, this orthosis holds potential for rehabilitation treatment in cases of stroke where the manual function is compromised. The CAD model of the orthopedic device and its additive manufacturing demonstrated efficacy in designing and fabricating a customized assistive device.

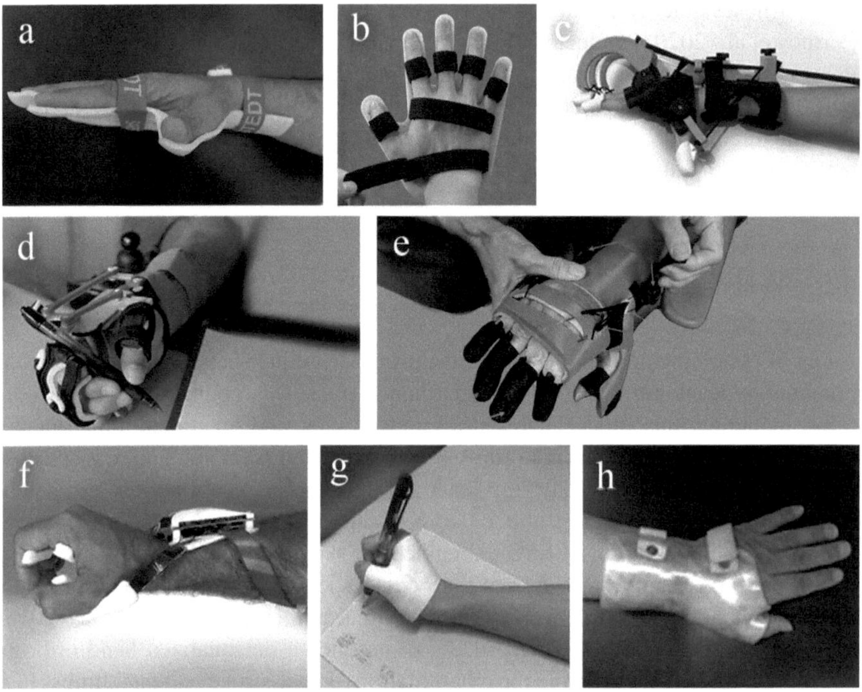

Fig. 4.2 Examples of hand orthoses produced by additive manufacturing. The image credits belong to the respective authors of the following studies: (**a**) Baronio et al. (2016); (**b**) Wang et al. (2018); (**c**) Chang et al. (2018); (**d**) Portnova et al. (2018); (**e**) Dudley et al. (2021); (**f**) Zhang et al. (2019); (**g**) Chu et al. (2020) and (**h**) Chae et al. (2020)

Portnova et al. (2018) developed a wrist-activated hand orthosis model using PLA and ABS materials for three individuals with spinal cord injuries (Fig. 4.2d). The orthosis exhibited favorable characteristics such as lightweight design and received high scores in comfort, aesthetics, ease of fabrication, and customization. The tests highlighted the orthosis's ability to enhance user performance during tasks and increase grip strength. The hand orthosis showed promise to improve functionality and quality of life for individuals with spinal cord injuries. Dudley et al. (2021) assessed the functional and neuromuscular changes resulting from using an orthosis specifically developed for a 67-year-old volunteer who had experienced a stroke 17 years prior (Fig. 4.2e). The volunteer exhibited satisfactory functional outcomes and obtained high scores on usability questionnaires, particularly in volume, weight, fit, and ease of use. The patient had increased extensor muscle activation in the affected limb and decreased flexor muscle activation compared to the patient's performance without using the device.

Zhang et al. (2019) compared the effects of two types of wrist orthoses on the treatment of flexor spasticity in 40 patients with chronic stroke (Fig. 4.2f). One type of orthosis was fabricated using the conventional method with thermoplastic material, while the other was produced through additive manufacturing using a photosensitive resin. The second orthosis yielded superior therapeutic outcomes, such as a significant reduction in wrist spasticity and swelling and improvements in wrist motor function and passive extension range of motion. The 3D scanning technology enabled the orthoses to adapt more precisely to the curves of the human body, providing patients with greater comfort even after prolonged periods of use. Toth et al. (2020) developed a personalized active orthosis using the SLS process and shape memory alloy nitinol for six patients with post-stroke spasticity (Fig. 4.2g). The aim was to enhance hand functionality enabling the patients to perform daily living and self-care tasks. Utilizing personalized orthosis yielded highly beneficial results, significantly improving hand functionality. This improvement allowed for completing everyday activities and self-care tasks that were previously challenging due to spastic paralysis. Additionally, the developed orthosis showcased positive characteristics such as easy assembly, lightweight design, and cost-effectiveness, making it a viable solution for home use.

Chu et al. (2020) developed a thumb orthosis model to treat a person with carpometacarpal osteoarthritis and pain (Fig. 4.2g). The volunteer underwent a usability questionnaire in which aspects such as fit, comfort, and efficiency were assessed during functional tasks. The results revealed a high satisfaction level reported by the volunteer. The orthosis model proved to be an effective solution for addressing the specific thumb conditions of the individual in question. The usability evaluation conducted via the questionnaire showed that the device achieved positive outcomes in terms of proper fit, comfort during use, and efficiency in performing functional tasks. Chae et al. (2020) developed personalized devices for three patients with neuropathy (Fig. 4.2h). Overall, the patients expressed satisfaction with the aesthetics and fit of the printed orthoses. The study emphasizes the importance of customization in manufacturing devices for patients with peripheral nerve injuries. By considering the individual condition of each patient, the devices were able to meet the specific needs of everyone. These findings underscore the relevance of the personalized approach in orthosis manufacturing for patients with neuropathy resulting from peripheral nerve injury.

4.6 Orthoses for the Treatment of Wrist Injuries and Fractures

A study by Palousek et al. (2014) presented a methodology for producing wrist immobilization orthoses utilizing ABS material. The approach incorporated digital technologies, including 3D scanning, reverse engineering, and additive manufacturing (Fig. 4.3a). Although the orthosis was not subjected to clinical testing, the pilot study showcased the promising application of digital technologies in orthosis fabrication, enabling data capture without direct contact and expediting the patient's reintegration into their social environment. This methodology is an

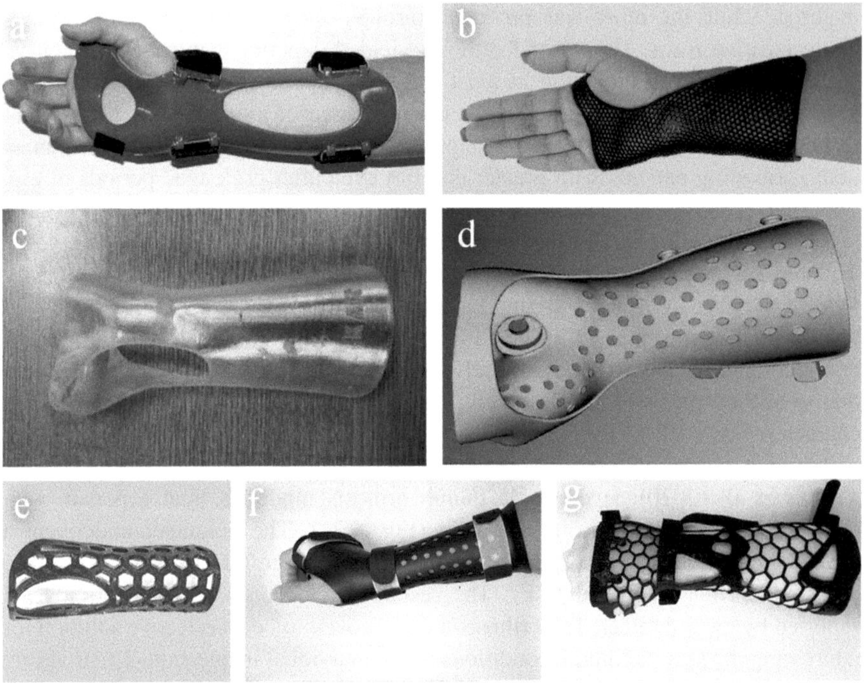

Fig. 4.3 Examples of orthoses for treating wrist injuries and fractures produced by additive manufacturing. The image credits belong to the respective authors of the following studies: (**a**) Palousek et al. (2014); (**b**) Cano (2017); (**c**) Kim et al. (2018); (**d**) Lee et al. (2019); (**e**) Popescu et al. (2020); (**f**) Chen et al. (2017) and (**g**) Yan et al. (2019)

innovative approach to orthosis manufacturing, harnessing the potential of digital technologies. While the pilot study lacked clinical trials, preliminary findings underscored the potential of digital technologies in orthosis production, setting the stage for developing more effective, personalized, and expeditiously integrated devices for patients. This approach holds significant implications for rehabilitation, enhancing individuals' quality of life and facilitating their social engagement.

Cano (2017) conducted a study describing the parameterization and fabrication of a thermo moldable orthosis for wrist immobilization utilizing PLA material through the MEX process (Fig. 4.3b). The orthosis design was based on measurements derived from a repository of hand images, and the author demonstrated the feasibility of production through additive manufacturing. In a study by Kim et al. (2018), a wrist orthosis was produced using the MEX printing process on TPU material and compared to a commercial wrist orthosis (Fig. 4.3c). The printed orthosis scored higher on two items of a comprehensive patient satisfaction questionnaire comprising 20 factors. Participants expressed high satisfaction with the orthosis design and its comfortable fit owing to the adherence of TPU material to the skin. Furthermore, the printed orthosis was lighter than the commercial counterpart. The study revealed that the 3D-printed and commercial wrist orthoses effectively alleviated wrist pain among patients. Moreover, no significant difference between the two groups was observed in the patient's ability to perform functional tasks. However, the 3D-printed wrist orthosis garnered superior ratings due to its design, lightweight nature, and comfort provided by the TPU material.

Lee et al. (2019) devised a customized orthosis model using 3D printing techniques for a 19-year-old male with right-sided hemiparesis resulting from a traumatic subdural hematoma after a fall. The orthosis was equipped with a connector to accommodate accessories, enabling the patient to perform daily activities such as eating, writing, and typing (Fig. 4.3d). The patient expressed contentment with the orthosis fit, finding it comfortable and user-friendly. In a study conducted by Popescu et al. (2020), the feasibility of rapid production of personalized flat hand orthoses using PLA material was analyzed (Fig. 4.3e). The production process involved an application that received data from 11 hand measurements provided by the physician, and these values were utilized to calculate orthosis parameters and automatically generate the design. The study demonstrated a significant reduction in production time and cost compared to traditional approaches, as it eliminated the need for hand scanning, manual data processing, and orthosis design. The 3D printing time for the orthosis was shorter in this case due to its flat structure.

Chen et al. (2017) conducted an initial experiment involving the treatment of forearm radius fractures in 10 patients using custom orthoses printed with polypropylene and polyamide materials (Fig. 4.3f). All patients preferred the custom fit feature, facilitating comfortable contact with the injured arm. Velcro straps integrated into the orthosis allowed adjustments to accommodate forearm swelling during the early inflammatory phase of the fracture. Throughout the seven-week test period, the printed orthosis fulfilled the orthopedic requirements for fracture treatment by maintaining bone alignment and immobilizing the forearm during the healing process. There were no instances of reduction loss or orthosis breakage during the treatment. In 2020, Chen et al. expanded this evaluation to include 60 patients and conducted finite element method simulations to assess the application of the orthosis above. The orthosis effectively preserved bone alignment and immobilized the forearm during healing. None of the patients experienced a secondary reduction, and no orthosis breakage occurred. The clinical effectiveness assessment questionnaire, completed by the surgeon for each patient, revealed high scores, as did the wrist functional assessment.

Yan et al. (2019) developed an optimized orthosis design for treating forearm radius fractures, reducing material usage by 40% compared to traditional designs (Fig. 4.3g). The porous structure of the orthosis allows the affected limb's skin to breathe, minimizing associated complications. Additionally, it facilitates cleaning and enables physicians to monitor the treated area. Finite element analysis simulations demonstrated that the maximum displacement of the structure met medical requirements for treating this type of fracture, with a value of ≤ 2 mm.

4.7 Concluding Remarks

The findings of the studies above emphasize the significance of additive manufacturing in the production of customized orthoses, which offer adequate therapeutic support for patients. The utilization of 3D printing enables the creation of tailor-made devices, leading to promising clinical outcomes and enhancements in the quality of life for affected individuals. Furthermore, these results underscore the promising potential of additive manufacturing with smart memory materials for developing personalized and functional orthoses for post-stroke patients with spasticity. This approach provides a viable and cost-effective alternative to enhance the well-being of these individuals, facilitating their reintegration into daily activities.

However, it is essential to acknowledge specific challenges and limitations associated with the application and accessibility of additive manufacturing in producing customized orthoses. Notably, utilizing this manufacturing method necessitates technical knowledge and acquiring new equipment. Currently, 3D printers need to be more utilized in clinical settings, despite the availability of various models in national and international markets. The promotion of this technology within clinical contexts is hindered by the requirement for expertise in 3D modeling, knowledge of biomechanics applied to design, and proficiency in 3D printing. Consequently, additive manufacturing can be daunting and inaccessible for occupational therapists, physicians, and orthopedic technicians due to needing more training and familiarity (Baronio et al. 2016; Lunsford et al. 2016; Li & Tanaka 2018; Portnoy et al. 2020). Moreover, significant investment is required in professional 3D modeling software, suitable materials that cater to specific applications, and high-quality 3D printers. The use of low-cost 3D printers can result in limitations in geometric definition, such as the replication of intricate details like holes and textures and may even lead to part fractures (Bourell 2012; Palousek et al. 2014; Wang et al. 2018).

Additionally, when producing custom orthoses through additive manufacturing, factors such as part size, finish quality, material selection, and printer quality influence production time and manufacturing cost (Miclaus et al. 2017). Industrial injection molding emerges as an efficient alternative, offering superior finish quality (e.g., texture, well-organized, and robust microscopic structure) and faster orthosis production (Dawoud et al. 2016; Lay et al. 2019). However, this process requires skilled personnel to operate the injection molding equipment and incurs high costs for molds, materials, and equipment, making it less viable for small-scale production.

These factors collectively impede the widespread adoption of additive manufacturing as the primary method for orthosis production. The initial investment in machinery, materials, and software, the need for modeling and 3D printing (or staff training) professionals, and the variability of printing time and cost pose challenges. Most studies in 3D printing for assistive technology devices have been conducted in developed countries where 3D printers are more accessible and commonly found in clinics, hospitals, and universities. Lastly, the cost aspect must be considered, encompassing the investment in machinery, materials, and skilled professionals for modeling and printing tasks. Generally, 3D printing in upper limb orthosis production enhances product comfort and aesthetics, facilitating patients' reintegration into their social environment. Furthermore, the presented models have demonstrated efficacy in fulfilling their intended purpose, withstanding post-processing and daily use without flaws or breakage, and garnering positive evaluations from patients, orthosis handlers, and orthopedic surgeons.

References

Bandyopadhyay, A., Bose, S., Das, & S. (2015). 3D printing of biomaterials. MRS bulletin, 40, 108–115.
Baronio, G., Harran, S., & Signoroni, A. (2016). A critical analysis of a hand orthosis reverse engineering and 3D printing process. Applied bionics and biomechanics, 2016.
Bourell, D. (2012, October). Evaluation of a digitised splinting approach with multi-material functionality using Additive Manufacturing Technologies. In Twenty-Third Annual International Solid Freeform Fabrication Symposium. University of Texas at Austin.
Cano, A. P. D. (2017). Parametrização e produção de órtese termomoldável para imobilização de punho produzida por manufatura aditiva. 102p. Trabalho de Conclusão de Curso de Engenharia Biomédica (TCC)-Universidade Federal de São Paulo.
Chae, D. S., Kim, D. H., Kang, K. Y., Kim, D. Y., Park, S. W., Park, S. J., & Kim, J. H. (2020). The functional effect of 3D-printing individualized orthosis for patients with peripheral nerve injuries: Three case reports. Medicine, 99(16).
Chang, K., Chang, J. H., Huang, M. W., & Lee, L. Y. (2018, April). Innovative orthosis for phalanx extension neurofacilitation (iOPEN)—development of a 3D-printed hand orthosis for chronic stroke patient. In 2018 IEEE International Conference on Applied System Invention (ICASI) (pp. 1175–1177). IEEE.
Chen, R. K., Jin, Y. A., Wensman, J., & Shih, A. (2016). Additive manufacturing of custom orthoses and prostheses—A review. Additive manufacturing, 12, 77–89.
Chen, Y. J., Lin, H., Zhang, X., Huang, W., Shi, L., & Wang, D. (2017). Application of 3D–printed and patient-specific cast for the treatment of distal radius fractures: initial experience. 3D Printing in Medicine, 3, 1–9.
Chen, Y., Lin, H., Yu, Q., Zhang, X., Wang, D., Shi, L., ... & Zhong, S. (2020). Application of 3D-printed orthopedic cast for the treatment of forearm fractures: finite element analysis and comparative clinical assessment. BioMed Research International, 2020.
Chu, C., Wang, I. J., Sun, J. R. & Liu, C. H. (2020). Customized designs of short thumb orthoses using 3D hand parametric models. Assistive Technology, 1–8.
Dawoud, M., Taha, I., & Ebeid, S. J. (2016). Mechanical behaviour of ABS: An experimental study using FDM and injection moulding techniques. Journal of manufacturing Processes, 21, 39–45.
Dudley, D. R., Knarr, B. A., Siu, K. C., Peck, J., Ricks, B., & Zuniga, J. M. (2021). Testing of a 3D printed hand exoskeleton for an individual with stroke: A case study. Disability and Rehabilitation: Assistive Technology, 16(2), 209–213.
Francisco, N. P. F. (2004). Avaliação das características de três materiais de baixo custo utilizados na confecção de órtese para estabilização de punho. Dissertação - (Mestrado em Engenharia Biomédica) - Instituto de Pesquisa e Desenvolvimento, Universidade do Vale do Paraíba.
Freeland, A. E., Hardy, M. A., & Singletary, S. (2003). Rehabilitation for proximal phalangeal fractures. Journal of Hand Therapy, 16(2), 129–142.
Gebhardt, A. (2011). Understanding additive manufacturing. Hanser Publications.
Gibson, I., Rosen, D. W., Stucker, B. Khorasani, M., Rosen, D., Stucker, B., & Khorasani, M. (2021). Additive manufacturing technologies. Cham, Switzerland: Springer.

Graham, J., Wang, M., Frizzell, K., Watkins, C., Beredjiklian, P., & Rivlin, M. (2020). Conventional vs 3-dimensional printed cast wear comfort. Hand, 15(3), 388–392.
Han, J. J., & Huang, H. (2011). Preparation and characterization of biodegradable polylactide/thermoplastic polyurethane elastomer blends. Journal of Applied Polymer Science, 120(6), 3217–3223.
Herr, H. (2009). Exoskeletons and orthoses: classification, design challenges and future directions. Journal of neuroengineering and rehabilitation, 6, 1–9.
Jacobs, M. A., & Austin, N. (2003). Splinting the hand and upper extremity: principles and process. Lippincott Williams & Wilkins.
Kadioglu, E., Aydin, H. E., Kaya, I., Aydin, N., & Sahin, M. C. (2019). Manufacturing and application of personal hand and finger splint with three dimensional printer technology following hand and finger trauma.
Kalita, B., Narayan, J., & Dwivedy, S. K. (2021). Development of active lower limb robotic-based orthosis and exoskeleton devices: a systematic review. International Journal of Social Robotics, 13, 775–793.
Kim, S. J., Kim, S. J., Cha, Y. H., Lee, K. H., & Kwon, J. Y. (2018). Effect of personalized wrist orthosis for wrist pain with three-dimensional scanning and printing technique: a preliminary, randomized, controlled, open-label study. Prosthetics and orthotics international, 42(6), 636–643.
Lay, M., Thajudin, N. L. N., Hamid, Z. A. A., Rusli, A., Abdullah, M. K., & Shuib, R. K. (2019). Comparison of physical and mechanical properties of PLA, ABS and nylon 6 fabricated using fused deposition modeling and injection molding. Composites Part B: Engineering, 176, 107341.
Lee, K. H., Kim, D. K., Cha, Y. H., Kwon, J. Y., Kim, D. H., & Kim, S. J. (2019). Personalized assistive device manufactured by 3D modelling and printing techniques. Disability and Rehabilitation: Assistive Technology, 14(5), 526–531.
Li, J., & Tanaka, H. (2018). Rapid customization system for 3D-printed splint using programmable modeling technique–a practical approach. 3D printing in medicine, 4(1), 1–21.
Lunsford, C., Grindle, G., Salatin, B., & Dicianno, B. E. (2016). Innovations with 3-dimensional printing in physical medicine and rehabilitation: a review of the literature. Pm&r, 8(12), 1201–1212.
Miclaus, R., Repanovici, A., & Roman, N. (2017). Biomaterials: Polylactic acid and 3D printing processes for orthosis and prosthesis. Materiale plastice, 54(1), 98–102.
Nam, H. S., Seo, C. H., Joo, S. Y., Kim, D. H., & Park, D. S. (2018). The application of three-dimensional printed finger splints for post hand burn patients: a case series investigation. Annals of Rehabilitation Medicine, 42(4), 634–638.
O'brien, V. H., & Thurn, J. (2013). A simple distal radioulnar joint orthosis. Journal of Hand Therapy, 26(3), 287–290.
Oksman, K., Skrifvars, M., & Selin, J.F. (2003). Natural fibers as reinforcement in polylactic acid (PLA) composites. Composite Science Technology, 1317–1324.
Palousek, D., Rosicky, J., Koutny, D., Stoklásek, P., & Navrat, T. (2014). Pilot study of the wrist orthosis design process. Rapid prototyping journal, 20(1), 27–32.
Popescu, D., Zapciu, A., Tarba, C., & Laptoiu, D. (2020). Fast production of customized three-dimensional-printed hand splints. Rapid Prototyping Journal, 26(1), 134–144.
Portnova, A. A., Mukherjee, G., Peters, K. M., Yamane, A., & Steele, K. M. (2018). Design of a 3D-printed, open-source wrist-driven orthosis for individuals with spinal cord injury. PloS one, 13(2), e0193106.
Portnoy, S., Barmin, N., Elimelech, M., Assaly, B., Oren, S., Shanan, R., & Levanon, Y. (2020). Automated 3D-printed finger orthosis versus manual orthosis preparation by occupational therapy students: Preparation time, product weight, and user satisfaction. Journal of Hand Therapy, 33(2), 174–179.
Sarı, M. İ., Şahin, İ., Gökçe, H., & Öksüz, Ç. (2020). Ring orthosis design and production by rapid prototyping approach. Journal of Hand Therapy, 33(2), 170–173.

Spaulding, S. E., Yamane, A., McDonald, C. L., & Spaulding, S. A. (2019). A conceptual framework for orthotic and prosthetic education. Prosthetics and Orthotics International, 43(4), 369–381.

Toth, L., Schiffer, A., Nyitrai, M., Pentek, A., Told, R., & Maroti, P. (2020). Developing an anti-spastic orthosis for daily home-use of stroke patients using smart memory alloys and 3D printing technologies. Materials & Design, 195, 109029.

Wang, K., Shi, Y., He, W., Yuan, J., Li, Y., Pan, X., & Zhao, C. (2018). The research on 3D printing fingerboard and the initial application on cerebral stroke patient's hand spasm. Biomedical engineering online, 17, 1–14.

Wong, J. Y. (2015). On-site 3D printing of functional custom mallet splints for Mars analogue crewmembers. Aerospace medicine and human performance, 86(10), 911–914.

Wong, K. V., & Hernandez, A. (2012). A review of additive manufacturing. International scholarly

Yan, W., Ding, M., Kong, B., Xi, X., & Zhou, M. (2019). Lightweight splint design for individualized treatment of distal radius fracture. Journal of medical systems, 43, 1–10.

Zhang, Z., Demir, K. G., & Gu, G. X. (2019). Developments in 4D-printing: a review on current smart materials, technologies, and applications. International Journal of Smart and Nano Materials, 10(3), 205–224.

Part II
Computational Systems Applied to Health

Chapter 5
Mobile Health Solution Through Machine Learning and Sensors in the Detection of Falls Associated with Aging

Priscyla Waleska Simões, Fellipe Soares de Oliveira, Camila Carvalho da Silva, Pablo Deoclecia dos Santos, and Harki Tanaka

Abstract This chapter presents a Mobile Health solution with the objective of detecting falls in the elderly from a system composed of an application, a backend module for processing the data provided by the application and sensors, and an ML model based on the algorithm Random Forest. The application was developed for the Android operating system using the Java development language. A model was developed using the Random Forest algorithm that considers smartphone accelerometer data, using the Scikit-Learn library version 0.24.2 of the Python language and the Visual Studio Code tool version 1.45. The data registered in the application and those collected by the sensors are managed in a backend module (in the cloud). The application uses the accelerometer and GPS sensors to capture the data and send it to a server for processing, fall detection, and sending alerts (SMS, email, and call). The Random Forest model showed an accuracy of 87.1% (95% CI: 86.9%–87.3%). The results are promising for achieving good accuracy metrics.

Keywords Digital health · Mobile health · Machine learning · Fall detection · Sensors

5.1 Introduction

Population aging presents itself as a global public health problem associated with the increase in the life expectancy of the population (Gafner et al. 2021). Between 2008 and 2015, the global incidence of falls was 30%; in 2015, the worldwide medical cost associated with falls exceeded US$50 billion (Shu and Shu 2021). In this context, some studies have been developed in monitoring and preventing falls, seeking to offer greater agility of alerts and help (Shu and Shu 2021); some of this research includes the detection of falls through thresholds (Usmani et al.

P. W. Simões (✉) · F. S. de Oliveira · C. C. da Silva · P. D. dos Santos · H. Tanaka
Graduate Program in Biomedical Engineering, Federal University of ABC – UFABC Al. da Universidade, São Bernardo do Campo, SP, Brazil
e-mail: priscyla.simoes@ufabc.edu.br

C. B. Lombello, P. A. da Ana (eds.), *Current Trends in Biomedical Engineering*,
https://doi.org/10.1007/978-3-031-38743-2_5

2021) and/or machine learning (ML) (Usmani et al. 2021) associated with the use of sensors.

In projects that use ML, supervised learning has been more documented through models developed from decision trees, random forests (RF), and support vector machines (Vallabh and Malekian 2018; Byeon 2020), in conjunction with a wearable device (Ma et al. 2019; Yacchirema et al. 2019; Er and Tan 2020; Usmani et al. 2021). Tree-based algorithms, such as random forests, process large amounts of data well and are flexible in both classification and regression tasks, in addition to the possibility of parallel processing, which improves their performance (Chowdhury et al. 2019).

Despite documented methods for detecting falls, most tests are performed in controlled environments and data outside this environment is scarce (Gafner et al. 2021; Usmani et al. 2021), sensor combination can also improve accuracy and reduce uncertainties (Usmani et al. 2021). However, the differential of fall detection systems is that their tools are able to restore autonomy to the patient at some level (Holden et al. 2020).

Thus, considering the high rates and prevalence of falls and their effects on the population, as well as the cost to health systems around the world, the objective of this chapter is to present a mHealth strategy aimed at detecting falls in the elderly. The following sections will be presented as: methods, results, discussion, and conclusion.

5.2 Methods

Figure 5.1 presents the system architecture. The user attaches a smartphone on their waist with the application installed and configured; then, data from the accelerometer (x, y, and z axis) and Global Positioning System (GPS) sensors (latitude and longitude) of the cell phone are collected and sent to a cloud server (backend).

The accelerometer data is processed in a model developed from the Random Forest algorithm that seeks to detect the fall and to enable alerts and help. In case the user needs help, or if the user does not activate the application (within 10 s), email and short message service (SMS) alert notifications are sent to previously registered emergency contacts, in addition, a call is made from the user's cell phone to the first emergency contact.

The application was developed for the Android operating system using the Java development language, this choice was made because it is the most used system in the world (Dysart 2014), and it is also a development standard of the operating system manufacturer (Google) (Google 2021). The development environment was Android Studio version 4.2.1, which is also the default suggested by Google.

The detection of falls is performed by the app through the smartphone's accelerometer sensor. Accelerometer-type sensors in smartphones are triaxial, contain X, Y, and Z axes, and are measured in acceleration force in m/s^2 (Lima et

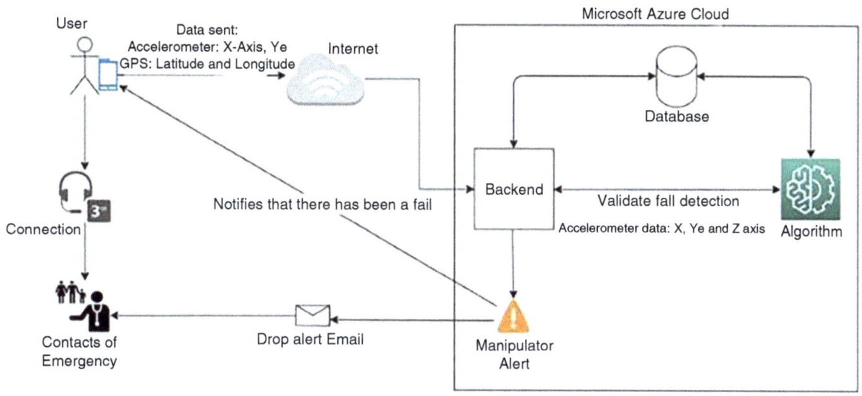

Fig. 5.1 Architecture of the fall detection system

al. 2013). The smartphone's GPS was also used to determine the user's location (Dassau et al. 2009) at the time of the fall (with the latitude and longitude information collected from the GPS), and this way, it sends the alert to the emergency contacts with the exact position of the user.

Subsequently, the data (comprising data acquired through the sensors) are sent to the backend module (in the cloud); that receives the data coming from the application through the HTTPS protocol (Kaleb et al. 2019), and REST standard (Zhou et al. 2014). Data from accelerometer and GPS sensors are sent to the backend module using the MQTT protocol (Sahlmann et al. 2020), when the user uses the application, information from the accelerometer sensor (X, Y, and Z axis) and GPS (latitude and longitude) are constantly collected at a rate of 10 Hz and sent through a queue (MQTT) to the server, which is responsible for processing such data along with the algorithm and detecting the fall. MQTT is a protocol widely used in IoT, due to data optimization (Sahlmann et al. 2020).

The backend module works on an Azure cloud server which is a free and secure platform (Shanahan et al. 2014), with security protocols and adherent to MQTT (Simms and Daggett 2012; Liu et al. 2014; Mate et al. 2015; Dufresne et al. 2018). Data transferred between the application and the cloud is securely exchanged, using the Advanced Encryption Standard (AES) algorithm to encrypt the data (Hossain et al. 2021). SQL server was used as a database, due to its high performance, robustness, and security (Kim et al. 2012; Shanahan et al. 2014; Mate et al. 2015), and for being the standard used in the Azure cloud (Antonopoulos et al. 2020).

For the training of the Random Forest algorithm, a public dataset aimed at the detection of falls called SisFall was used (Sucerquia et al. 2017), it is widely cited in the literature (Alizadeh et al. 2021) and it presents a larger sample when compared to other datasets (Casilari et al. 2017); it also offers drop data from more than one accelerometer sensor. For the present research, accelerometer and outcome data (presence or absence of fall) were used in the model. The Random Forest algorithm

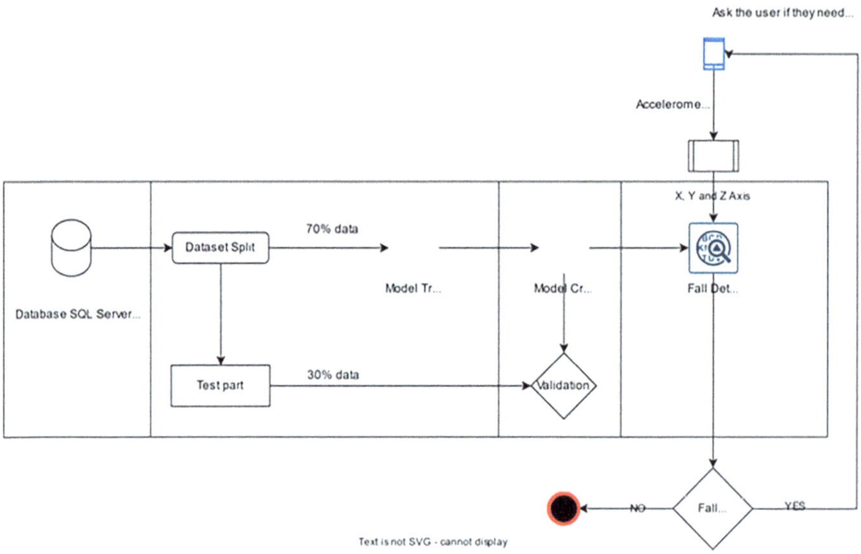

Fig. 5.2 Random Forest algorithm training flow

was chosen for its good accuracy measurements (from 83.0% to 95.0%) in related studies (Dogan and Hossain 2019; Kim et al. 2019; Zurbuchen et al. 2020).

The machine learning model was developed in Python language using the Scikit-Learn library version 0.24.2, and Visual Studio Code tool version 1.45. The choice of the Python language and the Scikit-Learn library was due to its wide use in academia, as it is a free tool, in addition to being a multiplatform language with an easy-to-use syntax (Pryss et al. 2020).

Figure 5.2 presents the fall detection flow. The collection and storage of SisFall data are performed automatically when the backend module is started, that is, if the model is not trained. The model is trained at the time of application startup, and, when the training is completed, the application starts to receive requests.

It was considered 70% of the data for training and 30% for testing (Ramos-Pollán et al. 2010). With the model created, it is possible to start monitoring (processing the data of the X, Y, and Z axes of the accelerometer).

The parameterization of the Random Forest algorithm was performed based on the values: n_estimator = 1000, criterion = gini, max_depth = None, min_samples_split = 2, and min_samples_leaf = 1.

The features that were used to train the algorithm (accelerometer sensor X-axis, Y-axis, and Z-axis) were selected by predictive relevance (Jahanjoo et al. 2017; Sucerquia et al. 2018). In training the model, the number of trees in the forest ranged from 100 to 1000 and the accuracy from 85.0% to 87.3% (95%CI: 90.7%–99.3%). The model generated from the RF algorithm obtained an accuracy of 87.1% (95%CI: 86.9%–87.3%). In the experimental tests ($n = 100$), the use of a smartphone fixed to a wooden structure was considered to simulate the fall.

5.3 Results and Discussion

The present research resulted in a Digital Health solution composed of an application and a backend module that processes the information that the application sends, that is, registration data (from the user himself or from emergency contacts) and information from the accelerometer and GPS for fall detection and alert.

With active monitoring (Fig. 5.3a), the application starts capturing the accelerometer and GPS data by sending it to the RF model for processing, and, in the event of a fall, by notifying the emergency contacts (Fig. 5.3b).

The first screen that the user has access to when starting to use the application is the login screen (Fig. 5.4a), which requests the email and password information to

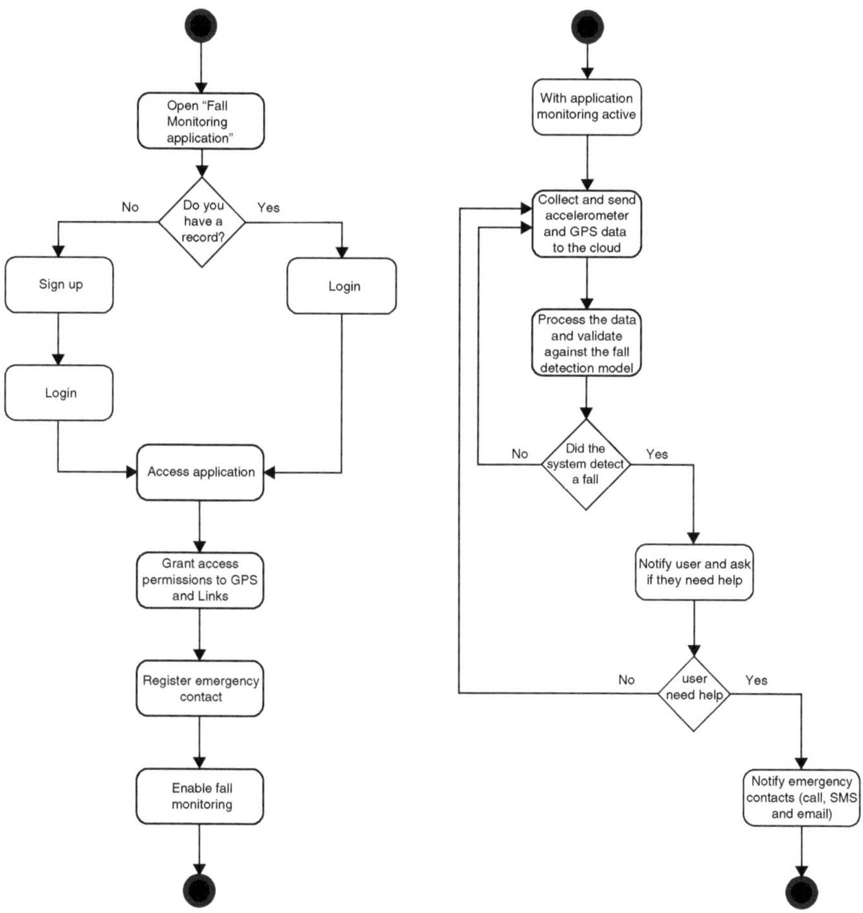

Fig. 5.3 Flowchart of access to the application and activation of fall monitoring (**a**), and monitoring, detection of falls, and notification of emergency contacts (**b**)

Fig. 5.4 App interface

access the app, or, in the "Register" option, registration is done for later access to the app (Fig. 5.4b). In Fig. 5.4b, information to access the app is requested (name, email, password, and mobile number). It also presents the privacy statement for the user acceptance in which personal information is collected and accessed by the smartphone. Upon acceptance of the terms, the registration is completed, and the app is started (Fig. 5.4c).

On the screen shown in Fig. 5.4c, the user can activate and deactivate monitoring; to activate at least one emergency contact must be registered. As soon as the user activates the fall detection, the application starts monitoring, by collecting data from the accelerometer sensor (X, Y, and Z axis at a frequency of 10 Hz) and the GPS (latitude and longitude), and sends them to the cloud (backend module) through the MQTT protocol. In the backend module, such information is processed and validated against the RF model. Upon a drop presented by the RF model, the alert is issued (Fig. 5.4f).

Figure 5.4d, f contain the functionalities related to emergency contacts and data that are managed in the cloud (backend module). When a fall is detected, a notification (alert) is displayed on the user's smartphone (Fig. 5.4f); if the user indicates that they do not need help, the flow ends. If the user selects that they need help or does not interact with the Help option within 10 s, a call is made to the first registered emergency contact, and an alert and help notifications are sent by email and SMS to all emergency contacts, with the information of the fall. The alert also presents the user's location as obtained via GPS.

Figure 5.5 shows the accelerometer events during the fall; on the Y axis of the graph, the acceleration data in m/s^2 (accelerometer X, Y, and Z axes) are presented, and on the X axis the time in seconds of the test duration. In the experimental tests (n = 100), an accuracy of 96.0% was observed (95%CI: 92.2%–99.8%).

For fall monitoring to be effective, it is important that the system offers alerts as suggested by some studies (Gokalp et al. 2018; Hantsoo et al. 2018; Urrea and Venegas 2020). Our research, therefore, corroborates this issue by sending alerts through e-mail, SMS, and voice calls (in the event of a fall).

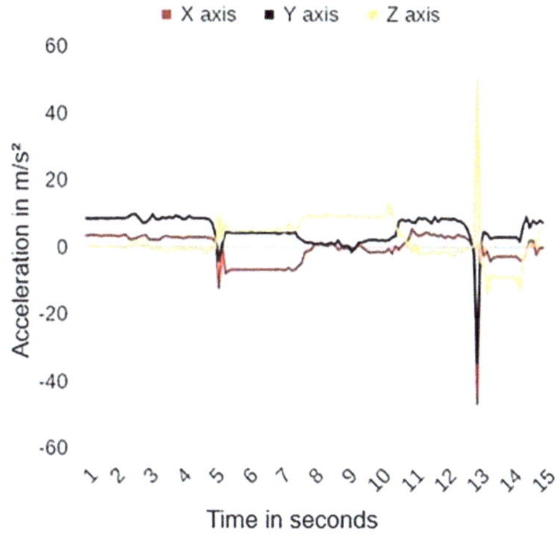

Fig. 5.5 Monitoring data during the fall

Some studies suggest WhatsApp (Walwema 2020) as decision support by allowing different formats of content such as alerts (text, audio and voice, and video call). In our study, the voice call is made to the first emergency contact registered in the application, and e-mail and SMS are sent to the others. Both SMS and e-mail were used in our research due to the possibility that the first emergency contact may reject the call, this way, SMS and e-mail alerts seek to speed up the notification of the fall and subsequent assistance. Collaborating with this issue, a study carried out in two cities in the United States (New York and Boston) between 2015 and 2018 showed that alerts reduced falls by 15% (Dykes et al. 2020). This feature was also observed in other studies (Snooks et al. 2014; Ploegmakers et al. 2022).

The waist region and the lower back (Özdemir 2016) are the most cited for monitoring device positioning (99.96% sensitivity), followed by the lumbar (Schwickert et al. 2013) as it is the center of the human body, and upon the occurrence of the event (fall), the waist moves toward the ground and therefore possibly presents a greater sensitivity compared to other regions of the human body (Howcroft et al. 2013). Therefore, in the tests conducted in our research, the smartphone was positioned on a wooden structure 90 cm from the floor, corresponding to the waist height of a human being of 170 cm.

Although threshold-based fall detection is an option (Yang et al. 2016; Sucerquia et al. 2017; Wang et al. 2018), it presented many false positives (Fakhrulddin et al. 2019; Hussain et al. 2019), thus, ML has shown promising results. Our study used the ML Random Forest algorithm in fall detection, compared to other studies (Yacchirema et al. 2018), the fall detection model developed from the RF algorithm reached an accuracy of 87.1% in the experimental test (95%CI: 86.9%–87.3%), a result superior to the other studies mentioned in this chapter.

In this chapter, a Digital Health solution was presented for the detection of falls in the elderly from a system composed of an application for the Android operating system, a backend module for processing the data provided by the application and sensors, and an ML model based on the Random Forest algorithm. Thus, it can be concluded that the results are promising due to the good accuracy presented.

Funding Declaration/Acknowledgments This study was financed in part by the Coordenação de Aperfeiçoamento de Pessoal de Nível Superior – Brasil (CAPES) – Finance Code 001, by São Paulo Research Foundation (FAPESP) under grant #2020/09838-0 (BIOS – Brazilian Institute of Data Science), and Federal University of ABC.

This chapter is based on the masters degree thesis of the Fellipe Soares de Oliveira on Biomedical Engineering at UFABC.

References

Alizadeh J, Bogdan M, Classen J & Fricke C (2021) Support Vector Machine Classifiers Show High Generalizability in Automatic Fall Detection in Older Adults. Sensors (Basel, Switzerland) 21.

Antonopoulos P, Arasu A, Singh K, Eguro K, Gupta N, Jain R, Kaushik R, Kodavalla H, Kossmann D, Ogg N, Ramamurthy R, Szymaszek J, Trimmer J, Vaswani K, Venkatesan R & Zwilling M. *Azure SQL Database Always Encrypted.* (2020).

Byeon H (2020) Is the Random Forest Algorithm Suitable for Predicting Parkinson's Disease with Mild Cognitive Impairment out of Parkinson's Disease with Normal Cognition? International journal of environmental research and public health 17.

Casilari E, Santoyo-Ramón JA & Cano-García JM (2017) Analysis of Public Datasets for Wearable Fall Detection Systems. Sensors (Basel, Switzerland) 17.

Chowdhury AR, Chatterjee T & Banerjee S (2019) A Random Forest classifier-based approach in the detection of abnormalities in the retina. Medical & biological engineering & computing 57:193–203.

Dassau E, Jovanovic L, Doyle FJ, 3rd & Zisser HC (2009) Enhanced 911/global position system wizard: a telemedicine application for the prevention of severe hypoglycemia–monitor, alert, and locate. Journal of diabetes science and technology 3:1501–1506.

Dogan JC & Hossain MS. in *2019 IEEE International Conference on Smart Computing (SMARTCOMP).* 434–438.

Dufresne J, Bowden P, Thavarajah T, Florentinus-Mefailoski A, Chen ZZ, Tucholska M, Norzin T, Ho MT, Phan M, Mohamed N, Ravandi A, Stanton E, Slutsky AS, Dos Santos CC, Romaschin A, Marshall JC, Addison C, Malone S, Heyland D, Scheltens P, Killestein J, Teunissen C, Diamandis EP, Siu KWM & Marshall JG (2018) The plasma peptidome. Clin Proteomics 15:39–39.

Dykes PC, Burns Z, Adelman J, Benneyan J, Bogaisky M, Carter E, Ergai A, Lindros ME, Lipsitz SR, Scanlan M, Shaykevich S & Bates DW (2020) Evaluation of a Patient-Centered Fall-Prevention Tool Kit to Reduce Falls and Injuries: A Nonrandomized Controlled Trial. JAMA network open 3:e2025889.

Dysart TJ. *Enforcing the GNU general public licence: copyright or contract?*, Oxford University, UK, (2014).

Er PV & Tan KK. in *Assistive Technology for the Elderly* (eds Nagender Kumar Suryadevara & Subhas Chandra Mukhopadhyay) 81–105 (Academic Press, 2020).

Fakhrulddin SS, Gharghan SK, Al-Naji A & Chahl J (2019) An Advanced First Aid System Based on an Unmanned Aerial Vehicles and a Wireless Body Area Sensor Network for Elderly Persons in Outdoor Environments. Sensors (Basel, Switzerland) 19:2955.

Gafner SC, Allet L, Hilfiker R & Bastiaenen CHG (2021) Reliability and Diagnostic Accuracy of Commonly Used Performance Tests Relative to Fall History in Older Persons: A Systematic Review. Clinical interventions in aging 16:1591–1616.

Gokalp H, De Folter J, Verma V, Fursse J, Jones R & Clarke M (2018) Integrated Telehealth and Telecare for Monitoring Frail Elderly with Chronic Disease. Telemedicine journal and e-health: the official journal of the American Telemedicine Association 24:940–957.

Google. *Arquitetura de Plataforma Android*, https://developer.android.com/guide/platform (2021).

Hantsoo L, Criniti S, Khan A, Moseley M, Kincler N, Faherty LJ, Epperson CN & Bennett IM (2018) A Mobile Application for Monitoring and Management of Depressed Mood in a Vulnerable Pregnant Population. Psychiatric services (Washington, D.C.) 69:104–107.

Holden RJ, Campbell NL, Abebe E, Clark DO, Ferguson D, Bodke K, Boustani MA, Callahan CM & Brain Health Patient Safety L (2020) Usability and feasibility of consumer-facing technology to reduce unsafe medication use by older adults. Res Social Adm Pharm 16:54–61.

Hossain FS, Sakib TH, Ashar M & Ferdian R (2021) A dual mode self-test for a stand alone AES core. PLoS One 16:e0261431.

Howcroft J, Kofman J & Lemaire ED (2013) Review of fall risk assessment in geriatric populations using inertial sensors. Journal of NeuroEngineering and Rehabilitation 10:91.

Hussain F, Basit Umair M, Ehatisham-Ul-Haq M, Pires IM, Valente T, Garcia NM & Pombo N (2019) An Efficient Machine Learning-based Elderly Fall Detection Algorithm. arXiv e-prints:arXiv:1911.11976.

Jahanjoo A, Tahan MN & Rashti MJ. in *2017 3rd International Conference on Pattern Recognition and Image Analysis (IPRIA).* 90–95.

Kaleb K, Vesztrocy AW, Altenhoff A & Dessimoz C (2019) Expanding the Orthologous Matrix (OMA) programmatic interfaces: REST API and the OmaDB packages for R and Python. F1000Research 8:42.

Kim I, Jung J-Y, Deluca TF, Nelson TH & Wall DP (2012) Cloud computing for comparative genomics with windows azure platform. Evol Bioinform Online 8:527–534.

Kim K, Yun G, Park SK & Kim DH. in *2019 41st Annual International Conference of the IEEE Engineering in Medicine and Biology Society (EMBC).* 4611–4614.

Lima M, Silva F, Fabrin A, Pazoti M & Garcia Júnior J (2013) UTILIZAÇÃO DOS RECURSOS DO IOS PARA MONITORAR PESSOAS DE TERCEIRA IDADE NA PRÁTICA DE ATIVIDADES F´iSICAS. Colloquium Exactarum 5.

Liu Y-L, Shih C-T, Chang Y-J, Chang S-J & Wu J (2014) Performance enhancement of a web-based picture archiving and communication system using commercial off-the-shelf server clusters. Biomed Res Int 2014:657417–657417.

Ma C, Shimada A, Uchiyama H, Nagahara H & Taniguchi R-I (2019) Fall detection using optical level anonymous image sensing system. Optics & Laser Technology 110:44–61.

Mate S, Köpcke F, Toddenroth D, Martin M, Prokosch H-U, Bürkle T & Ganslandt T (2015) Ontology-based data integration between clinical and research systems. PLoS One 10:e0116656–e0116656.

Özdemir AT (2016) An Analysis on Sensor Locations of the Human Body for Wearable Fall Detection Devices: Principles and Practice. Sensors 16.

Ploegmakers KJ, Medlock S, Linn AJ, Lin Y, Seppälä LJ, Petrovic M, Topinkova E, Ryg J, Mora MaC, Landi F, Thaler H, Szczerbińska K, Hartikainen S, Bahat G, Ilhan B, Morrissey Y, Masud T, Van Der Velde N & Van Weert JCM (2022) Barriers and facilitators in using a Clinical Decision Support System for fall risk management for older people: a European survey. European geriatric medicine 13:395–405.

Pryss R, Schlee W, Hoppenstedt B, Reichert M, Spiliopoulou M, Langguth B, Breitmayer M & Probst T (2020) Applying Machine Learning to Daily-Life Data From the TrackYourTinnitus Mobile Health Crowdsensing Platform to Predict the Mobile Operating System Used With High Accuracy: Longitudinal Observational Study. J Med Internet Res 22:e15547.

Ramos-Pollán R, Guevara-López MÁ & Oliveira E. (2010) in *Progress in Pattern Recognition, Image Analysis, Computer Vision, and Applications.* (eds Isabelle Bloch & Roberto M. Cesar) 517–524 (Springer Berlin Heidelberg).

Sahlmann K, Clemens V, Nowak M & Schnor B (2020) MUP: Simplifying Secure Over-The-Air Update with MQTT for Constrained IoT Devices. Sensors (Basel, Switzerland) 21.

Schwickert L, Becker C, Lindemann U, Maréchal C, Bourke A, Chiari L, Helbostad JL, Zijlstra W, Aminian K, Todd C, Bandinelli S, Klenk J, For The FC & The FMDCG (2013) Fall detection with body-worn sensors. Zeitschrift für Gerontologie und Geriatrie 46:706–719.

Shanahan HP, Owen AM & Harrison AP (2014) Bioinformatics on the cloud computing platform Azure. PLoS One 9:e102642–e102642.

Shu F & Shu J (2021) An eight-camera fall detection system using human fall pattern recognition via machine learning by a low-cost android box. Scientific reports 11:2471.

Simms AM & Daggett V (2012) Protein Simulation Data in the Relational Model. J Supercomput 62:150–173.

Snooks HA, Carter B, Dale J, Foster T, Humphreys I, Logan PA, Lyons RA, Mason SM, Phillips CJ, Sanchez A, Wani M, Watkins A, Wells BE, Whitfield R & Russell IT (2014) Support and Assessment for Fall Emergency Referrals (SAFER 1): cluster randomised trial of computerised clinical decision support for paramedics. PLoS One 9:e106436.

Sucerquia A, López JD & Vargas-Bonilla JF (2017) SisFall: A Fall and Movement Dataset. Sensors (Basel, Switzerland) 17:198.

Sucerquia A, López DJ & Vargas-Bonilla FJ (2018) Real-Life/Real-Time Elderly Fall Detection with a Triaxial Accelerometer. Sensors 18.

Urrea C & Venegas D (2020) Automatized follow-up and alert system for patients with chronic hypertension. Health informatics journal 26:2625–2636.

Usmani S, Saboor A, Haris M, Khan MA & Park H (2021) Latest Research Trends in Fall Detection and Prevention Using Machine Learning: A Systematic Review. Sensors (Basel, Switzerland) 21.

Vallabh P & Malekian R (2018) Fall detection monitoring systems: a comprehensive review. Journal of Ambient Intelligence and Humanized Computing 9.

Walwema J (2020) The WHO Health Alert: Communicating a Global Pandemic with WhatsApp. Journal of Business and Technical Communication 35:35–40.

Wang F-T, Chan H-L, Hsu M-H, Lin C-K, Chao P-K & Chang Y-J (2018) Threshold-based fall detection using a hybrid of tri-axial accelerometer and gyroscope. Physiological Measurement 39:105002.

Yacchirema D, Puga J, Palau C & Esteve M (2018) Fall detection system for elderly people using IoT and Big Data. Procedia Computer Science 130:603–610.

Yacchirema D, De Puga JS, Palau C & Esteve M (2019) Fall detection system for elderly people using IoT and ensemble machine learning algorithm. Personal and Ubiquitous Computing 23:801–817.

Yang L, Ren Y & Zhang W (2016) 3D depth image analysis for indoor fall detection of elderly people. Digital Communications and Networks 2:24–34.

Zhou W, Li L, Luo M & Chou W. in *2014 28th International Conference on Advanced Information Networking and Applications Workshops*. 358–365.

Zurbuchen N, Bruegger P & Wilde A. *A Comparison of Machine Learning Algorithms for Fall Detection using Wearable Sensors*. (2020).

Chapter 6
Managing the Future of Healthcare: The Importance of Health Information Management

Fernanda Nascimento Almeida, Alexandre Davi Santos Dias, and Pedro Henrique Pires Da Silva

Abstract Health Information Management (HIM) is the practice of organizing, analyzing, and managing health data in electronic form. HIM professionals are responsible for ensuring that health records are complete, accurate, and accessible to authorized personnel. This involves the use of data management techniques and databases, as well as information systems designed specifically for healthcare settings. One of the key challenges in HIM is the sheer volume of health data generated on a daily basis. HIM professionals must be skilled in handling large amounts of data while maintaining data privacy and security. Machine learning is also becoming an important tool in HIM, allowing for the automated analysis of health data to identify patterns and trends. Another critical aspect of HIM is the use of provenance data, which refers to information about the origin and history of a piece of data. Provenance data is crucial in healthcare as it allows for tracking changes made to health records over time, ensuring their accuracy and integrity. Overall, HIM is a crucial component of modern healthcare, providing the necessary infrastructure for the efficient and effective management of health data.

Keywords Health information management · Machine learning · Electronic health data · Data privacy and security · Provenance data

Conflict of interest statement: None.

F. N. Almeida (✉) · A. D. S. Dias · P. H. P. Da Silva
Graduate Program in Biomedical Engineering, Federal University of ABC—UFABC, Al. da Universidade, São Bernardo do Campo, SP, Brazil

Bioinformatics and Health Informatics Group—BHIG, Center for Engineering, Modeling and Applied Social Sciences—CECS, Federal University of ABC, São Bernardo do Campo, SP, Brazil
e-mail: fernanda.almeida@ufabc.edu.br

C. B. Lombello, P. A. da Ana (eds.), *Current Trends in Biomedical Engineering*,
https://doi.org/10.1007/978-3-031-38743-2_6

6.1 Introduction

From the 1990s, the tasks of in silico experiments began to be disseminated and performed on a large scale by the scientific community. This caused an increase in the demand for computational resources that could meet the great demand for data generated by these experiments. As a result, the need to manage data has become and is becoming increasingly fundamental to extract more efficient data analysis, especially for those coming from heterogeneous sources or with variable quality (Stevens et al. 2007; Discala et al. 2000; Davidson et al. 1995; Karp 1995).

A common solution to this demand lies in the functionalities made possible by the areas of scientific databases and data warehousing, which offer support to different projects, mainly management, organization, maintenance, analysis, and data availability. However, the growing interest in the use of databases, sometimes naive and sometimes mistaken, resulted in the propagation of inaccurate representations of the same data. The area of data management and analysis has contributed to the increasing advances in research on the elucidation of problems related to important biological and biomedical events, guaranteeing their application in the most varied sectors, in addition to acting in the design of strategies that may help in the development of therapeutic products or even in the presentation of effects on the cure or mitigation of a given disease (Hogeweg 2011).

The development of decision support systems in healthcare has long been advocated as they can promote consistent guideline-based diagnosis, supporting clinical reasoning at the point of care. According to the literature, decision support systems in medicine are conceptualized as all software that helps physicians in solving problems. However, the wider acceptance of these systems in clinical practice has been much slower, despite the availability of many commercial products. Concerns with their use remain about the performance of the diagnosis and the perception of a lack of transparency in the first-generation systems (so-called because they are focused on less structured problems and because they combine analytical techniques with traditional functions of accessing and retrieving information), which present evidence-based knowledge as a "black box" that generates clinical recommendations. These concerns about the quality of evidence and the effort required in maintaining the long-term sustainability of the underlying evidence base that supports such systems have led to the search for second-generation tools. Such tools were created with the purpose of supporting a more dynamic and iterative cycle of creation and updating of evidence using a technical infrastructure developed from the perspective of the Learning Health System (Friedman et al. 2010).

The Learning Health System community envisions all participants in the health system (clinician, patient, researcher, insurer, etc.) as both a producer and a consumer of data. Its main function is the notion of routine capture, transformation, and dissemination of data and the resulting knowledge. Clinical studies, analytics, quality improvement initiatives, decision support, and other scenarios can all be tied to the routes that data is taking through the Learning Health System. Trust information associated with the data needs to be available at each step of these use cases to support auditability and transparency.

Currently, it has become increasingly evident the importance of knowing the data derivation processes, not only its origin, and thus determining the quality and, mainly, guaranteeing that the information extracted from a database is, in fact, reliable. Therefore, storing the data and the respective processes that generated it is an important tool to help different groups of data users to have more clarity about the authenticity, legitimacy, and reliability of an analysis (Almeida et al. 2010; Moreau et al. 2008; Stevens et al. 2007).

In this sense, data management and analysis are critical areas for many sectors, and healthcare is no different. Therefore, this chapter provides an overview of how data management and analysis have become increasingly important in the healthcare field. It discusses some of the key challenges that the healthcare industry faces in relation to data management and analysis, as well as the solutions and technologies being used to overcome these challenges. The chapter also highlights the benefits that can be obtained through efficient data management in healthcare, such as improving the quality of care, reducing costs, and enabling professionals in the field to make more informed and accurate decisions. By the end of this chapter, the readers will have a clearer understanding of the importance of data management and analysis in healthcare and the critical role these fields play in delivering high-quality healthcare.

6.2 Data and Information in the Health Area

Data is understood as a sequence of quantifiable symbols, purely syntactic in nature, lacking inherent meaning but replicable without noticing any difference from the original. On the other hand, information refers to the organized arrangement of data that conveys meaning and understanding within a specific context for analysis or study. However, information cannot be reduced solely through logical or mathematical theories. Nonetheless, information can be represented through data. Within organizations, whether public or private, Information Systems (ISs) are commonly used to collect, store, retrieve, and disseminate information for specific purposes (Jannuzzi et al. 2014).

In the context of Brazilian healthcare, there are Health Information Systems (SISs) that can be defined as a set of interconnected components that collect, process, store, and distribute information to support decision-making processes and assist in the control of health organizations. Regarding the Unified Health System (SUS) in Brazil, these SISs serve several purposes, including serving as a foundation for resource management, organizing the production of information tailored to the needs of different levels, guiding the development of systems specific to the diverse operational units within the health system, contributing to the professional development in healthcare, fostering collective health awareness, and promoting a more compassionate relationship between institutions and citizens.

According to a document released in 2015 by the Brazilian Ministry of Health on the perspective of using information systems in the SUS, several problems were identified, including fragmentation, bias in billing, and challenges in data capture, as follows:

1. Fragmentation: The SUS currently consists of several independent SISs that lack integration, leading to the potential repetition of information and the creation of inaccurate indicators.
2. Billing bias: Systems like the SIA (Outpatient Information System)/SIH (Hospital Information System), originally designed for controlling payment of hospitalizations and procedures in SUS-integrated establishments, are now being used for epidemiological analysis. However, some institutions may have provided or continue to provide false data in order to manipulate the funding they receive from the government.
3. Information capture: Even today, healthcare facilities rely on paper records for capturing health information. When the data is later digitized, manual typing can introduce errors and biases. Additionally, the paper-based record capture and storage model can result in losses throughout the process.

The Brazilian Ministry of Health has proposed several strategies to address the aforementioned problems. One of these strategies is the implementation of the National Policy on Informatics and Health Information (PNIIS). The objective of this policy is to promote the innovative, creative, and transformative use of information technology to improve work processes in healthcare. The ultimate goal is to establish a coordinated National Information System that generates information for citizens, management, professional practice, knowledge generation, and social control. By doing so, the PNIIS aims to achieve measurable improvements in the efficiency and quality of healthcare services, while expanding access, equity, comprehensiveness, and humanization of health services, thus contributing to the overall improvement of the population's health situation (MS/Brasil 2015).

Interoperability is one of the key aspects emphasized in the PNIIS to overcome the problem of fragmentation. Interoperability in healthcare refers to the ability of different Health Information Systems (HISs) with diverse technologies and software applications to communicate, exchange data, and utilize shared information (Lehne et al. 2019). One of the significant challenges in achieving interoperability between HISs is ensuring the legibility and comprehensibility of health data, which often represent concepts rather than objects. To address this, the PNIIS recommends following the Catalog of Information Standards defined in Ordinance No. 2073/2011. Some of the mentioned standards include the TISS (Exchange of Information on Supplementary Health) standards established by the National Supplementary Health Agency of Brazil and the ICD (International Classification of Diseases) (MS/Brasil 2015). TISS is a comprehensive standard that encompasses codes and mandatory terms for the electronic exchange of healthcare data for beneficiaries of health plans and insurance. It covers various aspects, such as medication codes, types of care, and reasons for discharge. It is also designed to be compatible with other systems, such as the SIGTAP (*Management System for the Table of Procedures, Medications,*

Orthotics, Prostheses, and Special Materials). The ICD, established by the World Health Organization, provides a standardized cataloging system for diseases and health-related problems.

6.2.1 Database Modeling and Software Engineering

Database is a collection of related data and their relationship may not be something trivial, so its modeling is something of vital importance for its construction. By database modeling, it is understood to be the abstraction process where only the essential elements are emphasized, that is, it describes which elements will be part of the database.

An electronic medical record, for example, is not just a database, but a broad system to support medical and administrative decisions within a health unit. A way to develop it in a coherent way would be through the application of practices and technologies of database management systems and Software Engineering that have a systematic and disciplined approach in order to manage the entire software construction process (Pressman 2006). From the moment you plan to develop software, you should select the most appropriate life cycle model (Gonçalves et al. 2016). Within Software Engineering, there are different models of processes intended to represent the activities involved in software determination; among them the Waterfall Model can be broken down according to the activities below:

- Analysis and definition of requirements—objectives, functions, and restrictions are defined with the help of customers and users and serve as a specification of the system, indicating what must be implemented.
- Design of systems and software—it involves the description of the system and the software in terms of abstract units and their relationships, indicating how the software should be implemented.
- Implementation and unit testing—software units must be individually coded and tested.
- Integration and system testing—units are integrated and tested.
- Delivery, operation, and maintenance—the system is installed and put into operation. Maintenance involves correcting errors and evolving the system to meet new requirements.

Therefore, like other engineering disciplines, Software Engineering aims to support the process of developing, maintaining, and improving technologies through well-defined methods and techniques. With regard to data modeling for the development of the physical database, it is essential to study and implement data normalization techniques, which apply rules to all tables in the database in order to avoid failures in the project, such as data redundancy and a mix of different subjects in the same table. This tends to speed up the time for entering and processing data in the system. Another important point to be discussed about this type of system is data security, which covers access control to the system and the anonymization of patients' personal records, for example. In the latter case, in particular, it should always be used, as in the case of studies in public health and by research bodies. In this sense, it is strongly recommended to study anonymization techniques and standards that guarantee the irreversibility of the process or that establish a high degree of difficulty, to the point of inhibiting its reversal due to time and computational cost.

6.2.2 Data Provenance

Currently, it has become natural to ask questions about the reliability of a given set of consulted data. This makes it more and more evident the importance of knowing the data derivation processes, not only its origin but also, thus, determining the quality and mainly guaranteeing that the information extracted from a database is in fact reliable. Therefore, storing the data and the respective processes that generated it is an important tool to help different groups of data users to have more clarity about the authenticity, legitimacy, and reliability of an analysis (Almeida et al. 2016; Moreau et al. 2011; Stevens et al. 2007).

With the evolution of technology, information systems emerged capable of recording the origin of data, that is, the origin of each information inserted in a system. These systems allow for better data traceability, increasing the reliability and security of the information used in patient care. When applied to healthcare decision support systems, this requirement of trust translates into the ability to readily demonstrate the clinical reasoning performed in a clinical encounter, along with the recommendation received. In addition to supporting the auditability of the process, this feature should also promote transparency and traceability, from the recommendation to the rules applied to produce the recommendation. Given the arguments presented, the importance of obtaining explanations about the steps and rules (or reasoning) employed to arrive at an answer or knowledge becomes increasingly evident. As a result, understanding what manipulations were made and where the information was sourced from becomes crucial. In this sense, recovering the origin of data (its provenance), to ensure reproducibility in scientific research, in addition to enabling data auditing, is a task that can help in the process of making important decisions for different areas of research (Silva et al. 2003; Álvarez et al. 2006).

Provenance data can also be used to validate results of in silico experiments, recording the rules and reasoning used in the process of deriving a data set, known as knowledge provenance. This terminology is analogous to data source (Silva et al. 2003). This approach includes information, as if it were evidence, about the process by which knowledge arrived in a knowledge base. This process can include all of the reasoning used to generate conclusive information. However, "provenance" and "provenance of knowledge" have the same concerns and motivations, differing in terms of the purpose of the record that will be captured.

In the area of information technology, the source of data is responsible for maintaining information about all means used to obtain a result. In order to obtain explanations about how results were generated, it is necessary to know how, when, where, and why the original data were obtained, as well as all the means that modified it until the final result. This allows experiments to be reproduced and results to be analyzed and interpreted throughout their history (Moreau et al. 2011; Stevens et al. 2007).

However, for the results obtained to be reliable, it is necessary that both the original data and the means that modified it are stored in an organized way and that these data are able to bring quality and reliable information for the analysis (Paula et al. 2013). In the field of health, the origin of data in electronic medical records and decision support systems requires standardization so that these data are meaningful for use in scientific research and systems decision support (Galvao and Ricarte 2011). However, until now, there is no agreement regarding the data that should be captured and the way in which they should be stored and then processed in order to obtain useful information for use by health professionals (Johnson et al. 2014). In addition, in results that prove to be doubtful, provenance can be a qualitative tool, as it can be analyzed which data were used, identifying whether there are faulty data at the source that led to the propagation of errors.

One of the biggest challenges today to be transposed in the origin of data is to complete a model that can meet all the necessary requirements so that the entire history of a data is stored reliably and that it is possible to qualify it as effective for the various applications. This is also the challenge of provenance management in projects related to the health area since the programs to be used depend on the type of analysis performed, which generates a great diversity of these programs, indicating the need for flexible managers.

6.3 The Importance of Computerizing Health Data

Generating information is intrinsic to human beings. In the current digital age, we are constantly generating new information, and technological advancements have enabled this information to be captured, stored, and analyzed more efficiently than in the last century. Additionally, besides the information generated by our interactions as users, such as with smartphones, computers also rely on data from intelligent technologies that capture information about our existence and interactions with the environment, for example, sensors (Guha and Kumar 2018).

In order to understand the importance of Health Information Systems, it is important to establish how they can help professionals in managing patient information. Historically, scientific data were used mainly as a way to corroborate hypotheses and confirm beliefs about phenomena. However, the advancement of data collection, management, and analysis technologies has provided the opportunity for these data to be used as tools for building knowledge, for example, by assisting in predicting phenomena and aiding decision-making, not only in the medical environment but also in sectors such as business and consumer behavior.

Traditionally, the health sector has lagged behind other segments when it comes to the use of data, mainly due to resistance to changes by health professionals, who are used to making decisions based on clinical experience. Additionally, investors have under-invested in information technology due to uncertain returns and the difficulty of consolidating and standardizing data due to aging technologies. Furthermore, the nature of healthcare can make it difficult to share data due to privacy concerns (McGraw and Mandl 2021). Thus, finding ways to manipulate and deliver information efficiently and, consequently, dissemination and creation of knowledge can be an important differential in the delivery of health services (Siqueira 2005).

In Siqueira (2005), a definition for system is given as "we understand as systems a set of interdependent and interacting elements that form an organized whole and aims to transform inputs into outputs." The inputs, in turn, are the elements captured by the system and utilized in the output generation, which similarly can be defined as the input elements modified by the system processes. For a system to work, it needs three raw materials—data, information, and knowledge (Siqueira 2005).

In the field of health, Health Information Systems (HISs) can be instruments that, through the collection and processing of data from health services, for example, clinics and hospitals, help in the construction of knowledge for better decision-making (Coelho and Chioro 2021). HISs are computational systems that aim to enable environments that are free, or attenuated, from the use of physical documents in all aspects of the life of a health institution, for example, clinical, administrative, and financial.

However, the data stored by the health sector are still underused, and there is a difficulty in recovering these data and in understanding the care flow related to the patient. Additionally, outdated data regarding clinical treatment, difficulty in generating reliable indicators, and medical records with excessive records can be observed. Consequently, a large volume of difficult-to-use data are generated, which makes the information precarious to guide the management decision-making process (Cavalcante 2011).

When we observe the specialized literature, reports, and recommendations of Health Conferences, there is a consensus on the importance of using information as a way of evaluating health policies. Towards the end of the last century, technological advancements started to promote the use of large-scale databases that could be managed, with the expectation of achieving better access to clinical information and transitioning from efficacy studies to effectiveness studies. Moreover, this move aimed to encompass a more inclusive universe of analysis concerning access to information.

Still on the challenges inherent in the use of data, it is worth noting that we are currently living in the era of large volumes of data and this has generated a series of challenges for professionals who work with data. One of the main challenges is the ability to process, store, and analyze this data efficiently and effectively, using appropriate technologies to deal with the large amount of information generated daily. In addition, privacy and data security issues must also be addressed, ensuring that information is used ethically and responsibly. Even so, the benefits that the use of data can bring to companies and society are undeniable, as long as these challenges are adequately overcome. In this sense, the Machine Learning area offers several techniques that can be used to develop algorithms for analyzing medical data, including those stored in electronic medical records, with the aim of providing insights and support in clinical decision-making. For example, Machine Learning algorithms can be used to analyze large sets of medical data and identify patterns or trends that would be difficult or impossible to identify manually. These patterns can be used to make predictions about a patient's health and help clinicians make informed decisions about treatment and care needed.

6.4 Machine Learning and the Application in Healthcare

Machine Learning (ML) emerged as an alternative to statistical inference to improve a computer's ability to perform pattern identification (Friedman et al. 2001). Supervised learning is one of the best known paradigms of ML. From a set of data, the model infers a function that corresponds to a response variable (Friedman et al. 2001). The chosen base model dictates the structure of the final prediction function. The idea of supervised learning is that from a sample the predictive function is "trained" and this function can predict the response of future samples. This area of study is old and dates back to the 1960s when Frank Rosenblatt proposed a classifier model known as the "perceptron" (Rosenblatt 1958). One of the first applications in biology was in Stormo et al. (1982), in which, given an amino acid sequence, the perceptron was used to recognize transcription initiation patterns of *Escherichia coli* genes. This is just one example of the use of Machine Learning within the field of Biological Sciences.

Classifiers are a group of learning algorithms supervised, and the objective is to estimate a binary dependent variable from a data set of independent variables. Within clinical medicine, classifier models are incorporated as models to support clinical management (Tarca et al. 2007; Abreu et al. 2016; Johnson et al. 2016; Huang et al. 2017), optimizing decision-making, decisions within the medical clinic, and the outpatient setting.

Patient mortality prediction models were the first applications of supervised learning in clinical health (Johnson et al. 2016). From a set of variables of a patient, the model tries to estimate the patient's mortality within a period of time.

The APACHE II system is one of the first successful attempts to apply a mortality predictor within the clinic. Through logistic regression techniques in several physiological measures, this system tries to predict the risk of death of a patient recently admitted to an intensive care unit. Efforts are currently focused on optimizing these models, the review (Johnson et al. 2016) describes current efforts to update mortality classifiers within the intensive care unit, and according to this chapter, many of the applied models are neural networks.

The work presented by Chen et al. (2014) describes development of a neural network for predicting mortality in patients with lung cancer. Neural networks is a generic term for a group of prediction algorithms (Friedman et al. 2001), usually represented in a structure of nodes (dubbed neurons) and "layers." Simply put, the basic neural network consists of two steps. First, the independent variable data is fed to the network through a data input layer, and each node performs a weighted linear combination of its input signals. The resulting output signal is sent to an "output layer." Subsequently, the flow of information is reversed, and the neural network uses the prediction error and the real value of the dependent variable to optimize the parameters of the linear combination. For a more in-depth exploration of this topic, it is recommended to read (Friedman et al. 2001).

Mortality prediction models were developed focusing on mainly cancer patients (Abreu et al. 2016; Chen et al. 2014), where attempts are made to predict the evolution of the cancer pathology after a certain time interval. Incorporating genetic data into these models is an idea explored in the literature (Huang et al. 2017), but a problem that arises from this approach is the selection of specific genes and explanations of the model and observed pathology. In the article presented by Guyon et al. (2002), based on DNA microarray data, they use a specific model called Support Vector Machine (SVM) to select a gene substrate that best explains the leukemia condition, its results demonstrated superior performance to methodologies of the time.

The choice of model for surveying the decision function, as in statistical inference, involves raising mathematical hypotheses about the distribution of variables within the population and sampling characteristics (Friedman et al. 2001). Cancer models using genetic information have benefits when based on SVM, as this algorithm has little performance degradation for a small volume of data (Tarca et al. 2007; Johnson et al. 2016; Abreu et al. 2016). From the choice of the model, it is necessary to use techniques for data treatment (Friedman et al. 2001), and this phase of the analysis is called preprocessing of data. The preprocessing phase ranges from the normalization of gene expression data to the selection of variables that will be included for model training. Data preprocessing is essential for developing statistical learning models.

Gene expression data has the problem that the number of possible dimensions (in this case the expression of each gene) exceeds the number of entries, which leads to a phenomenon called overfitting, in which the predictor system outperforms the set of data, data used to build it, but it is not capable of making generalizations to new data (Friedman et al. 2001). This problem reflects the "bias-variance tradeoff" paradigm, in which we can opt for either a model that performs well for current data, but does not generalize to new samples or a model that performs poorly but has a stable performance with new samples. Therefore, the variable selection process is extremely important for the gene selection problem, since a microarray assay has information on thousands of genes for a few patients who specifically have the condition (Guyon et al. 2002; Chen et al. 2014; Abreu et al. 2016). Variable selection processes can be classified into filtering methods, which use statistical tests to discover correlations between variables and population effects, and "wrapping methods," which use machine learning models to select variables (Guyon et al. 2002; Chen et al. 2014). Selecting the appropriate method for variable selection is also another crucial decision for optimizing the evaluation metric. Consider a binary classification problem, where a feature vector x describes each individual in the sample and is associated with a corresponding response variable y, which can only take on the values -1 or 1. In other words, for a sample of m individuals described by n characteristics, we have the following data set:

$$(y_1, y_1), ..., (y_1, y_1), \in \mathbb{R}^n \times \{-1, +1\}. \tag{6.1}$$

The SVM model solves the binary classification problem by estimating a hyperplane that separates the space of independent variables into two spaces, each representing a value of y. Equation (6.2) represents the equation of this hyperplane.

$$(w.x) + b = 0. \tag{6.2}$$

The vector w is a vector of real values, and b is a constant. Theoretically, many hyperplanes can separate the space of variables, but only one can maximize the distance between the plane and the points of two different classes. Then,

$$(w.x_i) + b \geq 1, se(y_1) = 1, \tag{6.3}$$

$$(w.x_i) + b \leq 1, se(y_1) = -1. \tag{6.4}$$

Multiplying the value of the binary class y_1, we can transform equations (6.3) and (6.4) into just one:

$$y_i[(w.x_i) + b] \geq 1, i = 1, ..., m. \tag{6.5}$$

The Euclidean distance d from a point x to the hyperplane (w, b) is given by

$$d(w, b; x) = \frac{|w.x + b|}{||w||}. \tag{6.6}$$

The distance p from the hyperplane (w, b) to two parallel hyperplanes, called margin, is then defined and the objective is to maximize the equation:

$$p(w, b) = \min_{\{xi, yi=1\}} d(w, b; x_i) + \min_{\{xj, yj=-1\}} d(w, b; x_j). \tag{6.7}$$

Substituting Eq. (6.6) into (6.7) and performing simplifications, we can arrive at the equation

$$p(w, b) = \frac{2}{||w||}. \tag{6.8}$$

Therefore, the optimal hyperplane that separates the space of variables x into two classes is the one that minimizes

$$\phi(w) = \frac{||w||}{2}. \tag{6.9}$$

The formulation of Eq. (6.9) is simplistic, as it describes an ideal case of a hyperplane that manages to perfectly separate linear data into two classes. Another important point to emphasize is that minimizing Eq. (6.9) maximizes the distance between the hyperplane and its margins. Figure 6.1 graphically represents what has been explained so far to facilitate the visualization of the problem.

The explanation so far is sufficient for the Support Vector Classifier (SVC), and however binary classification problems are not even linearly separable, that is, there is not always a hyperplane that completely separates the two classes. The SVM incorporates in Eq. (6.5) a term that represents the prediction error (ε_i), that is, the model starts to "tolerate" that some samples are classified incorrectly, as long as this results in an improved model. Incorporation of this error term results in Eq. (6.10).

$$y_i[(w.x_i) + b] + \varepsilon_i \geq 1, i = 1, ..., m. \tag{6.10}$$

The algebraic development of Eq. (6.6) is carried out using Lagrange multipliers, and we arrive at the following function:

$$\min_{w,b,\varepsilon} \frac{1}{2} w^T w + C\Sigma_{i=1}^{m} \varepsilon_i. \tag{6.11}$$

Function (6.11) needs to be minimized while respecting the restriction imposed by function (6.10). The constant C is a hyperparameter of the model, adjusted according to the empirical result, and it controls how much the SVM model can

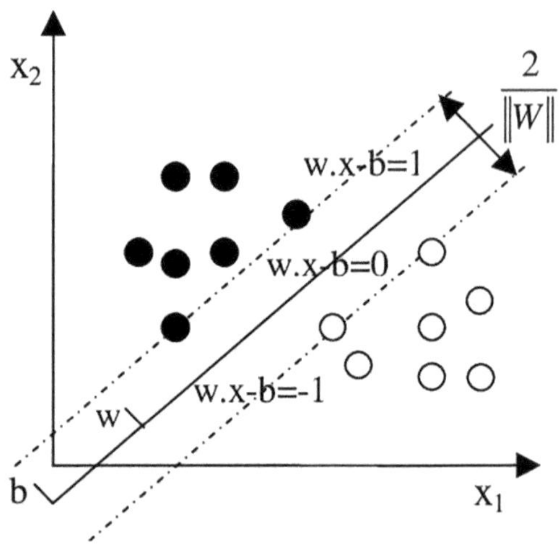

Fig. 6.1 Maximum-margin hyperplane and margins for an SVM trained with samples from two classes. Samples on the margin are called the support vectors (Islam et al. 2010)

err during the optimization phase of terms w and b. Another strategy applied is the use of a kernel function, which calculates the projection of the variables w in a space with more dimensions without transforming them.

Unlike neural networks, SVM generates a convex space, avoiding the local maximum problem, and this is one of the main advantages and reasons for using SVM instead of neural networks.

6.5 Conclusion

Health Information Management (HIM) plays a critical role in modern healthcare by organizing, analyzing, and managing electronic health data. HIM professionals ensure the completeness, accuracy, and accessibility of health records while also safeguarding data privacy and security. They possess expertise in data management techniques, databases, and specialized information systems designed for healthcare settings. Additionally, HIM professionals are increasingly utilizing Machine Learning techniques to automate data analysis, identify patterns and trends, and support decision-making processes.

One of the primary challenges faced by HIM professionals is the sheer volume of health data generated on a daily basis. Effectively managing this data requires a high level of precision and expertise to ensure privacy and security. Since health data is sensitive and personal, HIM professionals must prioritize protecting it from unauthorized access, disclosure, or misuse. Compliance with regulations and standards, such as the Health Insurance Portability and Accountability Act (HIPAA) in the United States, is vital in governing the collection, use, and storage of health data.

In the Brazilian context, ensuring the integrity of health records through provenance data management is of utmost importance. This allows HIM professionals to track changes made to health records over time, ensuring accuracy and keeping information up to date. The National Policy on Informatics and Health Information (PNIIS) implemented by the Brazilian Ministry of Health serves as a strategy to overcome challenges in health information management. The PNIIS promotes the innovative and transformative use of information technology, aiming to create an articulated National Information System that produces information for citizens, management, professional practice, knowledge generation, and social control. Interoperability, as advocated in the policy, addresses the problem of fragmentation by enabling different Health Information Systems (HISs) to communicate and exchange data seamlessly. Standards such as the TISS (Exchange of Information on Supplementary Health) and ICD (International Disease Code) are recommended for interoperability, ensuring legible and understandable health data representation.

HIM professionals play a pivotal role in ensuring the completeness, accuracy, and security of health records. They must stay updated on regulations, standards, and best practices in health information management while effectively handling large amounts of data. Machine Learning is increasingly becoming a valuable tool for HIM, enabling automated analysis, identification of patterns and trends, and informed decision-making. Provenance data management is vital for maintaining the integrity of health records and ensuring accuracy over time. With the advancements in telemedicine and digital health technologies, the role of HIM becomes even more critical in safeguarding the accuracy and security of health data exchanged over the Internet. Collaborating with vulnerable populations, such as rural and indigenous communities, is essential to ensure comprehensive and accurate health records, promoting equitable access to quality healthcare.

To conclude, health information management is an indispensable component of modern healthcare, providing the necessary infrastructure for efficient and effective management of health data. HIM professionals are instrumental in guaranteeing the completeness, accuracy, and accessibility of health records, while maintaining privacy and security. These professionals also must stay updated on regulations and best practices in the field while effectively handling data. The incorporation of Machine Learning techniques enhances data analysis capabilities, empowering HIM professionals to make informed decisions and improve patient care. Provenance data management ensures the integrity of health records, tracking changes, and preserving accuracy. The continuous advancement of HIM is essential in a data-driven healthcare landscape, as it plays a significant role in delivering high-quality healthcare and enhancing the overall health situation of the population in Brazil.

References

Abreu, Pedro Henrique et al. (2016). "Predicting breast cancer recurrence using machine learning techniques: a systematic review". In: *ACM Computing Surveys (CSUR)* 49.3, pp. 1–40.

Almeida, F N et al. (2016). "A provenance model based on declarative specifications for intensive data analyses in hemotherapy information systems". In: *Future Generation Computer Systems* 59, pp. 105–113. ISSN: 0167-739X. https://doi.org/10.1016/j.future.2015.09.019.

Almeida, F.N. et al. (July 2010). "Aplicando Procedência deDados para Análise dos Perfis de Doadores de Sangue". In: *IV Brazilian E-Science Workshop and XXX CSBC*.

Álvarez, Sergio et al. (2006). "Applying Provenance in Distributed Organ Transplant Management". In: *Provenance and Annotation of Data*. Ed. by Luc Moreau and Ian Foster. Berlin, Heidelberg: Springer Berlin Heidelberg, pp. 28–36. ISBN: 978-3-540-46303-0.

Cavalcante R. Nagata F. M., Silva P. (2011). "Sistemas de Informaçãao em Saúde: possibilidades e desafios". In: *Revista de Enfermagem da Universidade Federal de Santa Maria* 1, pp. 290–299. https://doi.org/10.5902/217976922580.

Chen, Yi-Cheng, Wen-Chyuan Ke, and Hua-Wen Chiu (2014). "Risk classification of cancer survival using ANN with gene expression data from multiple laboratories". In: *Computers in biology and medicine* 48, pp. 1–7.

Coelho Neto G. C., Chioro A. (2021). "Afinal, quantos Sistemas de Informação em Saúde de base nacional existem no Brasil?" In: *Caderno de Saúde Pública* 37, p. 7. https://doi.org/10.1590/0102-311X00182111.

Davidson, S.B., C. Overton, and P Buneman (1995). "Challenges in integrating biological data sources". In: *J Comput Biol* 2, pp. 557–72.

Discala, Claude et al. (Jan. 2000). "DBcat: a catalog of 500 biological databases". In: *Nucleic Acids Research* 28.1, pp. 8–9. ISSN: 0305–1048. https://doi.org/10.1093/nar/28.1.8. https://academic.oup.com/nar/article-pdf/28/1/8/9895206/280008.pdf. https://doi.org/10.1093/nar/28.1.8.

Friedman, Charles P., Adam K. Wong, and David Blumenthal (2010). "Achieving a Nationwide Learning Health System". In: *Science Translational Medicine* 2.57, pp. 57cm29–57cm29. https://doi.org/10.1126/scitranslmed.3001456. eprint: https://www.science.org/doi/pdf/10.1126/scitranslmed.3001456. https://www.science.org/doi/abs/10.1126/scitranslmed.3001456.

Friedman, Jerome, Trevor Hastie, and Robert Tibshirani (2001). *The Elements of Statistical Learning*. Vol. 1. Springer Series in Statistics 10. New York: Springer.

Galvao, Maria Cristiane Barbosa and Ivan Luiz Marques Ricarte (Dec. 2011). "O prontuário eletrônico do paciente no século XXI: contribuições necessárias da ciência da informação". In: *InCID: Revista de Ciência da Informação e Documentação* 2.2, pp. 77–100. https://doi.org/10.11606/issn.2178-2075.v2i2p77-100. https://www.revistas.usp.br/incid/article/view/42353.

Gonçalves, Rodrigo Franco et al. (Apr. 2016). "Uma abordagem sistêmica do processo de produçãao em engenharia web, na fase de concepção". In: *Production* 26.2, pp. 402–416.

Guha, S. and S. Kumar (2018). "Emergence of Big Data Research in Operations Management, Information Systems, and Healthcare: Past Contributions and Future Roadmap". In: *Prod Oper Manag* 27, pp. 1724–1735. https://doi.org/10.1111/poms.12833.

Guyon, Isabelle et al. (2002). "Gene selection for cancer classification using support vector machines". In: *Machine learning* 46.1-3, pp. 389–422.

Hogeweg, P (2011). "The Roots of Bioinformatics in Theoretical Biology". In: *PLoS Comput Biol* 7.3. https://doi.org/10.1371/journal.pcbi.1002021.

Huang, Mao-Wen et al. (2017). "SVM and SVM ensembles in breast cancer prediction". In: *PloS one* 12.1.

Islam, Md. Saiful, Abdullah Al Mahmud, and Md. Rafiqul Islam (2010). "Machine Learning Approaches for Modeling Spammer Behavior". In: *Information Retrieval Technology*. Ed. by Pu-Jen Cheng et al. Berlin, Heidelberg: Springer Berlin Heidelberg, pp. 251–260. ISBN: 978-3-642-17187-1.

Jannuzzi, Celeste Aída Sirotheau Corrêa, Orandi Mina Falsarella, and Cibele Roberta Sugahara (Dec. 2014). "Sistema de informação: um entendimento conceitual para a sua aplicação nas organizações empresariais". In: *Perspectivas em Ciência da Informação* 19.4, pp. 94–117. http://dx.doi.org/10.1590/1981-5344/1927

Johnson, Alistair EW et al. (2016). "Machine learning and decision support in critical care". In: *Proceedings of the IEEE* 104.2, pp. 444–466.

Johnson, K.E. et al. (2014). "How the Provenance of Electronic Health Record Data Matters for Research: A Case Example Using System Mapping". In: *Generating Evidence & Methods to improve patient outcomes* 2(1), p. 4. https://doi.org/10.13063/2327-9214.1058.

Karp, P (1995). "A strategy for database interoperation". In: *J Comput Biol* 2, pp. 573-86.

M, Lehne et al. (Aug. 2019). "Why digital medicine depends on interoperability". In: *NPJ Digit Med* 2–79. https://doi.org/10.1038/s41746-019-0158-1.

McGraw, Deven and Kenneth D. Mandl (Jan. 2021). "Privacy protections to encourage use of health-relevant digital data in a learning health system". In: *NPJ Digital Medicine* 4. https://doi.org/10.1038/s41746-020-00362-8.

Moreau, Luc, Ben Clifford, et al. (2011). "The Open Provenance Model core specification (v1.1)". In: *Future Generation Computer Systems* 27.6, pp. 743–756. ISSN: 0167–739X. https://doi.org/10.1016/j.future.2010.07.005. https://www.sciencedirect.com/science/article/pii/S0167739X10001275.

Moreau, Luc, Paul Groth, et al. (Apr. 2008). "The Provenance of Electronic Data". In: *Commun. ACM* 51.4, pp. 52–58. ISSN: 0001–0782. https://doi.org/10.1145/1330311.1330323. https://doi.org/10.1145/1330311.1330323.

MS/Brasil, Ministério da Saúde (2015). *Sistemas de Informação da Atenção à Saúde: Contextos Históricos, Avanços e Perspectivas no SUS/Organização Pan-Americana da Saúde—Brasília*. Secretaria de Atenção à Saúde. Departamento de Regulação, Avaliação e Controle. ISBN: 978-85-62258-10-7.

Paula, R. de et al. (2013). "Provenance in bioinformatics workflows". In: *BMC Bioinformatics* 14(Suppl 11). http://doi.org/10.1186/1471-2105-14-S11-S6.

Pressman, R. S (2006). *Engenharia de Software*. Vol. 6. McGraw-Hill, 720p.

Rosenblatt, Frank (1958). "The perceptron: a probabilistic model for information storage and organization in the brain". In: *Psychological review* 65.6, p. 386.

Silva, P.P.D., D.L. McGuinness, and R. McCool (2003). "Knowledge Provenance Infrastructure". In: *In Proceedings of IEEE Data Eng. Bull*, pp. 26–32.

Siqueira, Marcelo Costa (2005). "Gestão Estratégica da Informação". In: *Brasport Livros e Multimídia Ltda*, pp. 1–2.

Stevens, Robert, Jun Zhao, and Carole Goble (May 2007). "Using provenance to manage knowledge of In Silico experiments". In: *Briefings in Bioinformatics* 8.3, pp. 183–194. ISSN: 1467-5463. https://doi.org/10.1093/bib/bbm015. https://academic.oup.com/bib/article-pdf/8/3/183/575406/bbm015.pdf. https://doi.org/10.1093/bib/bbm015.

Stormo, Gary D et al. (1982). "Use of the 'Perceptron' algorithm to distinguish translational initiation sites in E. coli". In: *Nucleic acids research* 10.9, pp. 2997–3011.

Tarca, Adi L et al. (2007). "Machine learning and its applications to biology". In: *PLoS computational biology* 3.6, e116.

Chapter 7
Hemodynamic Modeling of Supraventricular Arrhythmias Using an Integrated Numerical Approach

João Loures Salinet, Ítalo Sandoval Ramos de Oliveira, John Andrew Sims, and João Lameu

Abstract The heart rhythm is controlled by electrical activity. Disorder in electrophysiology leads to cardiac arrhythmias (CA), which can be grouped into supraventricular, atrial, and ventricular arrhythmias. Among atrial arrhythmias, the most common is atrial fibrillation (AF). AF is the main cause of ischemic events and myocardial infarction. The risk of ischemic events in AF patients is about five times higher than in healthy individuals. About 1/3 of hospitalizations related to CA are due to AF, with an average annual cost of USD$5450/patient. Persistent cases of AF alter atrial contraction increasing their rigidity, and resulting in prothrombotic zones due to blood stagnation, especially in the left atrial appendage. Due to the complexity of hemodynamics in AF, computational fluid dynamics (CFD) is a useful complementary tool, providing patient-specific information related to thrombogenic risks. The proposed integrated model coupled realistic electrophysiological activity from 3D AF mapping, inducing local mechanical effects in the blood flow. Also, the coupling of a simplified coagulation cascade model allowed the direct prediction of clot stability. The main hemodynamic characteristics were well represented and the integrated model could be customized to patient-specific data providing a more realistic predictive tool for thrombogenesis risk assessment.

Keywords Cardiac arrhythmia · Thrombogenesis · Electrophysiology · Atrial fibrillation · Computational fluid dynamics

---

J. L. Salinet (✉) · Í. S. R. de Oliveira
Graduate Program in Biomedical Engineering, Federal University of ABC – UFABC, São Bernardo do Campo (SP), Brazil

Center for Engineering, Modeling and Applied Social Sciences (CECS), HEartLab – Biomedical Engineering, Federal University of ABC – UFABC, São Bernardo do Campo (SP), Brazil
e-mail: joao.salinet@ufabc.edu.br

J. A. Sims · J. Lameu
Center for Engineering, Modeling and Applied Social Sciences (CECS), HEartLab – Biomedical Engineering, Federal University of ABC – UFABC, São Bernardo do Campo (SP), Brazil

C. B. Lombello, P. A. da Ana (eds.), *Current Trends in Biomedical Engineering*,
https://doi.org/10.1007/978-3-031-38743-2_7

7.1 Introduction

Blood circulation results from the beating heart, which provides the mechanical driving force (pressure gradient) throughout the systemic and pulmonary circulation. The systemic circulation, related to the left side of the heart, provides oxygenated blood to the peripheral tissues and organs. The rhythmic heartbeat has critical dependence upon preceding electrical activations, and the disruption of this orderly electrophysiology leads to cardiac arrhythmias (Tse 2016). Cardiac arrhythmias encompass a broad spectrum of disorders of heart function disorders, including heart rate and rhythm abnormalities and it is generally categorized into bradyarrhythmia (heart rate below 60 bpm) and tachyarrhythmia (heart rate above 100 bpm) (Desai and Hajouli 2022). They can also be grouped by the location of abnormal electrical activation into supraventricular (originated in the atria) and ventricular (originated in the ventricles). Cardiac arrhythmias generally lead to the deterioration of mechanical function and consequently to hemodynamic-related outcomes, including increase in thrombogenic risks, which are the main cause of cerebral ischemia, myocardial infarction, and venous thromboembolism (Dueñas-Pamplona et al. 2021).

7.2 Supraventricular Arrhythmias

As we have seen, supraventricular tachycardias (SVTs) are characterized by the increase of atrial rates as higher as 100 beats per minute at rest. SVTs can be caused by several mechanisms, acting individually or simultaneously in the atrium, which can also affect the mechanical and hemodynamical functions of the heart (Page et al. 2016; Marques et al. 2020a). The three most common atrial arrhythmias are atrial tachycardia (AT), atrial flutter (AFL), and atrial fibrillation (AF).

AT can be defined as a rapid regular atrial rhythm, originated from ectopic focus which can be localized either in the right or left atrium, but not in the sinoatrial (SA) node. The ectopic mechanism is characterized by its regular high-frequency automaticity, generating a centrifugal wave that propagates from its regions close to the focus to its neighboring areas, in a ratio of 1:1 (Madeiro et al. 2018; Marques et al. 2020a).

In contrast to AT, the fast-regular atrial activations in AFL are triggered and maintained by a macro reentrant circuit propagating into a circular reentry circuit, at rates of 300 bpm or above, around an anatomical structure or functional substrate, such as the tricuspid valve or scar tissues. The flutter is more likely to occur in the right atrium than the left atrium, with the typical AFL occurring with the reentry wavefront propagating in a counter-clockwise direction starting from the right atrium, to its anterolateral wall, then to the istmo cavotricuspid and returning through the interatrial septum (Madeiro et al. 2018). The ventricles do not respond in a 1:1 ratio during AFL, due to the atrioventricular node refractoriness, slowing down its rate. It has been shown that AFL increases the risks of thromboembolic events (Page et al. 2016).

Atrial fibrillation (AF) is the most common sustained cardiac arrhythmia in clinical practice (Salinet et al. 2021; Rahman et al. 2014), with a prevalence of around 1% of the worldwide population, which is expected to increase 2.5-fold in 50 years (Salinet 2014). About 1/3 of hospitalizations related to cardiac arrhythmias are due to AF, giving an average annual cost per patient of USD$5450 (Kim et al. 2011; Wodchis et al. 2012). In this arrhythmia, the atria activated in an electrically uncoordinated manner, with the tissue not responding in a ratio of 1:1, resulting in the maintenance of the AF through fibrillatory conduction. AF can result from several different underlying mechanisms, including ectopic foci, multiple, random propagating wavelets, and localized reentrant activity with fibrillatory conduction, which can occur individually from patient to patient, and coexist simultaneously, favoring a complex arrhythmia, and with consequences that extend to mechanical deterioration, associated fibrosis and loss of atrial muscle mass which affects the ability of the atria to pump blood effectively (Salinet et al. 2021). As the mechanical pumping ability of the atria is compromised in AF, the atria do not empty effectively between contractions and this can lead to the formation of blood clots and thromboembolic events (frequently in the left atrial appendage, LAA), increasing the risk of stroke fivefold and doubling mortality (Salinet et al. 2021; Wang et al. 2003).

AF is classified as paroxysmal or persistent/permanent. Paroxysmal AF is defined as short duration: episodes of the arrhythmia with spontaneous termination. Persistent/permanent AF is defined as continuous episodes with durations that can last more than a year, and with interruption being made by use of chemical or electrical intervention. AF results in altered patterns of atrial contractions, and in persistent cases, it causes increased stiffness of the left atrial (LA) wall, leading to the formation of blood stagnation zones and potential thrombus formation (García-Isla et al. 2018). Thromboembolic events are mainly related to three factors: hypercoagulability, endothelial lesions, and blood stasis, according to Virchow's triad (Watson et al. 2009). Thus, thrombogenesis occurs especially in the left atrial appendage (LAA) (Mill et al. 2021). The LAA is a small anatomical pocket remaining from the embryonic development of the LA (Al-Saady et al. 1999), located close to the left pulmonary veins (LVP) (Fang et al. 2021). In healthy patients, the LAA presents great contractility, preventing the formation of thrombi, in contrast to patients with reduced contractility due to AF (Dueñas-Pamplona et al. 2021). About 70–99% of thrombi in patients with non-valvular AF are formed in the LAA (Cresti et al. 2019; Lee et al. 2015; Wang et al. 2010). These thrombi cause about 90% of strokes in patients with non-valvular AF and about 60% of patients with mitral valve (MV) lesions (Dueñas-Pamplona et al. 2021; Al-Saady et al. 1999).

Advances in medical imaging techniques such as angiography, computed tomography (CT), and transesophageal echocardiography (TEE) have allowed a better understanding of atrial morphology and dynamics (López-Mínguez et al. 2014; Vedula et al. 2015). New techniques such as 4D magnetic resonance imaging (4D-

MRI) (Morales et al. 2021; Markl et al. 2015) and Blood Speckle Tracking (Nymes et al. 2020; Zhang and Yu 2007) are promising in hemodynamic characterization, however, it is still uncertain whether these have sufficient resolution to capture the mechanisms of thrombogenesis in conditions related to low blood velocity (Mill et al. 2021). In this scenario, CFD is a possible complementary tool for hemodynamical studies, being able to obtain additional patient-specific information related to thrombogenic risks.

7.3 Applied Example: Integrated Hemodynamic Modeling for Atrial Fibrillation

Due to the inherent and specific complexity of blood flow under different arrhythmias, especially in AF, along with the limitation of spatial-temporal resolution of medical imaging techniques, numerical modeling of hemodynamics provides an alternative to enhance the understanding of the underlying physics and its consequences in cardiovascular disease development. This section is focused on atrial fibrillation and its hemodynamical characterization.

7.4 Electrophysiological Activity

As said before, supraventricular tachyarrhythmias and mainly atrial fibrillation (AF) can have serious consequences in the body, involving hemodynamic issues. In order to study the effects of AF on the blood, first we need to know the physiological behavior of the atrial tissues. This can be obtained by simulation software such as OpenCarp, which simulates the electric potentials in myocytes and can determine its propagation through a given tissue.

The mechanism of AF, such as rotors and ectopic foci, can be simulated by: (i) inducing the stimulus, (ii) changing the conductivity and connectivity properties of the tissue (e.g., randomly disconnecting cells for simulating fibrotic tissue). The activation frequency of the cells can be obtained by detecting the cell activation, which can be performed with the wavelet method described by Marques et al. (2020b), calculated by the inverse of the interval between activations. As the electrical potentials are related to the myocyte contraction, the activation frequency can be used as a boundary condition in the hemodynamic simulation.

7.5 Hemodynamic Modeling by Computational Fluid Dynamics (CFD)

Computational fluid dynamics (CFD), also referred to as in silico fluid dynamics, is a robust numerical technique that provides excellent spatial-temporal detail in the evaluated domain, being a versatile tool for hemodynamic evaluation in different scenarios, including specific morphological alterations, inclusion of devices for left atrial appendage occlusion (Mill et al. 2021; Tzikas et al. 2016; Aguado et al. 2019), consideration of wall motion (Mill et al. 2021; Fang et al. 2021; Masci et al. 2019; Koizumi et al. 2015), and among others. The accuracy and reliability of the CFD model are directly related to the boundary conditions (BCs) imposed on the numerical solution, as well as other modeling considerations, including flow regimes (laminar, transition, and turbulent) and blood rheological behavior (Newtonian or Non-Newtonian). Most studies have considered the laminar flow regime (Masci et al. 2019; Bosi et al. 2018; Aguado et al. 2019; Koizumi et al. 2015). Turbulence modeling was evaluated by Dueñas-Pamplona et al. (2021) in patients with and without AF, by comparing the laminar model and LES (Large Eddy Simulation) approach, capable of detecting the transition between regimes. The authors observed that laminar flow prevails throughout the entire cycle, in agreement with in vivo observations (Vedula et al. 2015) and other numerical studies (Bosi et al. 2018; Aguado et al. 2019; Koizumi et al. 2015).

In terms of the BCs, it is necessary to define: (i) flow conditions at the inlets and outlets – pressure or velocity profiles; (ii) the behavior of the wall, rigid or mobile/deformable; and (iii) its interaction with the viscous fluid, the so-called no-slip condition. In an ideal scenario, patient-specific data (velocity/pressure profile and medical images of the region of interest) provide the necessary BCs for hemodynamic simulation, however, this is not always feasible. Some studies have evaluated different BCs and their effects on hemodynamic results in AF conditions (Dueñas-Pamplona et al. 2021; Mill et al. 2021). The main features of AF are the absence of organized atrial contraction at the end of ventricular diastole (i.e., absence of the P wave on the ECG), leading to the absence of the second peak wave in the blood flow through the mitral valve, along the occurrence of high-frequency wall fibrillation. The first characteristic can be considered in BC by idealizing the velocity profile in the mitral valve, based on data from healthy patients, as studied by Koizumi et al. (2015) and Bosi et al. (2018), if hemodynamic data from AF patients are not available.

The consideration of high-frequency fibrillation presents a greater modeling challenge, since so far there are no techniques available for image acquisition that provide dynamic deformation details in a realistic way (Masci et al. 2019). Some studies have considered rigid atrial walls, justified as the worst scenario of AF (Masci et al. 2019; Bosi et al. 2018; Aguado et al. 2019; Zhang and Gay 2008), however, this consideration may be an incorrect approximation of wall behavior, since there is a certain degree of passive contraction in the atrium (Dueñas-Pamplona et al. 2021). Fibrillation idealizations have been proposed using dynamic meshes considering predefined oscillation frequencies and amplitudes (Masci et al. 2019; Koizumi et al. 2015) or passive movement of the atrium based on the displacement of the mitral annulus (Mill et al. 2021; Koizumi et al. 2015). Other more complex models with higher computational cost have been proposed based on fluid–structure interaction (FSI) simulations (Fang et al. 2021; Zhang and Gay 2008), however, other simplifications are often necessary due to uncertainties in the characterization of atrial mechanical properties (wall thickness, elastic behavior and properties, orientation of muscle fibers, etc.) (Mill et al. 2021), mainly for AF cases. There is also the possibility of dynamic reconstruction of the geometries from medical images (e.g., computed tomography – CT or magnetic resonance imaging – MRI) obtained at different times of the cardiac cycle, applying specific spatial and temporal reconstruction algorithms, but these are limited to healthy patients (Masci et al. 2019; Koizumi et al. 2015).

Few studies considered the electromechanical effects of high-frequency AF (Masci et al. 2019; Koizumi et al. 2015). Koizumi et al. (2015) provided numerical studies of hemodynamic changes in the LA due to AF. A dynamic model of the LA, obtained by MRI from a healthy volunteer, was used as a reference for the development of two idealized AF models. Dynamic reconstruction of atrial volume was based on mitral valve dynamics. The first AF model (AF1) considered the absence of the second velocity profile through the mitral valve, while the second model (AF2) also considered the effects of fibrillation by a 7.58 Hz sine wave (five times faster than the healthy patient heartbeat) with an amplitude of 2.35% of the difference between the minimum and maximum volumes of the healthy LA. Koizumi et al. (2015) observed a lower volume of atrial emptying (around 10%) in the AF models compared to the healthy one, in addition to a considerable increase in the relative residence time of blood, mainly by the AF2 model, indicating blood stagnation. The authors observed a greater fluctuation in the velocity predicted by the AF2 model, which is not consistent with clinical observations. Thus, they recommended that more realistic conditions of fibrillation in different phases should be considered in the LA wall.

Masci et al. (2019) also accounted for fibrillation by a sinusoidal function, considering a frequency of 4 Hz multiplied by a random factor obtained by a uniform probability density function, varying between 0 and 1. The wave was modulated to a maximum amplitude of 0.1 mm, in order to simulate the reduced AF contractions and avoid convergence problems related to the poor quality of the numerical mesh elements. The authors concluded that even less complex left atrial appendage (LAA) morphologies may have low velocities and vorticities, characterizing favorable scenarios for thrombogenesis. The correlation between several geometric factors was related to low blood turnover rates, leading to recirculation zones even in simpler or less tortuous LAA morphologies. Masci et al. (2019) concluded that more realistic models of AF should be considered.

Few experimental studies have been carried out focusing on the movement of the atrial wall in AF conditions (De Vos et al. 2009; Limantoro et al. 2014). The literature reports great difficulty in the dynamic reconstruction of atrial structures in these conditions, due to the limitations of spatial and temporal resolution of current equipment (Masci et al. 2019). De Vos et al. (2009) and Limantoro et al. (2014) used transthoracic echocardiography and tissue Doppler velocimetry (tissue velocity imaging, TVI) of the atrial myocardium to quantify AF noninvasively in more than 200 patients: 12 patients (De Vos et al. 2009), 215 patients (Limantoro et al. 2014) with paroxysmal or permanent AF. De Vos et al. (2009) validated the measurements by TVI with electrocardiograms. Permanent AF was associated with lower cardiac wall movement velocities and higher fibrillation frequencies (De Vos et al. 2009;

Limantoro et al. 2014). Data on mean wall fibrillation velocity and cycle duration were reported, serving as a starting point for a more realistic wall motion modeling by CFD.

Spatial and temporal information from electrical activity mapping can be used as BCs in CFD simulations, providing a more realistic way of considering AF in atrial hemodynamics. These can be obtained by electro-anatomical mapping techniques or even by data obtained from electrophysiological simulations of the propagation of electrophysiological activity in 2D or 3D models. Considering patient-specific data (Rodrigo et al. 2017; Marques et al. 2020b), our research group proposed an integrated numerical model based on the coupling of electrophysiological activity to consider the mechanical effects of fibrillation on hemodynamics by a momentum method, which has a lower computational cost than FSI-type simulations.

7.6 Governing Equations of Blood Flow

The integrated hemodynamic model was based on computational fluid dynamic (CFD) simulations, considering non-Newtonian, incompressible, and transient blood flow. The mass and momentum conservations (Eqs. 7.1 and 7.2) were numerically solved by the finite volume method using ANSYS Fluent code:

$$\nabla \cdot \mathbf{U} = 0 \tag{7.1}$$

$$\rho \left(\frac{\partial \mathbf{U}}{\partial t} + \mathbf{U} \cdot \nabla \mathbf{U} \right) = -\nabla p + \nabla \cdot \boldsymbol{\tau} + \mathbf{f} \tag{7.2}$$

where ρ is the specific mass (kg m^{-3}) ($\rho = 1055$ kg m^{-3}), $\mathbf{U}$ is the velocity vector (m s^{-1}), p is the pressure (kg m^{-1} s^{-2}), $\boldsymbol{\tau}$ is the viscous stress tensor (kg m^{-1} s^{-2}), given by Eq. (7.3), and $\mathbf{f}$ represents addition momentum source terms (kg m^{-2} s^{-2}), given by Eq. 7.4:

$$\boldsymbol{\tau} = \eta(\dot{\gamma}) \left(\nabla \mathbf{U} + (\nabla \mathbf{U})^T \right) \tag{7.3}$$

where $\eta(\gamma)$ is the apparent dynamic viscosity (kg m^{-1} s^{-1}), modeled by the Generalized Newtonian approach, e.g., using Carreau–Yasuda or Casson model.

7.7 Integration of Electromechanical Effects into Hemodynamics

The local oscillatory mechanical effects inherent from AF electrophysiological patterns were considered by a momentum-based method, considering rigid atria walls with the addition of an extra momentum source in the blood momentum balance (**f** in Eq. 7.2). This additional force was considered by a local, relative inertial effect induced by the oscillatory wall motion, as given by Eqs. (7.4)–(7.6), illustrated by the vector field of **f** shown in Fig. 7.1d:

$$\mathbf{f} = \rho\, \phi \left(\frac{U_w - \mathbf{U}}{\Delta t} \right) \tag{7.4}$$

$$U_w = \frac{d_w^{t_i} - d_w^{t_i - 1}}{\Delta t} \text{ with } d_w = A \ \sin\left(2\pi f_{\mathrm{AF}} t\right) \tag{7.5}$$

$$\frac{\partial \varphi}{\partial t} + \mathbf{U} \cdot \nabla \varphi = \phi_{\mathrm{activ}} \tag{7.6}$$

where ϕ_{activ} is the activity indicator from electrophysiological data from EGM mapping, as detailed in Fig. 7.1a, b with $\phi = \phi_{\mathrm{activ}} = 1$ for active walls, $\phi = \phi_{\mathrm{activ}} = 0$ for non-active walls, $\phi = \varphi$ for the fluid domain. U_w is the wall velocity, calculated using Eq. 7.5 from the wall displacement, d_w, determined by an isotropic motion field, given by a first-order sinusoidal function with spatial-temporal variable fibrillation frequency, f_{AF}, as obtained from 3D electrophysiological mapping (Fig. 7.1a, b), and a prescribed amplitude, A (e.g., $A = 0.88$ mm, for persistent AF patients (Limantoro et al. 2014)). The methodology of the integrated numerical model is provided in Fig. 7.1.

7.8 Hemodynamic Indicators of Pro-thrombotic Zones

The hemodynamic indicators of pro-thrombotic zones are based on values of time-averaged wall shear stress (*TAWSS* in Eq. 7.7). Three *TAWSS*-derived indicators are often used in cardiovascular disease assessment by CFD studies: Oscillatory Shear Index (*OSI* in Eq. 7.7), a dimensionless parameter expressing the temporal variation of the *WSS* vector direction relative to the predominant flow direction. The *OSI* varies between 0 and 0.5, with the maximum value indicating a variation of 180° in the *WSS* direction, relative residence time (*RRT* in Eq. 7.8), estimating the residence time of a fluid particle in a given local, endothelial cell activation potential (*ECAP* in Eq. 7.8), and characterizing the local thrombogenic susceptibility.

$$\mathrm{TAWSS} = \frac{1}{T} \int_o^T |\mathrm{WSS}|\, \mathrm{d}t \qquad \mathrm{OSI} = \frac{1}{2} \left(1 - \frac{\left| \int_0^T \mathrm{WSS}\ \mathrm{d}t \right|}{\int_0^T |\mathrm{WSS}|\ \mathrm{d}t} \right) \tag{7.7}$$

$$\mathrm{RRT} = \left[(1 - 2 \cdot \mathrm{OSI})\, \mathrm{TAWSS} \right]^{-1} \ \mathrm{ECAP} = \frac{\mathrm{OSI}}{\mathrm{TAWSS}} \tag{7.8}$$

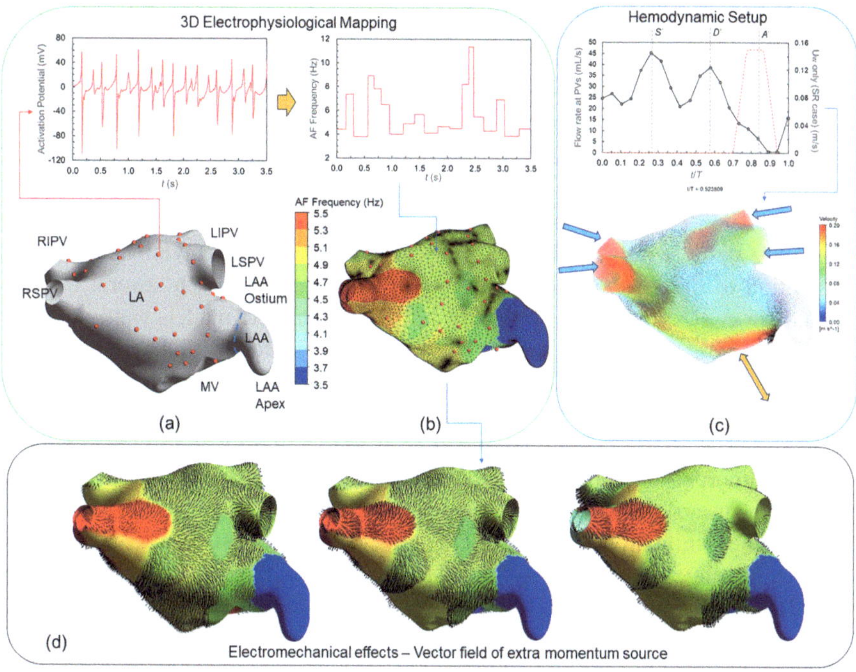

Fig. 7.1 Schematics of integrated numerical model: (**a**, **b**) Procedure for spatial-temporal map of electrophysiological activity, providing a 3D map boundary condition for the CFD setup: (**a**) 3D mapping of AF activity mimicking a 64-multiarray-electrode with EGMs with details of the left atrial portions (*LSPV* left superior pulmonary vein, *LIPV* left inferior pulmonary vein, *RSPV* right superior pulmonary vein, *RIPV* right inferior pulmonary vein, and *LAA* left atrial appendage), the left atrium geometry was obtained from open dataset (Roney et al. 2022); (**b**) determination of local AF frequency (f_{AF}), providing a 3D transient map as a BCs for the CFD model to account for the mechanical effects of oscillatory wall motion; (**c**) definition of inlet flow rates (black line with dots) at pulmonary veins (PVs) (Fernandez-Perez et al. 2012), red dotted lines: LAA wall velocity, prescribed only for healthy sinus rhythm case (Farese et al. 2019) in order to mimic the atrial contraction for healthy case, S', D', and A' indicate the systolic, diastolic, and atrial contraction instants of the cardiac cycle and T is the total cardiac flow cycle duration; and (**d**) transient sequence of the vector field of the extra momentum source, illustrating the local flow perturbation inherent from the oscillatory wall motion initiated by the random electrophysiological activity (Lameu et al. 2022)

where WSS is the instantaneous local wall shear stress, T is the total cardiac cycle duration, dt is the simulation timestep, and $\overline{\text{TAWSS}}$ is $TAWSS$ normalized by the global WSS average.

7.9 Simplified Biochemical Model for Direct Thrombi Prediction

The direct prediction of thrombogenesis is based on biochemical models by the solution of advection–diffusion–reaction transport equations of the biochemical and cellular species involved in the coagulation cascade. The complete network of the coagulation cascade results in a very high computational cost, mostly when simulating cases with complex geometries and flow fields (Taylor et al. 2016), in addition to the characteristic clot growth time that can reach several minutes (Menichini et al. 2016; Taylor et al. 2014, 2016). Complete models were proposed containing a large number of species (e.g., 23 species) (Bodnár and Sequeira 2008; Anand et al. 2003). Recently, simplified models have been evaluated by reducing the number of species involved in the cascade: four species (Menichini and Xu 2016; Menichini et al. 2016; Grande-Gutiérrez et al. 2021) and three species (Taylor et al. 2016), in order to reliably capture the main features of thrombogenesis. Another aspect is the use of artificially accelerated kinetics (Menichini and Xu 2016) in order to reduce the minimum time necessary to predict stable clots in a few cycles, usually 10–20 cardiac cycles or about 4–10 s (Menichini et al. 2016; Grande-Gutiérrez et al. 2021). This strategy is possible since prothrombotic regions are associated with relatively low-velocity zones, and thrombus growth results only in gradual changes in hemodynamics (Ngoepe and Ventikos 2016; Malaspinas et al. 2016; Peach et al. 2014).

In general, the coagulation process is divided into two pathways: intrinsic (or contact) and extrinsic (initiated by tissue factor). Both converge on a common path leading to the activation of factor X and generation of thrombin, fibrin, and, finally, the stable clot (polymeric fibrin) (Grande-Gutiérrez et al. 2021). The proposed models are usually based on vascular lesions, that is, on the extrinsic path, not being fully adequate for predicting thrombogenesis in cases of absence of vascular lesions as in the case of low shear regions of the blood, as observed in the LAA especially under AF. Grande-Gutiérrez et al. (2021), evaluating Kawasaki disease (KD) in coronary arteries, proposed a biochemical model of four species: deactivated platelets (PD), activated platelets (AP), adenosine diphosphate (ADP), and polyphosphate (PoliP), for conditions of thrombogenesis initiated in the absence of the extrinsic factor. The authors noticed, through experimental observations of MRI images a high thromboembolic risk related to KD aneurysms, due to the high concentration of activated platelets and agonist biochemical factors (e.g., ADP) in the pro-coagulation environment. The model was proposed based on the critical role of PolyP released from PA, initiating the intrinsic cascade (Choi et al. 2011; Müller et al. 2009) and the critical concentration of ADP related to platelet activation and aggregation (Hubbell and Mcintire 1986). Thus, the thrombogenesis model proposed by Grande-Gutierrez et al. (2021) is based on two contributions: mechanical activation, function of shear rate, and biochemical activation, function of the local concentration of PA and ADP, based on threshold values observed in several studies (Shadden and Hendabadi 2013; Bluestein et al. 1997; Hubbell and Mcintire 1986). A good agreement was observed in the confirmed regions of thrombosis and stagnation of the outflow for high concentration of PA and critical concentrations of ADP and PolyP (Grande-Gutiérrez et al. 2021). Few studies in AF (Qureshi et al. 2020; Wang et al. 2020) used this approach in the prediction of thrombi. Integrating a biochemical model of the coagulation cascade is an interesting addition to obtain a more detailed prediction of thrombogenesis in AF conditions, with emphasis on models based on hemodynamic factors (Menichini and Xu 2016; Menichini et al. 2016; Taylor et al. 2016; Grande-Gutiérrez et al. 2021). The biochemical model of Grande-Gutiérrez et al. (2021) was adapted and implemented in the proposed integrated numerical model (Lameu et al. 2022) including the volumetric indicator of stable clot, the aggregation intensity, as proposed by Taylor et al. (2016).

7.10 Left Atrial Hemodynamic Patterns: Healthy Sinus Rhythm *Versus* Atrial Fibrillation Case by Coupling 3D Electromechanical Activity

In order to illustrate the applicability of the proposed integrated hemodynamic model, blood flow patterns and the hemodynamic indicators were generated and compared for the cases of a healthy conditioned sinus rhythm (considering $T = 0.92$ s or 64 bpm of heart rate) and an AF case (3D Atrial Fibrillation, $T = 0.462$ s, equivalent to 130 bpm), given by the atrial fibrillation activity, which has 3D spatial-temporal variation, obtained from the LA phantom by simulating of rotor-type AF (Rodrigo et al. 2017). Figure 7.2 presents the *TAWSS*-based indicators. Figure 7.3 shows the transient-average blood velocity in cross-sectional planes after the complete cardiac flow cycle and also the shear rates at atrial contraction instant and the aggregation intensity, a volumetric indicator of stable clot, after 4.5 s of simulation.

From Fig. 7.2, we concluded that the proposed integrated models were capable of representing the major features of left atrial hemodynamics in healthy conditions and atrial fibrillation. In healthy conditions, there are no stagnation zones in the LAA, in contrast with its occurrence under AF, due to the lack of organized atrial contraction at the end of the cardiac flow cycle. In general, the 3D-EGM AF model predicted higher values of *TAWSS* than the SR model. This behavior can be attributed to the local flow perturbation inherent from the oscillatory effect of AF activity. However, in LAA, lower values of *TAWSS* were observed in the AF model (Fig. 7.2b). The consideration of the atrial contraction by the momentum method in the SR model provided local acceleration of blood (Fig. 7.2f) increasing the shear rate (Fig. 7.3c) and the *TAWSS* in LAA (Fig. 7.2b), reducing the blood stagnation, as expected for healthy conditions. The *OSI*, *RRT*, and *ECAP* fields highlight the recirculation zones in LAA (Fig. 7.2c, d) and also, in the middle section of the superior portion of LA (between left and right PVs). The 3D-AF model predicted

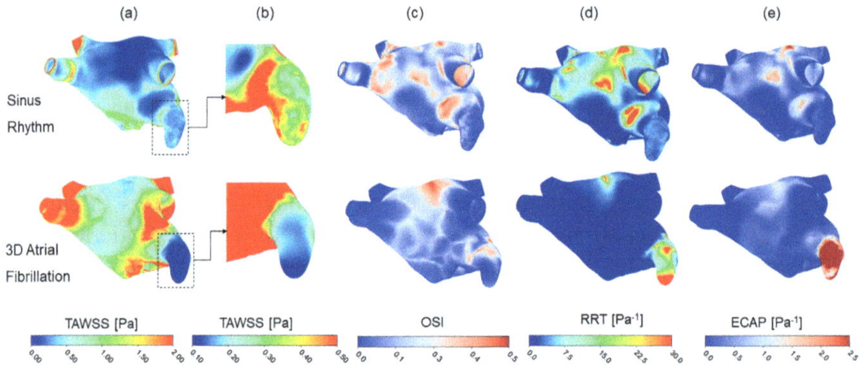

Fig. 7.2 CFD results for sinus rhythm (SR) and atrial fibrillation (AF) cases: (**a–e**) *TAWSS*-based hemodynamic indicators: (**a**) global *TAWSS*, (**b**) details of *TAWSS* field in the LAA, (**c**) *OSI*, (**d**) *RRT*, and (**e**) *ECAP* (Lameu et al. 2022)

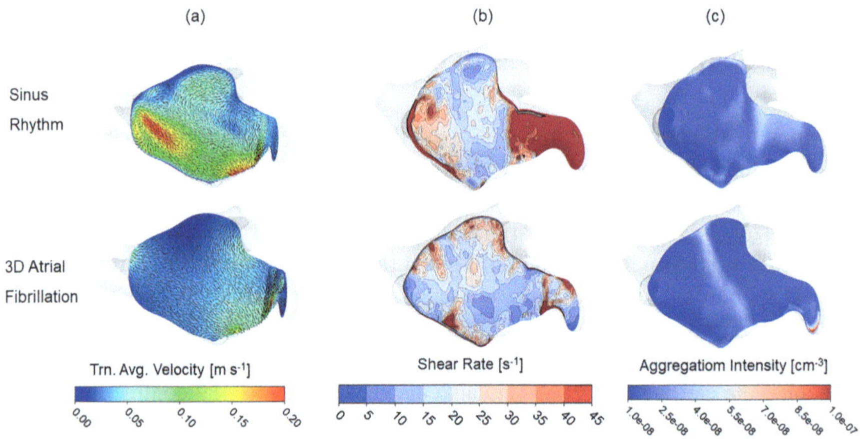

Fig. 7.3 CFD results for sinus rhythm (SR) and atrial fibrillation (AF) cases: (**a**) transient-average velocity field, (**b**) shear rate at atrial contraction instant, (**c**) aggregation intensity (Lameu et al. 2022)

zones with higher *OSI* near LAA ostium, where the blood fluid is accelerated due to the oscillatory perturbation from LA walls. The higher *RRT* and *ECAP* were observed from the LAA apex, a region where blood stagnation is expected.

The transient-average velocity field (Fig. 7.3a) illustrates this behavior. For the healthy case, high-velocity zones can be observed from the ejection of the blood from the PVs throughout the left atrium toward the mitral valve for the healthy case. In contrast, the AF results show a general lower velocity field, as a result of the local oscillatory perturbation from the wall fibrillation. The results agree with the expected hemodynamics, predicting stagnation zones inside the LAA under AF, which results in a pro-thrombotic environment. This was also in agreement with the aggregation intensity indicator, obtained from biochemical modeling of the coagulation cascade (Fig. 7.3c). Considering the experimental in vitro data from Taylor et al. (2014, 2016) and the accelerated kinetic factor from the Grande-Gutierrez et al. (2021) model, the threshold for a stable clot is about 1×10^{-7} cm^3 (Lameu et al. 2022), observed at the LAA apex only in 3D-AF case. The results from both models highlight the reliability of the proposed momentum-method approach, using the integrated data from electrophysiological activity to generate the mechanical effects and subsequent hemodynamic perturbation, which in future work could be used for customized patient-specific analysis. For improved visualization of hemodynamic patterns, animations of the blood velocity fields in the left atrium are provided for sinus rhythm (https://youtu.be/TZZcrbHtahA) and 3D atrial fibrillation (https://youtu.be/vx00V6b5Xd8).

7.11 Challenges and Future Perspectives

The proposed integrated model considers realistic electrophysiological activity from 3D AF mapping as a BCs to promote mechanical effects in the hemodynamic patterns. It represents a simplified approach to evaluate the outcomes from cardiac arrhythmia, regarding complex and higher computational cost FSI models. The applied example for an AF case showed the main features expected for this type of supraventricular arrhythmia. The additional coupling of a simplified biochemical model for coagulation cascade also allowed direct prediction of the formation of stable clots, which was only observed in the AF case, but not in the healthy reference case, as expected. The main hemodynamic characteristics were well represented by the proposed integrated model, including the wall shear stress-based indicators, often used in in silico analysis of cardiovascular pathology development. Finally, the proposed integrated model could be customized to patient-specific data, including real 3D anatomy from medical imaging, hemodynamic derived boundary conditions (flow rates and pressure waves), and also biochemical composition of blood, to provide a more realistic predictive tool for thrombogenesis risk assessment under different scenarios of cardiac arrhythmias.

Funding Declaration / Acknowledgments J. Salinet is supported by grant no. 2018/25606-2, São Paulo Research foundation (FAPESP). I. Sandoval is supported by Coordenacão de Aperfeiçoamento de Pessoal de Nível Superior – Brasil (CAPES). This study was financed in part by the Coordenação de Aperfeiçoamento de Pessoal de Nível Superior – Brasil (CAPES) – Finance Code 001.

References

Aguado AM, Olivares AL, Yagüe, C et al. (2019) In silico optimization of left atrial appendage Occluder implantation using interactive and modeling tools. Front Physiol 10:1–13. https://doi.org/10.3389/fphys.2019.00237

Al-Saady NM, Obel OA, Camm AJ (1999) Left atrial appendage: structure, function, and role in thromboembolism. Heart 82:547–554. https://doi.org/10.1136/hrt.82.5.547
Anand M, Rajagopal K, Rajagopal, KR et al. (2003) Model Incorporating some of the Mechanical and Biochemical Factors Underlying Clot Formation and Dissolution in Flowing Blood. J Theor Med, 5:183–218. https://doi.org/10.1080/10273660412331317415
Bluestein D, Niu L, Schoephoerster RT et al. (1997) Fluid mechanics of arterial stenosis: relationship to the development of mural thrombus. Annals Biomed Eng 25(2):344. https://doi.org/10.1007/BF02648048
Bodnár, T, Sequeira, A (2008) Numerical simulation of the coagulation dynamics of blood. Comput Math Meth Med 9(2):83–104. https://doi.org/10.1080/17486700701852784
Bosi GM, Cook A, Rai R et al. (2018) Computational Fluid Dynamic Analysis of the Left Atrial Appendage to Predict Thrombosis Risk. Front Cardiovasc Med 5:34. https://doi.org/10.3389/fcvm.2018.00034
Choi SH, Smith SA, Morrisey JH et al. (2011) Polyphosphate is a cofactor for the activation of factor XI by thrombin. Blood 118(26):6963–6970. https://doi.org/10.1182/blood-2011-07-368811
Cresti A, García-Fernández MA, Sievert H et al. (2019) Prevalence of extra-appendage thrombosis in non-valvular atrial fibrillation and atrial flutter in patients undergoing cardioversion: a large transoesophageal echo study. EuroIntervention 15(3):e225–e230. https://doi.org/10.4244/EIJ-D-19-00128
Desai DS, Hajouli S (2022) Arrhythmias, In: StatPearls [Internet], NIH Nation Library of Medicine, available at: https://www.ncbi.nlm.nih.gov/books/NBK558923/. Accessed March, 1st, 2022.
De Vos CB, Pison L, Pisters R et al. (2009) Atrial FibrillatoryWall Motion and Degree of Atrial Remodeling in Patients with Atrial Fibrillation: A Tissue Velocity Imaging Study. J Cardiovasc Electrophysiol 20(12):1374–1381, https://doi.org/10.1111/j.1540-8167.2009.01628.x
Dueñas-Pamplona J, García JG, Sierra-Pallares J et al. (2021) A comprehensive comparison of various patient-specific CFD models of the left atrium for atrial fibrillation patients. Comput Biol Med 133:104423, 2021. https://doi.org/10.1016/j.compbiomed.2021.104423
Fang R, Li Y, Zhang Y et al. (2021) Impact of left atrial appendage location on risk of thrombus formation. Biomech Model Mechanobiol 20(4):1431–1443. https://doi.org/10.1007/s10237-021-01454-4
Farese GE, Tayal B, Stöbe S et al. (2019) Regional disparities of left atrial appendage wall contraction in patients with sinus rhythm and atrial fibrillation. J Am Soc Echocard 32:755–762. https://doi.org/10.1016/j.echo.2019.01.016
Fernández-Pérez GC, Duarte R, de la Calle MC et al. (2012) Analysis of left ventricular diastolic function using magnetic. Radiol 4:295–305. https://doi.org/10.1016/j.rx.2011.09.018
García-Isla G, Olivares AL, Silva E et al. (2018) Sensitivity analysis of geometrical parameters to study haemodynamics and thrombus formation in the left atrial appendage. Int. J. Num. Methods Biomed. Eng. 2018, Vols. 34, 1–14. https://doi.org/10.1002/cnm.3100.
Grande-Gutiérrez N, Alber, M, Kahn AM et al. (2021) Computational modeling of blood component transport related to coronary artery thrombosis in Kawasaki disease. PLoS Comput Biol 17(9):e1009331. https://doi.org/10.1371/journal.pcbi.1009331
Hubbell JA, McIntire LV (1986) Platelet active concentration profiles near growing thrombi. A mathematical consideration. Biophys J 50(5):937–945. https://doi.org/10.1016/S0006-3495(86)83535-4
Kim MH, Johnston SS, Chu BC et al. (2011) Estimation of total incremental health care costs in patients with atrial fibrillation in the United States. Circ Cardiovasc Qual Outcomes 4(3):313–320. https://doi.org/10.1161/CIRCOUTCOMES.110.958165
Koizumi R, Funamoto K Hayase T et al. (2015) Numerical analysis of hemodynamic changes in the left atrium due to atrial fibrillation. J Biomech 48:472–478. https://doi.org/10.1016/j.jbiomech.2014.12.025

Lameu J, Oliveira ISR, Salinet J (2022) Thrombogenesis and Hemodynamics in Left Atrium Under Atrial Fibrillation. In: Computing in Cardiology 2022, 49, DOI: https://doi.org/10.22489/CinC.2022.097

Lee JM, Seo J, Uhm JS et al. (2015) Why is left atrial appendage morphology related to strokes? An analysis of the flow velocity and orifice size of the left atrial appendage. J Cardiovasc Electrophysiology 26(9):922–927. https://doi.org/10.1111/jce.12710

Limantoro I, de Vos CB, Delhaas T et al. (2014) Clinical correlates of echocardiographic tissue velocity imaging abnormalities of the left atrial wall during atrial fibrillation. Europace, 16,1546–1553. https://doi.org/10.1093/europace/euu047

López-Mínguez JR, González-Fernámndez R, Fernández-Vegas C et al. (2014) Comparison of imaging techniques to assess appendage anatomy and measurements for left atrial appendage closure device selection. J Invasive Cardio. 26(9):462–497.

Madeiro JPV, Pedrosa RC, Salinet J, Cortez PC. (2018) Classical and Modern Features for Interpretation of ECG signal. In: Madeiro JPV, Cortez P, Filho JMS et al. (Org.). Developments and Applications for ECG Signal Processing: Modeling, Segmentation, and Pattern Recognition. 1st Ed.: Academic Press, p. 1–210.

Malaspinas O, Turjman A, Sousa DR et al. (2016) A spatio-temporal model for spontaneous thrombus formation in cerebral aneurysms. J Theor Biol 394:68–76. https://doi.org/10.1016/j.jtbi.2015.12.022

Markl M, Lee DC, Carr ML et al. (2015) Assessment of left atrial and left atrial appendage flow and stasis in atrial fibrillation. J Cardiovasc Magn Reson. 17:M3. https://doi.org/10.1186/1532-429X-17-S1-M3

Marques VG, Rodrigo M, Guillem MS, Salinet J (2020a) Characterization of atrial arrhythmias in body surface potential mapping: A computational study. Comput Biol Med 127:103904. https://doi.org/10.1016/j.compbiomed.2020.103904

Marques VG, Rodrigo M, Guillem MS, Salinet J (2020b) A robust wavelet-based approach for dominant frequency analysis of atrial fibrillation in body surface signals. Physiol Meas 41(7):075004. https://doi.org/10.1088/1361-6579/ab97c1

Masci A, Barone L, Dedè L et al. (2019) The impact of left atrium appendage morphology on stroke risk assessment in atrial fibrillation: A computational fluid dynamics study. Front Physiol 9:1–11. https://doi.org/10.3389/fphys.2018.01938

Menichini C, Xu XY (2016). Mathematical modeling of thrombus formation in idealized models of aortic dissection: initial findings and potential applications. J Math Biol 73(5):1205–1226. https://doi.org/10.1007/s00285-016-0986-4

Menichini C, Cheng Z, Gibbs RGJ et al. (2016) Predicting false lumen thrombosis in patient-specific models of aortic dissection. J R Soc Interface 13:20160759. https://doi.org/10.1098/rsif.2016.0759

Mill J, Agudelo V, Olivares AL et al. (2021) Sensitivity Analysis of In-Silico Fluid Simulations to Predict Thrombus Formation after Left Atrial Appendage Occlusion. Mathematics 9(18):2304. https://doi.org/10.3390/math9182304

Morales X, Mill J, Delso, G et al. (2021). 4D Flow Magnetic Resonance Imaging for Left Atrial Haemodynamic Characterization and Model Calibration. In: et al. Statistical Atlases and Computational Models of the Heart. M&Ms and EMIDEC Challenges. STACOM 2020. Lecture Notes in Computer Science, 12592. Springer, Cham. https://doi.org/10.1007/978-3-030-68107-4_16

Müller F, Mutch NJ, Schenk WA et al. (2009) Platelet polyphosphates are proinflammatory and procoagulant mediators in vivo. Cell 139(6):1143–1156. https://doi.org/10.1016/j.cell.2009.11.001

Ngoepe MN, Ventikos Y (2016) Computational modelling of clot development in patient-specific cerebral aneurysm cases. J. Thromb. Haemost. 2016, Vols. 14, 262–272. https://doi.org/10.1111/jth.13220

Nyrnes SA, Fadnes S, Wigen MS et al (2020) Blood Speckle-Tracking Based on High-Frame Rate Ultrasound Imaging in Pediatric Cardiology. J Am Soc Echocardiogr 33(4):493–503. https://doi.org/10.1016/j.echo.2019.11.003

Page RL, Joglar JA, Caldwell MA et al. (2016) ACC/AHA/HRS guideline for the management of adult patients with supraventricular tachycardia: a report of the American College of Cardiology/American Heart Association Task Force on Clinical Practice Guidelines and the Heart Rhythm Society. Circulation. 2016 Apr 5;133(14):e471–505. https://doi.org/10.1161/CIR.0000000000000310

Peach TW, Ngoepe MN, Spranger K et al. (2014) Personalizing flow-diverter intervention for cerebral aneurysms: from computational hemodynamics to biochemical modeling. Int J Numer Method Biomed Eng 30:1387–1407. https://doi.org/10.1002/cnm.2663

Qureshi A, Darwish O, Dillon-Murphy D et al. (2020) Modelling Left Atrial Flow and Blood Coagulation for Risk of Thrombus Formation in Atrial Fibrillation. Proceedings of the Computing in Cardiology, Rimini, Italy, 13–16 September 2020. https://doi.org/10.22489/CinC.2020.219.

Rahman F, Kwan GF, Benjamin E (2014) Global epidemiology of atrial fibrillation. Nat Rev Cardiol 11:639–654. https://doi.org/10.1038/nrcardio.2014.118

Rodrigo M, Climent AM, Liberos A et al. (2017) Technical considerations on phase mapping for identification of atrial reentrant activity in direct and inverse-computed electrograms. Circ Arrhythmia Elec 10(9):e005008. https://doi.org/10.1161/CIRCEP.117.005008

Roney CH, Sim, I, Yu J et al. (2022) Predicting atrial fibrillation recurrence by combining population data and virtual cohorts of patient specific left atrial models. Circ Arrhythm Electrophysiol 15:e010253. https://doi.org/10.1161/CIRCEP.121.010253

Salinet JL (2014) Density Frequency Mapping of Human Intracardiac Persistent Atrial Fibrillation Electrograms. PhD Thesis. University of Leicester, UK, pp. 1–222.

Salinet J, Molero R, Schlindwein FS, et al. (2021) Electrocardiographic Imaging for Atrial Fibrillation: A Perspective from Computer Models and Animal Experiments to Clinical Value. Front Physiol 12:653013. https://doi.org/10.3389/fphys.2021.653013

Shadden SC, Hendabadi S (2013) Potential fluid mechanic pathways of platelet activation. Biomech Model Mechanobiol 12(3):467–474. https://doi.org/10.1007/s10237-012-0417-4

Taylor JO, Witmer KP, Neuberger T et al. (2014) In vitro quantification of time dependent thrombus size using magnetic resonance imaging and computational simulations of thrombus surface shear stresses. J Biomech Eng 136(7). https://doi.org/10.1115/1.4027613

Taylor JO, Meyer RS, Deutsch S et al. (2016) Development of a computational model for macroscopic predictions of device-induced thrombosis. Biomech Model Mechanobiol 15:1713–1731. https://doi.org/10.1007/s10237-016-0793-2

Tse G (2016) Mechanisms of cardiac arrhythmias. J Arrhythm 32(2): 75–81. https://doi.org/10.1016/j.joa.2015.11.003

Tzikas S, Shakir S, Gafoor, S et al. (2016) Left atrial appendage occlusion for stroke prevention in atrial fibrillation: multicentre experience with the AMPLATZER Cardiac Plug. EuroIntervention 11:1170–1179. https://doi.org/10.4244/EIJY15M01_06

Vedula V, George R, Younes L et al. (2015) Hemodynamics in the left atrium and its effect on ventricular flow patterns. J Biomech Eng 137:1–8. https://doi.org/10.1115/1.4031487

Zhang LT, Gay M (2008) Characterizing left atrial appendage functions in sinus rhythm and atrial fibrillation using computational models. J Biomech 41:2515–2523. https://doi.org/10.1016/j.jbiomech.2008.05.012

Zhang Q., Yu CM (2007) Atrial Function, In: Marwick TH, Yu, CM, Sun J. Myocardial Imaging: Tissue Doppler and Speckle Tracking. pp. 255–264, Blackwell Publishing, Massachusetts, USA, 2007.

Wang TJ, Massaro JM, Levy D et al. (2003) A risk score for predicting stroke or death in individuals with new-onset atrial fibrillation in the community: the Framingham heart study. JAMA. 290(8):1049–56. https://doi.org/10.1001/jama.290.8.1049

Wang Y, Di Biase L, Horton RP et al. (2010) Left atrial appendage studied by computed tomography to help planning for appendage closure device placement. J Cardiovasc Electrophysiology 21(9):937–982. https://doi.org/10.1111/j.1540-8167.2010.01814.x

Wang Y, Qiao YH, Mao YK et al. (2020) Numerical prediction of thrombosis risk in left atrium under atrial fibrillation. Math Biosci Eng 17(3):2348–2360. https://doi.org/10.3934/mbe.2020125

Watson T, Shatsila E, Lip GYH (2009) Mechanisms of thrombogenesis in atrial fibrillation: Virchow's triad revisited. Lancet 373:155–166. https://doi.org/10.1016/S0140-6736(09)60040-4

Wodchis WP, Bhatia RS, Leblanc K et al. (2012) A review of the cost of atrial fibrillation. Value Health 15(2):240–248. https://doi.org/10.1016/j.jval.2011.09.009

Part III
Medical Devices

Chapter 8
Principles of Tissue Engineering and Regenerative Medicine

Christiane Bertachini Lombello, Anand Oliveira Masson, Felipe Nogueira Ambrosio, Débora Carajiliascov Ferraraz, and Monica Helena Monteiro do Nascimento

Abstract Tissue engineering (TE) and regenerative medicine (RM) aim to restore or remodel damaged or diseased tissues by developing in vitro biological substitutes that can repair, replace, maintain, or enhance organ function. Tissue engineering is based on three essential components that are cells of variable sources, guiding tissue formation scaffolds, and microenvironment and its inductive factors. The structure, mechanical properties, and chemical composition of the scaffold can influence cell behavior, including proliferation, differentiation, extracellular matrix synthesis, and secretion. Bioreactors and bioprinting are also applied to improve the interaction between cells and scaffolds in order to achieve cell adhesion, proliferation, migration, extracellular matrix secretion, and differentiation, resulting in newly formed constructs or tissues. Translation from the laboratory to applied clinics must consider the integration of implanted constructs achieving functional restoration and regeneration of the target tissue. Tissue engineering and regenerative medicine hold important ongoing innovative developments with the potential to revolutionize healthcare, proponing treatments, beyond just repairing lost tissue function, but regenerating the original characteristics of tissues and organs.

Keywords Cell culture techniques · Growth factors · Regenerative medicine · Scaffolds · Tissue engineering

C. B. Lombello (✉) · F. N. Ambrosio · D. C. Ferraraz · M. H. M. do Nascimento
Graduate Program in Biomedical Engineering, Federal University of ABC – UFABC, São Bernardo do Campo, SP, Brazil
e-mail: christiane.lombello@ufabc.edu.br

A. O. Masson
Biomedical Engineering Graduate Program, University of Calgary, Calgary, AB, Canada

C. B. Lombello, P. A. da Ana (eds.), *Current Trends in Biomedical Engineering*,
https://doi.org/10.1007/978-3-031-38743-2_8

8.1 Introduction

Tissue engineering (TE) and regenerative medicine (RM) are rapidly advancing multidisciplinary fields with the primary aim to restore or regenerate the structure and function of damaged or diseased tissues in the human body. These sciences hold great potential for improving patient's quality of life and advancing healthcare (Caplan 2007; Berthiaume et al. 2011; O'Brien 2011; Nascimento et al. 2017; Freitas-Ribeiro et al. 2022).

Traditional treatments for trauma, chronic diseases, infectious pathologies, and other clinical conditions affecting the physiology and structure of the body involve various approaches such as drug therapies, physiotherapy, surgery, or even grafting. These approaches typically focus on functional improvement and tissue repair rather than true regeneration. Although they can enhance the patients' quality of life, this is often temporary and ineffective in the long run. Tissue and organ transplantation, while a viable option for restoring function, faces significant limitations including reduced availability, compatibility requirements, complex surgical procedures, risk of rejection and contamination, and logistical challenges (Hoffman et al. 2019). Aiming to address these limitations, the fields of TE and RM strive to enhance techniques and expand treatment options to a wide range of tissues and organs. The ultimate goal is to achieve biological reconstitution of the organism, enabling patients to regain their quality of life while ensuring the long-term stability of the results. The clinical adoption of tissue engineering-derived innovations also expands the corresponding technological market niche. Companies involved in research and development (R&D) in the fields of tissue engineering (TE) and regenerative medicine (RM), mostly referred together as TERM, have demonstrated sustained growth since the commercial approval of their first product, Integra™ Artificial Skin, by the FDA on March 1, 1996, highlighting the increasing recognition and demand for these technologies (Brockbank 2022; Lysaght and Reyes 2001; Hoffman et al. 2019; Kim et al. 2019).

8.2 History, Concepts, and Strategies

The term tissue engineering was introduced in 1988, as the topic of a workshop organized by the National Science Foundation, and the foundational publications that shaped the field emerged soon after in the 1990s (Vacanti and Vacanti 1991; Langer and Vacanti 1993). However, the scientific advancements leading to the establishment of this field are more recent, encompassing both basic research and clinical applications viewing regenerative medicine (Nerem 1992; Langer and Vacanti 1993; Vacanti 2006; Kaul and Ventikos 2015).

Tissue engineering is an interdisciplinary field of knowledge, that integrates concepts and techniques from various disciplines, including biomedical engineering, materials science, chemistry, biology, and medicine. It aims to restore or remodel

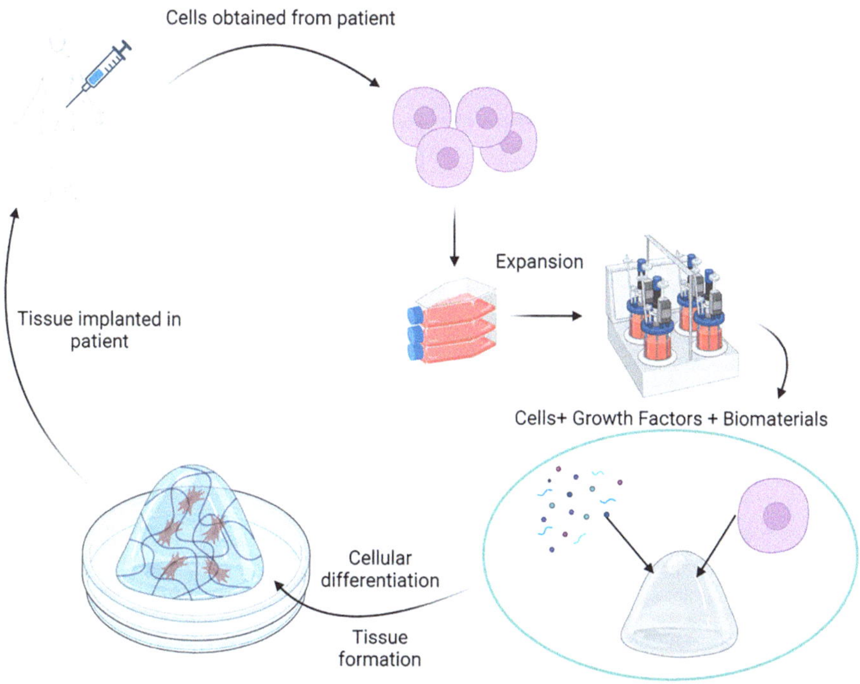

Fig. 8.1 Basic principles of tissue engineering (TE). Cells obtained from a patient can be used as an autologous source for different tissue engineering applications. The cells can be cultured on a two-dimensional proliferative system. Differentiation can be achieved with cell seeding onto a scaffold, inductive biomolecules, and specific culture conditions stimuli, such as bioreactor chambers. Newly formed tissues can be implanted on a damaged area to regenerate original composition and function. (Created with http://www.biorender.com)

damaged or diseased tissues by developing in vitro biological substitutes that can repair, replace, maintain, or enhance organ function (Nerem 1992; Langer and Vacanti 1993; Hoffman et al. 2019).

It is established that tissue engineering is primarily based on three essential components (Fig. 8.1): cells, which are responsible for tissue formation; scaffolds, which act as templates to guide tissue growth, by providing support for cell adhesion, proliferation, migration, and differentiation; and the microenvironment, and its inductive factors such as growth factors, which facilitate the regenerative process (Chan and Leong 2008; Zhang et al. 2009; O'Brien 2011; Khademhosseini and Langer 2016; Courtenay et al. 2018).

In recent decades, tissue engineering has undergone significant advancements (Fig. 8.2). For instance, the development of surgical techniques for clinical translation and the improvement of cryopreservation processes of newly formed tissues; moreover, progress on design engineering, involving the use of bioreactors, bioprinters, organs-on-a-chip systems, robotics; and the emergence of bioinformatics, leading to complex systems modeling (Khademhosseini and Langer 2016; Courte-

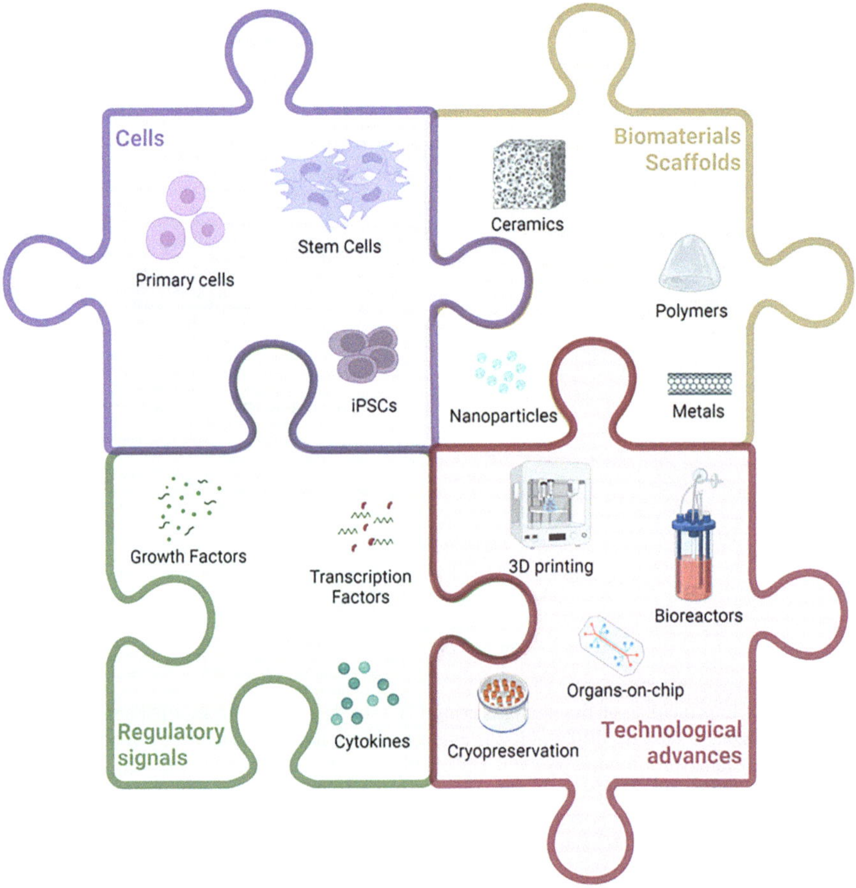

Fig. 8.2 Basic elements and technologies applied to tissue engineering and regenerative medicine. Cells, from different sources, interact with scaffolds and are subjected to biomolecules that act as regulatory signalization to cell behavior, for example, extracellular matrix synthesis. Technologies can be used to improve in vitro tissue formation (as bioreactors and 3D-bioprinters) and its clinical use (to correct organs-on-a-chip). (Created with http://biorender.com)

nay et al. 2018; Hoffman et al. 2019; Freitas-Ribeiro et al. 2022). Such technological advancements have enabled the integration of tissue engineering with cutting-edge strategies such as cellular therapy, gene therapy, and immunomodulation in the regenerative medicine field, further advancing therapeutic approaches for tissue and organ regeneration (Vacanti 2006; Daar and Greenwood 2007; Mason and Dunnill 2008; Mao and Mooney 2015; Hoffman et al. 2019). Importantly, regulatory affairs have been concomitantly strengthened to ensure the safety and efficacy of tissue engineering and regenerative medicine applications (Lee et al. 2010; ISO/TS21560: 2020; Beheshtizadeh et al. 2022).

Tissue engineering approaches involve seeding cells onto a scaffold, which provides a supportive framework for cell growth and tissue development in vitro. Additionally, biochemical factors can be used to promote cell proliferation and guide differentiation into a desired tissue type. Cells, which are required for this purpose (Mao and Mooney 2015), can be derived from the patient's own body (autologous) or obtained from other sources (homologous and xenogeneic). The use of autologous cells, as seen in Autologous Chondrocyte Implantation for the treatment of chondral lesions (Brittberg et al. 1994), is typically preferred. Reasons for that include the lower risk of immune rejection and ethical considerations related to cell source, however, donor site morbidity remains a challenge in clinical practice. Homologous cells, as seen in Apligraf (Falanga and Sabolinski 1999), or the less commonly xenogeneic cells (Wolf et al. 2019), appear as alternatives. Unfortunately, despite the wider availability of cells compared to autologous sources, increased risk of immune rejection and ethical concerns limit their therapeutic application.

Regardless of the tissue source, however, cells are typically isolated through mechanical and enzymatic methods and expanded in two-dimensional systems. At this stage, cells isolated from cartilage can be used in cell therapy applications, such as the FDA-approved Carticel, used for Autologous Chondrocyte Implantation (Brittberg et al. 1994; Kessler and Grande 2008). However, a more common approach is to inoculate the cultured cells onto three-dimensional scaffolds, biological or synthetic in nature, that mimic the physico-chemical properties of the target tissue. The resulting construct can be directly implanted into the target tissue or remain in culture for longer to allow cells to mature and develop tissue-like characteristics through the synthesis of extracellular matrix (Caplan 2007; Hoffman et al. 2019; Ashammakhi et al. 2022).

The structure, mechanical properties, and chemical composition of the scaffold can influence cell behavior, including proliferation, differentiation, extracellular matrix synthesis, and secretion, as elegantly shown by Folkman and Moscona in 1978 (Folkman and Moscona 1978; Howard et al. 2008; Vagaská et al. 2010). Therefore, in addition to carefully selecting cell type and source, the choice of scaffold is key for tissue engineering applications (Howard et al. 2008; Courtenay et al. 2018). In this regard, scaffolds that closely resemble the target tissue, known as biomimetic scaffolds, are preferable as they can more effectively guide cellular processes toward the desired tissue regenerative outcome. Ultimately, a comprehensive understanding of the properties of the scaffold and its similarities to the target tissue is necessary to improve the success rate of tissue engineering applications (Pina et al. 2019).

The biochemical microenvironment also plays an important role in the development of tissue-engineered constructs. Inductive molecules, such as growth factors and signaling molecules, are often incorporated into the scaffold to stimulate proliferation and direct subsequent cell differentiation (Lee et al. 2011; Ren et al. 2020). Recent advancements have also introduced innovative technologies such as bioreactors and bioprinters to enhance cell-scaffold interaction, extracellular matrix synthesis, and in vitro tissue formation (Tomasina et al. 2019; Fu et al. 2021), improving the structure and function of the final construct. Once implanted into the target tissue or organ, successful integration is the main objective in achieving functional restoration and regeneration of the affected tissue area (Hoffman et al. 2019; Ashammakhi et al. 2022).

8.3 Cells and In Vitro Culture

The human body is composed of four main tissue types: epithelial, connective, muscular, and nervous. Each tissue is composed of specific cells and structurally formed by the extracellular matrix (ECM) that these cells secrete and interact with Farley et al. (2012) and Lanza et al. (2020). Cells are responsible for producing, secreting, and assembling the different ECM molecules forming a macromolecular complex. In turn, the matrix modulates the tissue structure, physiology, and biomechanics. Generally, ECM components can be divided into three groups: fibrillar, corresponding to fibrillar collagens and elastic fibers; nonfibrillar, represented by glycosaminoglycans, proteoglycans, and noncollagenous structural glycoproteins; and microfibrils, formed by collagen VI and elastin-associated microfibrils (Farley et al. 2012; Theocharis et al. 2016).

Human cells are usually anchorage dependent, meaning that they require attachment to the ECM for proper function. Consequently, when culturing these cells in vitro, adhesion to a substrate is important for cell survival and growth, as well as to prevent apoptosis (Chan and Leong 2008). Cell-matrix adhesion is facilitated by ligands and often involves the interaction between integrins on the cell surface and fibronectin or other ECM components on the matrix. This interaction is known to trigger signal transduction events, leading to cytoskeletal remodeling, cell spreading, migration, differentiation, and among other cellular responses (Anselme 2000; Sage 2001).

As previously mentioned, in tissue engineering, it is necessary to first isolate the cells from the donor tissue. This can be accomplished through techniques like explant culture (using tissue fragments) or mechanical and enzymatic dissociation, which results in a cell suspension. The isolated cells are referred to as primary cells, and their phenotype and genotype are consistent with their tissue of origin (Falanga and Sabolinski 1999; Lanza et al. 2020). With time in culture (passages), these cells are able to retain most of the characteristics of the original tissue. However, they generally demonstrate a higher proliferation rate, being currently classified as continuous cell lines, as they can be maintained in culture for extended periods of time (Capes-Davis and Freshney 2010).

Maintenance of a controlled in vitro environment is also essential during cell culture. Factors such as carbon dioxide concentration, temperature, and nutrient levels must be rigorously regulated. The cell culture medium is responsible for providing the nutrition necessary for cellular development, as well as for maintaining appropriate pH and osmolarity levels required for cell survival. Additionally, the supplementation of cell media with animal-derived serum is a common practice, as it contains adhesion and growth factors, that promote cell adhesion and metabolic activity, as well as cell spreading, migration, proliferation, and differentiation (Brunner et al. 2010; Capes-Davis and Freshney 2010).

Early products developed in tissue engineering and regenerative medicine involved the use of homologous or autologous cells, as illustrated by Integra and Carticel products, respectively (Brittberg et al. 1994; Falanga and Sabolinski 1999). The use of human cells is favorable from a compatibility point of view, but the limited availability of autologous cells needs to be considered, as well as the risks of disease transmission associated with homologous cells. In all cases, cell expansion to a desired number prior to implantation in the damaged area is required. For Carticel, this process takes approximately 4 weeks and is crucial because chondrocytes are naturally low-proliferative cells. For treatment of the damaged cartilage, cells are implanted under periosteal coverage, as originally described by Brittberg et al. (1994) or by utilizing scaffolds to delineate the treated area and provide cellular support (Marlovits et al. 2006). However, the use of predifferentiated cells in regenerative medicine has certain restrictions. For instance, there is a limited number of in vitro cell passages without hindering the differentiated state of these cells. Additionally, cellular aging can occur, leading to a decreased cellular response and activity, including ECM synthesis and tissue neoformation impairment (Kessler and Grande 2008).

Stem cells can be obtained from different sources in the human body and present good proliferative (self-renewal) and cellular differentiation potential (potency), which makes them an interesting cell source alternative for tissue engineering and regenerative medicine (Ashammakhi et al. 2022; Fu et al. 2021). Under certain physiological or experimental conditions, these unspecialized cells can be induced to take on a specific fate/phenotype (differentiation), which can be identified morphologically and confirmed by detection of specific cell surface proteins. CD73, CD90, and CD105, for example, are markers of mesenchymal stem cells (MSCs) and CD34 and CD45 identify hematopoietic stem cells (Fig. 8.3) (La Noce et al. 2014; Verma et al. 2014; Mahla 2016; Los et al. 2018).

Wakitani et al. (1994) demonstrated the potential to use MSCs obtained from periosteum in cartilage repair. However, this technique shows the difficulty in maintaining the differentiated chondrogenic phenotype. MSCs for cartilage repair may be derived from endogenous, underlying bone, or exogenous sources, such as adipose tissue or blood-derived (Le et al. 2020). Wise et al. (2009) showed promising results with human MSCs onto an electrospun scaffold, with oriented fibers of 500 or 3000 nm diameter, made of polycaprolactone (PCL) to mimic articular cartilage superficial zone. But the challenge of the use of MSCs in cartilage regenerative medicine remains (Le et al. 2020).

Considering bone tissue engineering, Gómez-Barrena et al. (2020) presented results of an MSC obtained from bone marrow and seeded onto beta-tricalcium phosphate/hydroxyapatite scaffold with bone formation efficacy after 8 months follow-up. Furthermore, a study with autologous MSC obtained from aspiration of the anterior ilium used with beta-tricalcium phosphate scaffold, compared with scaffold alone, and applied to femoral defects in the revision of total hip arthroplasty.

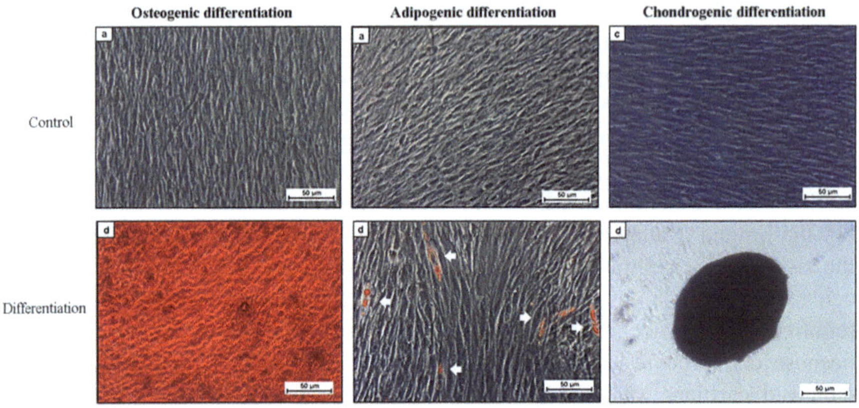

Fig. 8.3 Differentiation of mesenchymal stem cells (MSC) obtained from the pulp of deciduous teeth. (**a**, **c**, **e**) Control, cells cultivated with α-MEM medium. (**b**) Cells cultured with osteogenic differentiation-inducing medium. Calcium accumulation can be seen by the intense red color of cultures stained with alizarin red. (**d**) Cells cultured with adipogenic differentiation-inducing medium and stained with Oil red. The arrows show the intracellular lipids. (**f**) Cells cultured with chondrogenic differentiation-inducing medium, exhibiting micromass arrangement and stained with toluidine blue. Observation by light microscopy, with phase contrast (Ferraraz 2020)

The authors concluded that a scaffold containing MSC associated with cells is a safe and efficient alternative to treat bone defects once trabecular remodeling was observed in all patients, the scaffold alone did not achieve this performance (Šponer et al. 2016).

MSC has been applied to wound healing with promising results. Jin et al. (2011) showed that bone marrow-derived MSC cultivated on electrospun nanofibrous scaffold of collagen/poly(l-lactic acid)–co-poly(3-caprolactone) (Coll/PLLCL), and in the presence of inductive factors (epidermal growth factor (EGF) and 1,25-dihydroxyvitamin D3). This system could mimic skin extracellular matrix environments and for this reason was considered promising for skin tissue engineering. More recently, bioprinting techniques have been applied to skin tissue engineering and regenerative medicine. A human bilayered skin was 3D bioprinted with fibroblasts and keratinocytes, which mimics human skin (Cubo et al. 2016). Another printed construct was made with multiple layers of fibroblast on collagen hydrogel, with a top printed layer of melanocytes and keratinocytes. Epidermal and dermal layers were characterized resulting in biomimetic skin (2018).

As seen in the mentioned constructs, scaffolds play a central role in receiving cells, guiding tissue formation, and even more integration with adjacent original tissue.

8.4 Scaffolds

Scaffolds play a crucial role in tissue engineering and regenerative medicine applications. They provide the necessary structural support for cells and guide ECM synthesis, which is essential for tissue regeneration. Additionally, scaffolds are able to modulate cellular behavior and interactions (Folkman and Moscona 1978; Zhang et al. 2009; O'Brien 2011; Shimojo et al. 2020).

When selecting a scaffold for use in tissue engineering, several key factors need to be taken into consideration. First, the scaffold should possess a three-dimensional structure that is both biocompatible and bioactive. Additionally, the scaffold should facilitate cell adhesion and growth and upon implantation, it should not elicit a harmful inflammatory response, in order to prevent rejection. This can be facilitated by incorporating biochemical factors that promote such properties. Therefore, when designing or selecting a scaffold intended for tissue engineering, it is essential to evaluate its biocompatibility and to ensure its clinical safety.

To this end, biological evaluations are performed according to the ISO 10993-1 standard (2018). One of the initial assessments to be conducted is cytotoxicity testing, which aims at evaluating the in vitro cellular response to scaffolds and the biomaterials (Williams 1999) they are made of, as well as their potential toxic effects. Cytotoxicity assays include direct contact, indirect diffusion in agar, and direct or elution assays, which differ with regards to the way cells interact with the biomaterial under analysis, or in the case of elution assay, to biomaterials leachable (ISO10993-5 2009; Masson 2018) (Fig. 8.4). These assessments help in selecting the most suitable material for a desired application.

The surface characteristics of scaffolds, such as topography, chemistry, and surface energy, play an essential role in cell adhesion and spreading, thus affecting the interaction with cells and adjacent tissues (Chan and Leong 2008; O'Brien 2011; Pina et al. 2019; Shimojo et al. 2020). Considering that, it is important for the scaffold to mimic, at least in part, the ECM of the target tissue. Reproducing the mechanical properties, such as compression and tensile strength of the specific anatomical site wherein the scaffold will be inserted, is also important for the success of the regenerative process. Notably, the mechanosensitivity transmitted through the scaffolds can also influence the differentiation of stem cells (Chan and Leong 2008; Zhang et al. 2009).

Biological substitutes implanted in the body to restore the function of damaged tissues and organs can be classified into two groups: permanent or temporary implants. The first includes heart valves, intraocular lenses, and joint prostheses, which aim to replace damaged tissues indefinitely, retaining their characteristics over long periods. On the other hand, temporary implants, commonly used in tissue engineering and regenerative medicine, are designed to allow cells to produce their own matrix and gradually replace the implants over time, thereby promoting tissue remodeling. In this case, scaffolds must exhibit a degradation rate that is compatible with the tissue's regenerative process (Chan and Leong 2008; O'Brien 2011; Pina et al. 2019).

The microstructure of a scaffold is one of the determinants of its application. The three-dimensional structure of the scaffold promotes cell interaction, tissue neoformation, cell accommodation, and the synthesis of the extracellular matrix (Fig. 8.5). The efficient transport of nutrients and oxygen to cells within the scaffold, as well as the elimination of metabolites, is influenced by the dimensions and

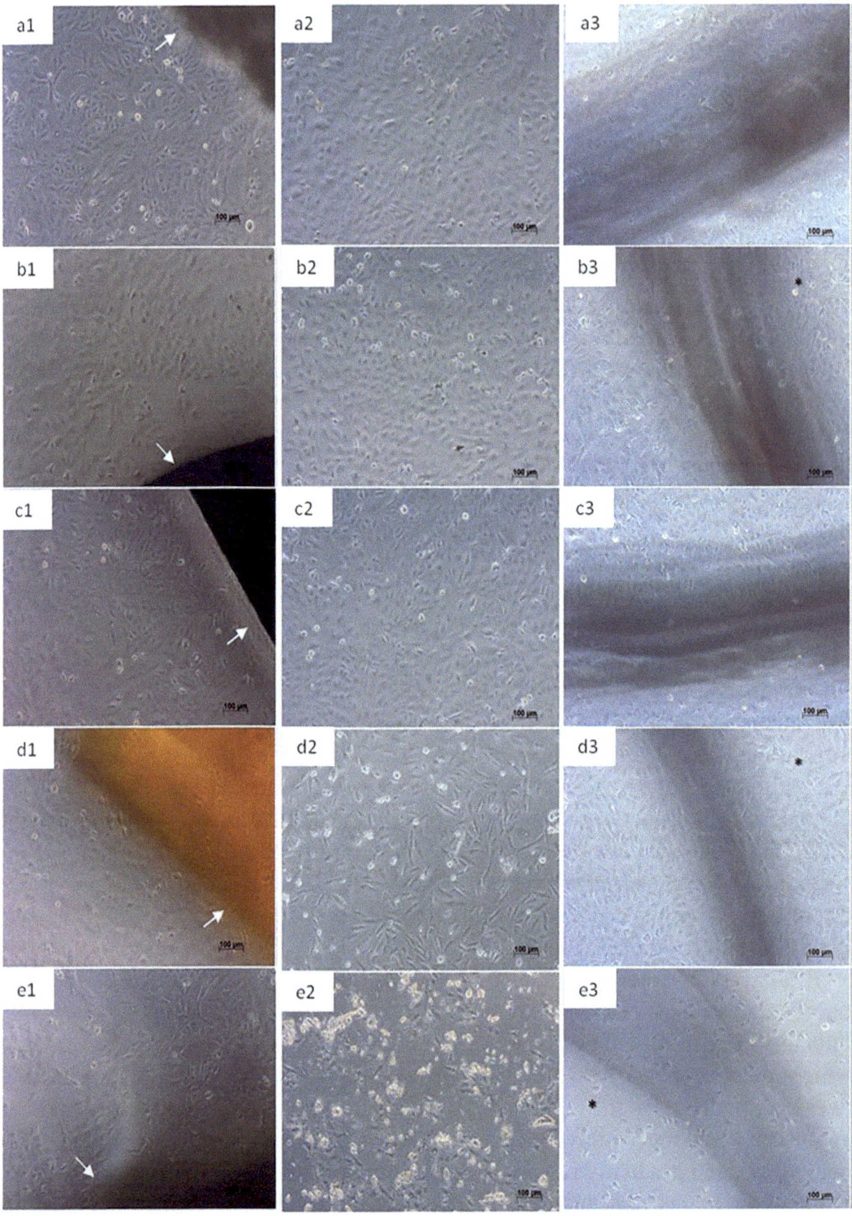

Fig. 8.4 Cytotoxicity of biomaterials. (**a**) Negative control: filter paper; test biomaterials: (**b**) ceramic, beta-tricalcium phosphate (BTCP); (**c**) metallic, stainless steel AISI 316L; (**d**) polymeric: gelatin crosslinked in 1% glutaraldehyde solution. (**e**) Positive control: latex. Cytotoxicity methods: (1) direct control; (2) elution; and (3) indirect contact, agar

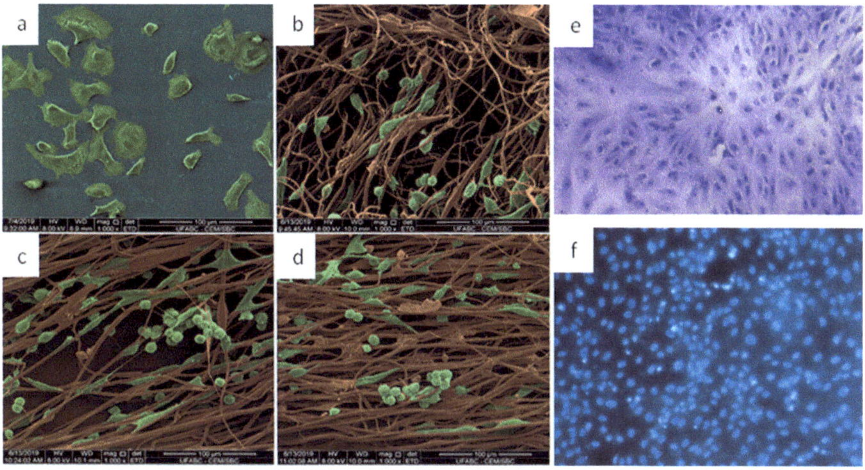

Fig. 8.5 Interaction of Vero cells with PLA scaffold. (**a–d**) scanning electron microscopy. (**a**) control cells on coverslips, (**b–d**) Cell interaction with electrospinning PLLA with parallel-oriented fibers (Lopes et al. 2019). (**e–f**) PLA obtained by casting, (**e**) Vero cells grown in culture plates in the presence of PLA sample, stained with toluidine blue; (**f**) Vero cells on PLA scaffolds, stained with DAPI (Nascimento and Lombello)

interconnectivity of pores. For in vitro techniques, pore sizes of 75–100 μm are recommended, as they enhance cellular interaction and matrix synthesis, while pore sizes of 200–500 μm are considered ideal for integration with native tissues and neovascularization (Roseti et al. 2017; Perić Kačarević et al. 2019; Krishani et al. 2023). Furthermore, scaffold manufacturing should be economically and technologically viable (O'Brien 2011; Krishani et al. 2023). Of note, the scaffold can also serve as a delivery vehicle or reservoir for controlled release of growth factors, which play a role in tissue remodeling and regeneration (Chan and Leong 2008; Lin et al. 2019).

Biomaterials used in scaffolds can be classified as either natural or synthetic. By definition, biomaterial is a "material intended to interface with biological systems to evaluate, treat, augment or replace any tissue, organ, or function of the body," converging with tissue engineering purpose. Natural biomaterials are derived from natural sources and exhibit properties that resemble those of the original tissues, often consisting of components forming the extracellular matrix. However, due to their origin, natural materials may exhibit greater variation in their makeup and are less amenable to modifications during the manufacturing process. Examples of widely used natural materials in tissue engineering include collagen, gelatin, and chitosan (Krishani et al. 2023). Conversely, synthetic biomaterials are artificially produced through the synthesis of selected components, thereby they can be designed to possess specific characteristics and manufactured on a large scale with little batch variation. Synthetic materials often have lower manufacturing costs and are more easily tuned compared to natural materials. Examples of synthetic biomaterials used as scaffolds include α-hydroxy acid polymers such as poly(ε-caprolactone) (PCL), polylactide (PLA) and polyglycolic acid (PGA) (O'Brien 2011; Pina et al. 2019). By combining natural and synthetic biomaterials, it is possible to enhance the properties of natural biomaterials, such as improving their mechanical properties, scalability, reproducibility, and processing capabilities (Pina et al. 2019; Krishani et al. 2023).

Polymers are the most widely employed in tissue engineering. Some examples include the natural polymers collagen, hyaluronic acid, chitosan, alginate, and synthetic poly-L-lactic/polyglycolic acids (PLLA/PGA) for cartilage and soft tissue-associated applications (Mao and Mooney 2015; Nascimento et al. 2017; Fu et al. 2021; Yari et al. 2022). An interesting study by Nascimento et al. (2017) developed and evaluated polymer-based scaffolds using hyaluronic acid (HA) and chitosan (CHI). The synthesis of the HA and CHI scaffold resulted in a biocompatible biomaterial with a high degree of swelling (300%). Evaluation of the cellular interaction with the polymer scaffold revealed that Vero cell line exhibited adhesion, spreading, and proliferation on the biomaterial for up to 12 days, highlighting its potential for tissue engineering applications (Nascimento et al. 2017). Wang et al. (2019) presented a used bioimpression to construct a bilayer membrane scaffold of PLGA (poly acid lactic-co-glycolic) with an alginate hydrogel layer to mimic epidermis and dermis, which can promote wound healing as skin substitute.

Other materials used include ceramics and metallic biomaterials. Ceramic scaffolds are used in bone tissue engineering, either alone or in combination with polymers, and commonly utilize materials such as calcium sulfate and calcium phosphates (CaPs) including hydroxyapatite and tricalcium phosphate (Vagaská et al. 2010; Mishra et al. 2016). Filipowska et al. (2014) showed results for a 3D porous bioactive composite scaffold made of PLGA and two different sol-gel-derived bioactive glasses (SBGs) of CaO-SiO2-P2O5, varying on the content of SiO2 and CaO. This scaffold was used to cultivate human MSC under different culture conditions (supplementation with recombinant human bone morphogenetic protein-2, rhBMP-2, or dexamethasone). Cell attachment and osteogenic differentiation were attained, and the scaffolds were fundamental to guide cell behavior, more than biomolecular induction. Moreover, nanostructured biomaterials have shown promising results as tissue engineering scaffolding. Hydroxyapatite (HA) scaffolds obtained from micro and nanostructured powder, resulted in 1.9 micron and 55 nm average grain size, respectively. Preosteoblast lineage cells (MC3T3) were cultured on the scaffolds, and no significant attachment difference comparing micro and nano-HA, however, greater cell number on nano-HA scaffold was observed (Smith et al. 2006). The same preosteoblastic lineage was used by Dos Anjos et al. (2019) to evaluate nanocarbonated hydroxyapatite (nCHA) cell interaction. No cytotoxicity was observed and cellular aggregates were observed on the scaffold surface, as a promising result for clinical application.

Metallic scaffolds have specific applications, particularly in bone associate applications, and can be used in combination with other biomaterials as composites. Examples include titanium (Ti), iron (Fe), and magnesium (Mg) (Ghassemi et al. 2018; Zerankeshi et al. 2022).

The emergence of bioreactors allowed for improvements in cell-scaffold interaction, facilitating the generation of three-dimensional tissues in vitro. One example is the use of bioreactors for cartilage synthesis in culture systems, which has shown promising results (Fu et al. 2021). By controlling parameters such as cell distribution on the scaffold, oxygen concentration, mechanical stimuli, hydrostatic pressure, compression forces, and shear stress, among others, bioreactors enable researchers to more closely mimic tissue conditions in vivo. In cartilage, oxygen concentration can influence cell proliferation and differentiation patterns, as cellular differentiation is dependent on low oxygen tension, similar to what is encountered in its natural tissue environment (Yasui et al. 2016). Min et al. (2018) applied bioreactor to culture MSC on alginate hydrogel scaffolds viewing cartilage in vitro formation. One of the bioreactor advantages is the homogeneous cell seed and culture, favoring cell-scaffold interactions. Oxygen tension was controlled by the culture bioreactor system, once it is known to directly influence chondrogenesis, and it was related that 3% oxygen pressure significantly enhanced production of sulfated glycosaminoglycan and collagen II, which are normally present as cartilage extracellular matrix elements sulfated glycosaminoglycan and collagen II), if compared to static culture system. So, the bioreactor improves cartilage in vitro formation and also allows scaling up the process, toward clinical application.

Ultimately, by considering the discussed biomaterial and scaffold characteristics and conducting comprehensive biological evaluations, researchers can develop scaffolds that not only support tissue regeneration but also minimize adverse reactions and maximize their therapeutic potential in tissue engineering applications.

8.5 Inductive Molecules and Growth Factors

Several factors or bioactive molecules, such as growth factors, are necessary for morphogenesis and tissue regeneration. Growth factors are multifunctional peptides that bind to cell surface receptors, activating signaling pathways and controlling gene expression (Marx 2004; Sánchez-González et al. 2012; Han et al. 2020). The action of these growth factors depends on their concentration and cell type, influencing cellular response and primarily affecting proliferation, differentiation, and extracellular matrix synthesis (Graziani et al. 2006; Drengk et al. 2009).

The transforming growth factor beta (TGF-β) superfamily constitutes mediators that stimulate proliferation in mesenchymal stem cells, fibroblasts, osteoblasts, and smooth muscle cells while inhibiting macrophage and lymphocyte proliferation (Marx et al. 1998). TGF-β also regulates extracellular matrix production and the mitogenic effect of other growth factors (Eppley et al. 2004).

Bone morphogenetic proteins (BMPs) are growth factors belonging to the TGF-β superfamily. Specifically, BMP-2 plays a crucial role in bone formation and repair, as well as in the differentiation of mesenchymal stem cells into osteoblasts, increasing alkaline phosphatase activity and osteocalcin synthesis (Eppley et al. 2004). BMP-2 has been used in combination with heparinized-Ti implants, along with recombinant human platelet-derived growth factor (rhPDGF), to enhance new bone formation (Lee et al. 2018). The use of genetically modified BMP-2 for controlled release in a fibrin matrix has also been reported, resulting in faster bone healing in the treatment of critical-size rat craniotomy (Schmoekel et al. 2005). Tissue engineering Infuse, a bone substitute product composed of a collagenous scaffold and BMP-2, is used to improve bone healing in spine surgeries, with comparable performance to autologous bone grafting, which is considered the gold standard for bone healing (James et al. 2016). In a recent study, stem cells combined with the human BMP-2 gene and encapsulated within hydrogel scaffolds were shown to differentiate into an osteogenic lineage, resulting in new bone formation (Lin et al. 2019).

BMP-2 also stimulates cartilage formation by enhancing prechondrogenic precursors, while BMP-4 can induce mesenchymal cells to differentiate into chondroprogenitor cells and promote their maturation into chondrocytes, thus favoring chondrogenesis (Hatakeyama et al. 2004; Giannoni and Cancedda 2006). BMP-7, also known as an osteogenic promoter, can be found in tissue engineering products associated with other biomaterials, such as collagen matrices, as in the case of OP1-Putty (rh-BMP-7), which enhances tissue regeneration (Mishra et al. 2016).

Other growth factors that have been studied for application in tissue engineering are the insulin-like growth factor 1 (IGF-1), PDGF, epidermal growth factor (EGF), and fibroblast growth factor (FGF). IGF-1 plays a significant role in the bone healing process, acting on mesenchymal stem cells and serving as an important modulator of cellular apoptosis. PDGF acts on various cell types, promoting proliferation and matrix synthesis. In combination with IGF, it contributes to bone restoration. EGF induces proliferation in certain cell types, such as mesenchymal stem cells, endothelial cells, fibroblasts, and keratinocytes. It stimulates endothelial chemotaxis and angiogenesis and regulates extracellular matrix production. Lastly, the FGF family consists of polypeptides involved in matrix synthesis, proliferation, differentiation, and survival of various cell types, including osteoblasts and chondrocytes (Eppley et al. 2004; Gaissmaier et al. 2008; Ren et al. 2020).

Growth factors can be used individually or combined with other molecules or even as a pool of molecules, such as in platelet-rich plasma (PRP). PRP is acquired by blood centrifugation, resulting in platelet concentration at levels higher than those found in circulating blood (approximately 2×10^5 platelets/μl) (Marx, 1998; Sánchez-González et al. 2012). Growth factors are present in platelets alpha granules, acting after cascade coagulation activation, on the endothelial and adjacent tissue reestablishment. The various types of platelet growth factors are also found in PRP, for example, epidermal growth factor (EGF), fibroblast growth factor (FGF), insulin-like growth factor 1 (IGF), platelet-derived growth factors (PDGF), transforming growth factors beta (TGF-β), and vascular endothelial growth factor (VEGF) (Eppley et al. 2004; Sánchez-González et al. 2012). There are various types of growth factors in PRP and the cells can respond to them as for attachment, spreading, migration, proliferation and differentiation depending on cell type, PRP concentration and microenvironment conditions, addressing tissue engineering and regenerative medicine objectives. Some approaches suggest PRP

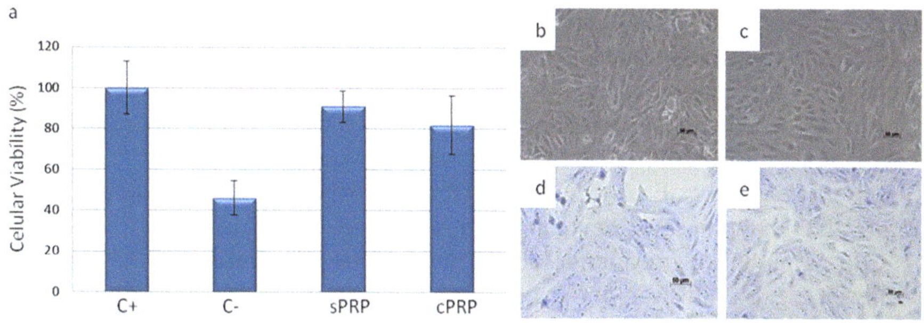

Fig. 8.6 Use of PRP as culture medium supplement (sPRP), in substitution to fetal bovine serum (C+), or without supplement (C-). It was used serum from PRP (sPRP) or culture medium incubated blood clot after PRP induction (cPRP). (**a**) The results showed no significant differences in cultured Vero cells, with MTT assay (3-(4,5-dimethylthiazol-2-yl)-2,5-dyphenyl tetrazolium bromide), absorbance read at 570nm (SpectraMax M5). (**b–e**). Morphological 5-day culture, light microscopy analysis with phase contrast (**b**, **c**) and stained with toluidine blue (**d**, **e**); sPRP (**b**, **d**); cPRP (**c**, **e**). (Ferraraz 2014)

use both as inducing cellular behavior, once supplement to culture medium, and also as a scaffold for cells in the tissue regeneration process (Ferraraz et al. 2019) (Fig. 8.6).

Growth factors can be added to cell cultures and/or to scaffolds, or even directly at the surgical site. Nevertheless, the use of bioactive factors chemically associated or physically encapsulated in polymeric matrices for controlled release has been more widely explored as it ensures prolonged local effects and dosage reduction, while also minimizing adverse systemic effects. For example, the use of poly-glycolic acid (PLG) for controlled release of VEGF (vascular endothelial growth factor) and PDGF has been mentioned to increase local vascularity in different tissues (Richardson et al. 2001; Lee et al. 2011; Lin et al. 2019).

Aside from growth factors, other molecules influence tissue engineering. For instance, cytokines, the small integrin-binding N-linked glycoprotein (SIBLING) family, highly expressed in mineralized tissues such as bone and dental tissue, plays a role in initiating and modulating cell differentiation and mineralization processes through cell-matrix interactions. They contribute to bone development, remodeling, and repair. In vivo, bioactive factors work synergistically to amplify a stimulus or in an antagonistic manner. Therefore, rather than relying on a single factor, a combination of factors is required for proper tissue formation or repair (Eppley et al. 2004; Ren et al. 2020).

8.6 3D Bioprinting for Tissue Engineering

3D bioprinting, also referred to as biofabrication, is a cutting-edge technology that combines cells, bioactive molecules, and biomaterials in a computer-aided printing process to construct organs and tissues layer by layer. As such, this technique has significant implications for the development of complex biological structures and the field of regenerative medicine (Guillemot et al. 2010; Tomasina et al. 2019). Bioprinting techniques can be characterized based on their working principle:inkjet-based, extrusion-based, and laser-based printing (Vijayavenkataraman et al. 2018; Tomasina et al. 2019). Each technique offers different resolution ranges, fabrication times, and limitations (Moroni et al. 2018; Tomasina et al. 2019).

Current bioprinting techniques allow the creation of cellular scaffolds, tissues, and even organs, however, the choice of bioprinting material is limited to fluid or powder form, since the structure of the biomaterial must be adjusted quickly so that there is stability to maintain the integrity of the impression. Cells can be included and printed on a cytocompatible material, with this, the subsequent construct will be immediately formed with a consistent and highly viable cell distribution throughout the entire bioprinting process (Ramos et al., 2020).

3D bioprinting offers a promising alternative for regenerating damaged or lost tissues in clinical practice (La Noce et al. 2014; Verma et al. 2014; Los et al. 2018). In this innovative approach, stem cells are integrated into a matrix along with appropriate growth factors, and then precisely deposited at the targeted site for regeneration, either through injection or placement (Ramos and Moroni 2020).

Researchers have been actively exploring the use of stem cells in 3D bioprinting to enhance cell viability and stimulate specific cell properties, including post-printing differentiation (Snyder et al. 2015). Over the past three decades, tissue engineering has successfully advanced the development of skin and cartilage therapies to replace functionally impaired tissues, as well as bone, liver, and heart tissue substitutes. In regenerative medicine, 3D bioprinting holds immense potential for fabricating complex tissues and organs suitable for transplantation, as it allows for the selection of cell types, materials, as well as growth and differentiation factors. However, bioprinting poses other technical challenges when compared to nonbiological printing, including software and hardware development (Murphy and Atala 2014; Zhao et al. 2015).

Therefore, 3D bioprinting emerges as a tool for generating biological tissues and organs with intricate structures and specialized functions, thus transforming the fields of tissue engineering and regenerative medicine, and holding great promise for clinical applications (Khatiwala et al. 2012; Ramos et al., 2020).

8.7 Conclusions and Future Perspectives

Tissue engineering and regenerative medicine have gradually evolved as promising approaches to address the needs of patients. These sciences hold great potential to restore or regenerate the structure and function of damaged tissues or organs and provide alternative approaches to improving patients' quality of life and well-being. Tissue engineering combines cells, scaffolds, and inductive factors to regenerate or replace damaged or diseased tissues. Meanwhile, regenerative medicine integrates tissue engineering with other strategies such as cell-based therapy, gene therapy, and immunomodulation to induce tissue/organ regeneration in vivo.

Recent advances in tissue engineering have led to the novel development and refinement of scaffolds. These scaffolds need to be biocompatible, promoting cell adhesion, differentiation, proliferation, and the production of tissue-specific molecules. The incorporation of signaling molecules, including growth factors, cytokines, and nonprotein chemical compounds, into scaffolds is commonly used to address tissue regeneration. These molecules can bind to cell surface receptors, activating signaling pathways that stimulate cell proliferation, differentiation, and extracellular matrix protein synthesis. Various cell types can be utilized in tissue engineering, including adult cells of homologous or autologous origin, mesenchymal stromal cells from the bone marrow, and embryonic stem cells capable of differentiating into multiple lineages in vitro. Stem cells, obtained from different sources in the body, possess excellent proliferative and differentiation capabilities, making them an interesting cell source for tissue engineering and regenerative medicine.

The emergence of 3D bioprinting technology has opened up new perspectives and promising applications in tissue engineering and regeneration. Further research into this cutting-edge technology is underway, aiming to improve the treatment and regeneration of tissues using scaffolds made from stem cells and biomaterials. Other areas associated with tissue engineering and regenerative medicine have witnessed significant progress in the last decades. These include the development of surgical techniques for clinical translation, improved cryopreservation processes for newly formed tissues, design engineering incorporating bioreactors, organ-on-a-chip systems, and robotics. These advancements combine multiple innovative technologies focused on fast-tracking the translation into clinical practice.

In summary, tissue engineering and regenerative medicine represent a promising scientific and technological field that combines diverse disciplinary approaches to treat, enhance or replace tissues, organs, and bodily functions, with clinically relevant results and groundbreaking technologies capable of improving patients' quality of life. As research continues, tissue engineering and regenerative medicine hold the potential to revolutionize the fields of medicine and surgery even further, beyond just repairing lost tissue function, but regenerating the original characteristics of tissues and organs.

References

Dos Anjos S, Mavropoulos E, Alves GG, Costa AM, de Alencar Hausen M, Spiegel CN, Longuinho MM, Mir M, Granjeiro JM, Rossi AM (2019) Impact of crystallinity and crystal size of nanostructured carbonated hydroxyapatite on pre-osteoblast in vitro biocompatibility. J Biomed Mater Res A 107(9):1965–1976

Anselme K (2000) Osteoblast adhesion on biomaterials. Biomaterials 21:667–681

Ashammakhi N, Ghavaminejad A, Tutar R, Fricker A, Roy I, Chatzistavrou X, Hoque Apu E, Nguyen Kl, Ahsan T, Pountos I, Caterson EJ (2022) Highlights on advancing frontiers in tissue engineering. Tissue Eng Part B Rev 28(3)633–664

Beheshtizadeh N, Gharibshahian M, Pazhouhnia Z, Rostami M, Zangi Ar, Maleki R, Azar HK, Zalouli V, Rajavand H, Farzin A, Lotfibakhshaiesh N, Sefat F, Azami M, Webster TJ, Rezaei N (2022) Commercialization and regulation of regenerative medicine products: Promises, advances and challenges. Biomed Pharmacother 153:113431

Berthiaume F, Maguire TJ, Yarmush ML (2011) Tissue engineering and regenerative medicine: history, progress, and challenges. Annu Rev Chem Biomol Eng 2:403–30

Brockbank KGM (2022) Tissue engineering constructs and commercialization. In: Madame Curie Bioscience Database [Internet]. Austin (TX): Landes Bioscience; 2000–2013.

Brittberg M, Lindahl A, Nilsson A, Ohlsson C, Isaksson O, Peterson L (1994) Treatment of deep cartilage defects in the knee with autologous chondrocyte transplantation. N Engl J Med 331(14):889–95

Brunner D, Frank J, Appl H, Schöffl H, Pfaller W, Gstraunthaler G (2010) Serum-free cell culture: the serum-free media interactive online database. Altex 27(1):53–62

Capes-Davis A, Freshney RI (2010) Culture of animal cells: a manual of basic technique and specialized applications. 8. ed. New Jersey: Wiley-Blackwell 732 p.

Caplan AI (2007) Adult mesenchymal stem cells for tissue engineering versus regenerative medicine. J Cell Physiol 213(2):341–7

Chan BP, Leong KW (2008) Scaffolding in tissue engineering: general approaches and tissue-specific considerations. Eur Spine J 17:S467–S479

Courtenay JC, Sharma RI, Scott JL (2018) Recent advances in modified cellulose for tissue culture applications. Molecules 23(3):654

Cubo N, Garcia M, Del Cañizo JF, Velasco D, Jorcano JL (2016) 3D bioprinting of functional human skin: production and in vivo analysis. Biofabrication 9(1):015006

Daar AS, Greenwood Hl (2007) A Proposed Definition Of Regenerative Medicine. J Tissue Eng Regen Med 1:179

Drengk A, Zapf A, Stürmer EK,Stürmer KM, Frosch KH (2009) Influence of platelet-rich plasma on chondrogenic differentiation and proliferation of chondrocytes and mesenchymal stem cells. Cells Tissues Organs 189(5):317–326

Eppley BL, Woodell JE, Higgins J (2004) Platelet quantification and growth factor analysis from platelet-rich plasma: implications for wound healing. Plast Reconstr Surg 114(6):1502–1508

Falanga V, Sabolinski MA (1999) bilayered living skin construct (APLIGRAF) accelerates complete closure of hard-to-heal venous ulcers. Wound Repair Regen 7:201–207

Farley A, Mclafferty E, Hendry C (2012) Cells, tissues, organs and systems. Nurs Stand 26(52): 40–45

Ferraraz DC (2014) Substituição do soro fetal bovino por plasma rico em plaquetas nas culturas celulares. [Unpublished undergraduate thesis on Biological Sciences] - Federal University of ABC (UFABC)

Ferraraz DC (2020) Avaliação dos efeitos da laserterapia em baixa intensidade na proliferação e diferenciação de células-tronco da polpa de dentes humanos decíduos esfoliados em arcabouços de dentina/osso. Thesis on Biotecnoscience (PhD) - Federal University of ABC (UFABC)

Ferraraz DC, Tatsui NH, Rodrigues LR, Lombello CB (2019) Platelet-rich plasma as supplement and scaffold for the culture of Vero cell line. Res Biomed Eng A:1–9

Filipowska J, Pawlik J, Cholewa-Kowalska K, Tylko G, Pamula E, Niedzwiedzki L, Szuta M, Laczka M, Osyczka AM (2014) Incorporation of sol-gel bioactive glass into PLGA improves mechanical properties and bioactivity of composite scaffolds and results in their osteoinductive properties. Biomed Mater 9(6):065001

Freitas-Ribeiro S, Reis RL, Pirraco RP (2022) Long-term and short-term preservation strategies for tissue engineering and regenerative medicine products: state of the art and emerging trends. PNAS Nexus 1(4):pgac212

Folkman J, Moscona A (1978) Role of cell shape in growth control. Nature 273:345–349

Fu L, Li P, Li H, Gao C, Yang Z, Zhao T, Chen W, Liao Z, Peng Y, Cao F, Sui X, Liu S, Guo Q (2021) The application of bioreactors for cartilage tissue engineering: Advances, limitations, and future perspectives. Stem Cells Int 2021:6621806

Gaissmaier C, Koh JL, Weise K (2008) Growth and differentiation factors for cartilage healing and repair. Injury 39:88–96

Ghassemi T, Shahroodi A, Ebrahimzadeh MH, Mousavian A, Movaffagh J, Moradi A (2018) Current Concepts in scaffolding for bone tissue engineering. Arch Bone Jt Surg 6(2):90–99

Giannoni P, Cancedda R (2006) Articular chondrocyte culturing for cell-based cartilage repair: needs and perspectives. Cells Tissues Organs 184(1):1–15

Gómez-Barrena E, Padilla-Eguiluz N, Rosset P, Gebhard F, Hernigou P, Baldini N, Rouard H, Sensebé L, Gonzalo-Daganzo RM, Giordano R, García-Rey E, Cordero-Ampuero J, Rubio-Suárez JC, García-Simón MD, Stanovici J, Ehrnthaller C, Huber-Lang M, Flouzat-Lachaniette CH, Chevallier N, Donati DM, Spazzoli B, Ciapetti G, Fleury S, Fernandez MN, Cabrera JR, Avendaño-Solá C, Montemurro T, Panaitescu C, Veronesi E, Rojewski MT, Lotfi R, Dominici M, Schrezenmeier H, Layrolle P (2020) Early efficacy evaluation of mesenchymal stromal cells (MSC) combined to biomaterials to treat long bone non-unions. Injury 51 Suppl 1:S63–S73

Guillemot F, Mironov V, Nakamura M (2010) Bioprinting is coming of age: report from the International Conference on Bioprinting and Biofabrication in Bordeaux (3B'09). Biofabrication 2(1):010201

Graziani F, Ivanovski S, Cei S, Ducci F, Tonetti M, Gabriele M (2006) The *in vitro* effect of different PRP concentrations on osteoblasts and fibroblastos. Clin Oral Impl Res 17(2):212–219

Han F, Wang J, Ding L, Hu Y, Li W, Yuan Z, Guo Q, Zhu C, Yu L, Wang H, Zhao Z, Jia L, Li J, Yu Y, Zhang W, Chu G, Chen S, Li B (2020) Tissue engineering and regenerative medicine: achievements, future, and sustainability in Asia. Front Bioeng Biotechnol.8:83

Hatakeyama Y, Tuan RS, Shum L (2004) Distinct functions of BMP4 and GDF5 in the regulation of chondrogenesis. J Cell Biochem 91:1204–1217

Hoffman T, Khademhosseini A, Langer R (2019) Chasing the paradigm: clinical translation of 25 years of tissue engineering. Tissue Eng Part A 25(9–10):679–687

Howard D, Buttery LD, Shakesheff KM, Roberts SJ (2008) Tissue engineering: strategies, stem cells and scaffolds. J Anatomy 213(1):66–72

ISO - International Organization for Standardization 10993-1 (2018) Biological evaluation of medical devices — Part 1: Evaluation and testing within a risk management process p.41

ISO - International Organization for Standardization 10993-5 (2009) Biological Evaluation of Medical Devices. Part 5: Tests for in Vitro Cytotoxicity p.34

ISO/TS - International Organization for Standardization (Technical Standard), 21560:2020. General requirements of tissue-engineered medical products, p.18

James AW, LaChaud G, Shen J, Asatrian G, Nguyen V, Zhang, X,Ting K, Soo C (2016). A review of the clinical side effects of bone morphogenetic protein-2. Tissue Eng. Part B Rev 22:284–297

Jin G, Prabhakaran MP, Ramakrishna S (2011) Stem cell differentiation to epidermal lineages on electrospun nanofibrous substrates for skin tissue engineering. Acta Biomater 7(8):3113–22

Kaul H, Ventikos Y (2015) On the genealogy of tissue engineering and regenerative medicine. Tissue Eng Part B Rev 21(2):203–217

Khademhosseini A, Langer R (2016) A decade of progress in tissue engineering. Nat Protoc 111775–1781

Khatiwala C, Law R, Shepherd B, Dorfman S, Csete M (2012) 3D cell bioprinting for regenerative medicine research and therapies. Gene Ther Regul 7(1)1230004

Kim YS Smoak MM, Melchiorri AJ, Mikos AG (2019) An overview of the tissue engineering market in the United States from 2011 to 2018. Tissue Eng Part A 25(1–2)1–8

Kessler MW, Grande DA (2008) Tissue engineering and cartilage. Organogenesis 4(1):28–32

Krishani M, Shin WY, Suhaimi H, Sambudi NS (2023) Development of Scaffolds from bio-based natural materials for tissue regeneration applications: A review. Gels 9(2):100

La Noce M, Paino F, Spina A, Naddeo P, Montella R, Desiderio V, De Rosa A, Papaccio G, Tirino V, Laino L (2014) Dental pulp stem cells: state of the art and suggestions for a true translation of research into therapy. J Dent 42(7):761–768
Langer R, Vacanti JP (1993) Tissue engineering. Science 260(5110):920–926
Lanza R, Langer R, Vacanti JP (2020) Principles of Tissue Engineering. Academic Press, 5th Ed. p.1678
Le H, Xu W, Zhuang X et al. (2020) Mesenchymal stem cells for cartilage regeneration. J Tissue Eng 11:2041731420943839
Lee K, Silva EA, Mooney DJ (2011) Growth factor delivery-based tissue engineering: general approaches and a review of recent developments. J R Soc Interface 8(55):153–170
Lee MH, Arcidiacono JA, Bilek AM, Wille JJ, Hamill CA, Wonnacott KM, Wells MA, Oh SS (2010) Considerations for tissue-engineered and regenerative medicine product development prior to clinical trials in the United States. Tissue Eng Part B Rev 16(1):41–54
Lee SH, Bae EB, Kim SE, Yun YP, Kim HJ, Choi J-W, Lee J-J, Huh J-B (2018). Effects of immobilizations of rhBMP-2 and/or rhPDGF-BB on titanium implant surfaces on osseointegration and bone regeneration. Coatings 8:17 https://doi.org/10.3390/coatings8010017
Lin H, Tang Y, Lozito TP, Oyster N, Wang B, Tuan RS (2019) Efficient in vivo bone formation by *BMP-2* engineered human mesenchymal stem cells encapsulated in a projection stereolithographically fabricated hydrogel scaffold. Stem Cell Res Ther 10:254
Lopes CC, Ambrosio FN, Lombello CB, Romani AP (2019) Estudo do alinhamento de fibras de arcabouços poliméricos de policaprolactona e blendas de policaprolactona/poli (ácido lático). In: 6° Workshop de Orgãos Artificiais,Biomateriais e Engenharia de Tecidos (OBI), 2019, São Paulo- SP. Anais 6° OBI, 2019.
Los M, Hudecki A, Wiechec E (2018) Stem cells and biomaterials for regenerative medicine. 1st ed. Academic Press, p.249
Lysaght MJ, Reyes J (2001) The growth of tissue engineering. Tissue Eng 7(5):485–493
Mahla RS (2016) Stem cells applications in regenerative medicine and disease therapeutics. Intern J Cell Biol 2016:6940283
Mao AS, Mooney DJ (2015) Regenerative medicine: Current therapies and future directions. Proc Natl Acad Sci USA 112(47):14452–14459
Marlovits S, Zeller P, Singer P, Resinger C, Vécsei V (2006) Cartilage repair: generations of autologous chondrocyte transplantation. Eur J Radiol 57(1):24–31
Mason C, Dunnill P (2008) A brief definition of regenerative medicine. Regen Med 3(1)1–5
Masson AO (2018) Estudo comparativo de metodologias de citotoxicidade aplicadas à biomateriais.Thesis on Biomedical Engineering (Master Degree) - Federal University of ABC (UFABC)
Marx RE, Carlson ER, Eichstaedt RM, Schimmele SR, Strauss JE, Georgeff KR (1998) Platelet-rich plasma: growth factor enhancement for bone grafts. Oral Surg Oral Med Oral Pathol Oral Radiol Endod 85(6): 638–46
Marx RE (2004) Platelet-rich plasma: evidence to suport its use. J Oral Maxillofac Surg 62(4):489–496
Min D, Lee W, Bae IH, Lee TR, Croce P, Yoo SS (2018) Bioprinting of biomimetic skin containing melanocytes. Exp Dermatol 27(5):453–459
Mishra R, Bishop T, Valerio IL, Fisher JP, Dean D (2016) The potential impact of bone tissue engineering in the clinic. Regen Med 11(6):571–87
Moroni L, Boland T, Burdick JA, De Maria C, Derby B, Forgacs G, Groll J, Li Q, Malda J, Mironov VA, Mota C, Nakamura M, Shu W, Takeuchi S, Woodfield TBF, Xu T, Yoo JJ, Vozzi G (2018) Biofabrication: a guide to technology and terminology. Trends Biotechnol 36(4):384–402
Murphy SV, Atala A (2014) 3D bioprinting of tissues and organs. Nat Biotechnol 32:773–785
Nascimento MHM, Ferreira M, Malmonge SM, Lombello CB (2017) Evaluation of cell interaction with polymeric biomaterials based on hyaluronic acid and chitosan. J Mater Sci: Mater Med 28(5):68
Nerem RM (1992) Tissue engineering in the USA. Med Biol Eng Comput 30:CE8–CE12
O'Brien FJ (2011) Biomaterials & scaffolds for tissue engineering. Materials today 14:88–95

Perić Kačarević Ž, Rider P, Alkildani S, Retnasingh S, Pejakić M, Schnettler R, Gosau M, Smeets R, Jung O, Barbeck M (2019) An introduction to bone tissue engineering. Int J Artif Organs 43:69–86

Pina S, Ribeiro VP, Marques CF, Maia FR, Silva TH, Reis RL, Oliveira JM (2019) Scaffolding strategies for tissue engineering and regenerative medicine applications. Materials (Basel) 12(11):1824

Ramos T, Moroni L (2020) Tissue engineering and regenerative medicine 2019: The Role of Biofabrication—A Year in Review. Tissue Engineering: Part C 26:91–106

Ren X, Zhao M, Lash B, Martino Mm, Julier Z (2020) Growth factor engineering strategies for regenerative medicine applications. Front Bioeng Biotechnol 7:469

Richardson TP, Peters MC, Ennett AB, Mooney DJ (2001) Polymeric system for dual growth factor delivery. Nat. Biotechnol 19:1029–1034

Roseti L, Parisi V, Petretta M, Cavallo C, Desando G, Bartolotti I, Grigolo B (2017) Scaffolds for Bone tissue engineering: State of the art and new perspectives. Mater. Sci. Eng. C 78:1246–1262

Sage EH (2001) Regulation of interactions between cells and extracellular matrix: a command performance on several stages. J Clin Invest 107(7):781–783

Sánchez-González DJ, Mendez-Bolaina E, Trejo-Bahena NI (2012) Platelet-rich plasma peptides: key for regeneration. Int J Pept 2012:532519

Schmoekel HG, Weber FE, Schense JC, Grätz KW, Schawalder P, Hubbell JA (2005) Bone repair with a form of BMP-2 engineered for incorporation into fibrin cell ingrowth matrices. Biotechnol Bioeng 89(3):253–62

Shimojo AAM, Rodrigues ICP, Perez AGM, Souto EMB, Gabriel LP, Webster T (2020). Scaffolds for tissue engineering: A State-of-the-Art review concerning types, properties, materials, processing, and characterization. In: Li, B., Moriarty, T., Webster, T., Xing, M. (eds) Racing for the Surface. Springer

Smith IO, McCabe LR, Baumann MJ (2006) MC3T3-E1 osteoblast attachment and proliferation on porous hydroxyapatite scaffolds fabricated with nanophase powder. Int J Nanomedicine 1(2):189–94

Snyder J, Rin Son A, Hamid Q, Wang C, Lui Y, Sun W (2015) Mesenchymal stem cell printing and process regulated cell properties. Biofabrication 7(4):044106

Šponer P, Filip S, Kučera T, Brtková J, Urban K, Palička V, Kočí Z, Syka M, Bezrouk A, Syková E (2016) Utilizing Autologous Multipotent Mesenchymal Stromal Cells and β-Tricalcium Phosphate Scaffold in Human Bone Defects: A Prospective, Controlled Feasibility Trial. Biomed Res Int 2016:2076061

Theocharis AD, Skandalis SS, Gialeli C, Karamanos NK (2016) Extracellular matrix structure. Adv Drug Deliv Rev 97:4–27

Tomasina C, Bodet T, Mota C (2019) Bioprinting vasculature: materials, cells and emergent techniques. Materials (Basel) 12:2701

Vacanti CA, Vacanti JP (1991) Functional organ replacement: The new technology of tissue engineering. Braverman MH, Tawes RL, eds. Surgical Technology Int'l, London: Century Press. p. 43–49

Vacanti CA (2006) The history of tissue engineering. J Cel Molec Med 10(3):569–576

Vagaská B, Bacáková L, Filová E, Balík K (2010) Osteogenic cells on bio-inspired materials for bone tissue engineering. Physiol Res 59(3):309–322

Verma K, Bains R, Bains VK, Rawtiya M, Loomba K, Srivastava SC (2014) Therapeutic potential of dental pulp stem cells in regenerative medicine: An overview. Dent Res J (Isfahan) 11(3):302–308

Vijayavenkataraman S, Yan WC, Lu WF (2018) 3D bioprinting of tissues and organs for regenerative medicine. Adv Drug Deliv Rev 132:296–332

Wakitani S, Goto T, Pineda SJ, Young RG, Mansour JM, Caplan AI, Goldberg VM (1994) Mesenchymal cell-based repair of large, full-thickness defects of articular cartilage. J Bone Joint Surg Am 76(4):579–92

Wang S, Xiong Y, Chen J, Ghanem A, Wang Y, Yang J, Sun B (2019) Three Dimensional Printing Bilayer Membrane Scaffold Promotes Wound Healing. Front Bioeng Biotechnol 7:348

Williams DF (1999). Dictionary of biomaterials. Liverpool: Liverpool University Press p.42

Wise JK, Yarin AL, Megaridis CM, Cho M (2009) Chondrogenic differentiation of human mesenchymal stem cells on oriented nanofibrous scaffolds: engineering the superficial zone of articular cartilage. Tissue Eng Part A 15(4):913–21

Wolf E, Kemter E, Klymiuk N, Reichart B (2019) Genetically modified pigs as donors of cells, tissues, and organs for xenotransplantation. Anim Front 9(3):13–20

Yari D, Ebrahimzadeh MH, Movaffagh J, Shahroodi A, Shirzad M, Qujeq D, Moradi A (2022) Biochemical Aspects of scaffolds for cartilage tissue engineering; from basic science to regenerative medicine. Arch Bone Jt Surg 10(3):229–244

Yasui Y, Chijimatsu R, Hart DA, Koizumi K, Sugita N, Shimomura K, Myoui A, Yoshikawa H, Nakamura N (2016) Preparation of Scaffold-free tissue-engineered constructs derived from human synovial mesenchymal stem cells under low oxygen tension enhances their chondrogenic differentiation capacity. Tissue Eng Part A, 22(5–6):490–500

Zerankeshi MM, Bakhshi R, Alizadeh R (2022) Polymer/metal composite 3D porous bone tissue engineering scaffolds fabricated by additive manufacturing techniques: A review. Bioprinting 25:e00191

Zhang X, Reagan MR, Kaplan DL (2009) Electrospun silk biomaterial scaffolds for regenerative medicine. Adv Drug Deliv Rev 61(12):988–1006

Zhao Y, Li Y, Mao S, Sun W, Yao R (2015) The influence of printing parameters on cell survival rate and printability in microextrusion-based 3D cell printing technology. Biofabrication 7(4):045002

Chapter 9
Natural Hydrogels for Drug Delivery Systems

Sônia Maria Malmonge, Juliana Kelmy Macário Barboza Daguano, Amanda Castro Juraski, Kaline do Nascimento Ferreira, and Marcos Antonio Sabino Gutierrez

Abstract Tissue engineering can be improved by the addition of molecules – therapeutic drugs, growth factors, cellular signaling, or binding molecules – that will facilitate cellular function and tissue regeneration. With that in mind, researchers have been exploring methods to have a better control over the release profile of drugs in order to enhance the speed, quantity, and quality of tissue regeneration. The addition of therapeutics in the scaffold can accelerate tissue regeneration process and offer the drug directly to the injured site, avoiding systemic effects. Drug delivery systems (DDS) also reduce the necessary dose to obtain the desired effect

The original version of the chapter has been revised. A correction to this chapter can be found at https://doi.org/10.1007/978-3-031-38743-2_16.

S. M. Malmonge
Graduate Program in Biomedical Engineering, Federal University of ABC – UFABC, São Bernardo do Campo, SP, Brazil

J. K. M. B. Daguano (✉)
Graduate Program in Biomedical Engineering, Federal University of ABC – UFABC, São Bernardo do Campo, SP, Brazil

Center for Information Technology Renato Archer, Campinas, SP, Brazil
e-mail: juliana.daguano@ufabc.edu.br

A. C. Juraski
Graduate Program in Biomedical Engineering, Federal University of ABC – UFABC, São Bernardo do Campo, SP, Brazil

Department of Chemical Engineering, Polytechnic School, University of São Paulo, São Paulo, SP, Brazil

K. d. N. Ferreira
Center for Information Technology Renato Archer, Campinas, SP, Brazil

M. A. S. Gutierrez
Center for Information Technology Renato Archer, Campinas, SP, Brazil

Simón Bolívar University, Caracas, Venezuela

C. B. Lombello, P. A. da Ana (eds.), *Current Trends in Biomedical Engineering*, https://doi.org/10.1007/978-3-031-38743-2_9

and amount of drugs needed to counter-effect adverse reactions. Hydrogels are among the most used materials for drug delivery systems (DDS). Hydrogels' DDSs main issue is that they usually display a burst release in the beginning, and drug load hardly lasts beyond a week. However, it is possible to promote a better control over drug release profile through the method of drug loading, hydrogel cross-linking, and even chemical or physical modifications to the surface of the material.

Keywords Natural hydrogels · Polysaccharides · Drug delivery · Tissue engineering

9.1 Introduction

Hydrogels are polymeric materials that have several hydrophilic domains. Their polymeric chains are cross-linked physically or chemically, which allows hydrogels to entrap water inside without dissolving. Hydrogels have a soft and rubbery consistency and low surface tension (Katime et al. 1996; Blanco et al. 2008). These same characteristics are observed in natural soft tissues, making hydrogels interesting materials for biomedical applications as scaffolds, because they must be able to promote and hold cell adhesion, usually through a compatible topography, meaning roughness values that will allow cell adhesion without sharp edges. Figure 9.1 shows some of the expected properties of a scaffold applied to tissue engineering and drug delivery; because most studied hydrogel scaffolds, especially natural hydrogels, offer these characteristics (Kesharwani et al. 2021).

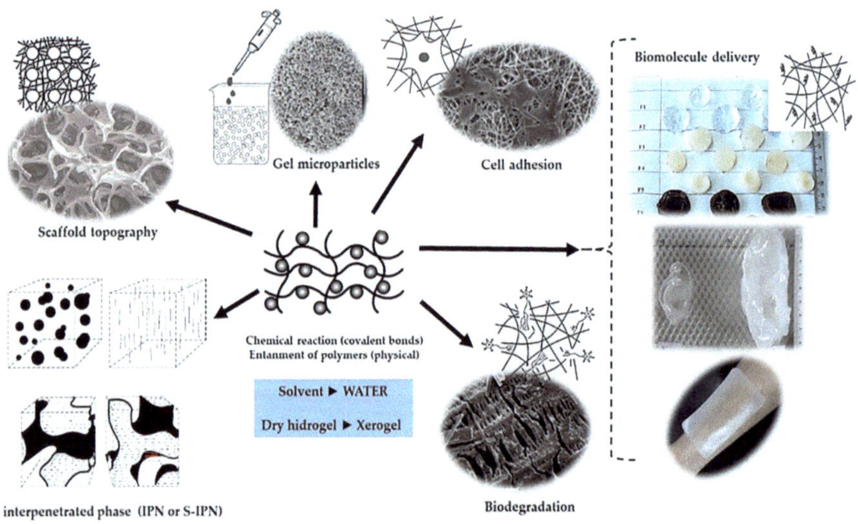

Fig. 9.1 The ideal hydrogels structures for tissue engineering and drug delivery is expected to be biocompatible, biodegradable, interpenetrate porous and networks, gels micro/nano-particles, plus able to encapsulate and deliver drugs, and biomolecules and cells at the repair site

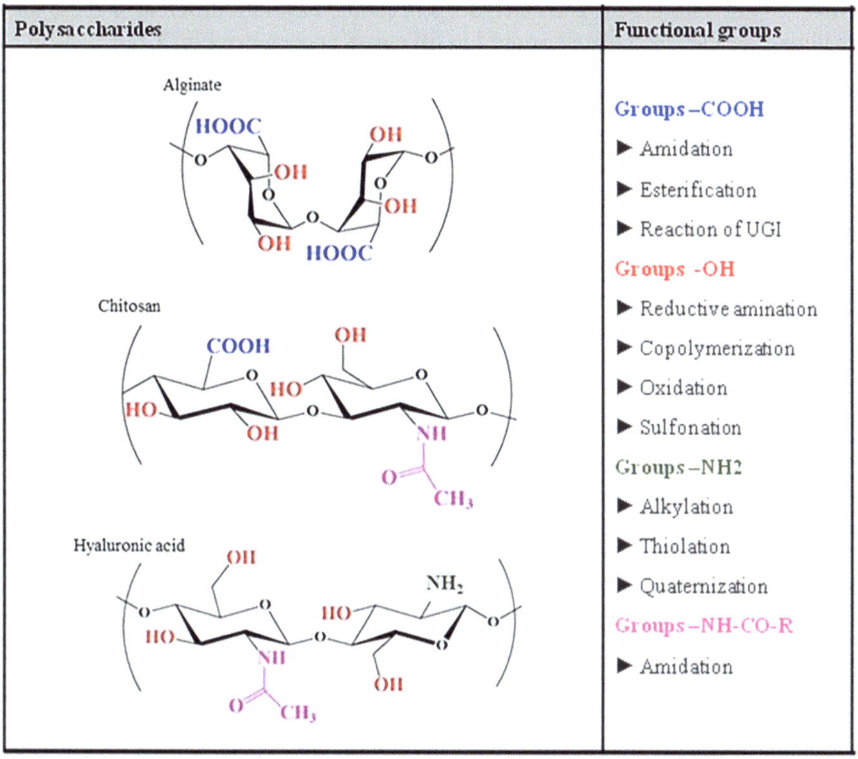

Fig. 9.2 Chemical structures of polysaccharides and chemical reactions that allow modifying the characteristics and properties of these polysaccharides according to their reactive functional groups

Synthetic hydrogels usually are not naturally reabsorbed, have the disadvantage of provoking more severe immunological reactions, and for this reason demand a second surgery to remove the scaffold after tissue repair. On the other hand, natural hydrogels provide structures similar to living tissues and induce a better cellular response due to their similarities with extracellular matrix (ECM), which avoids immunological complications. They also present biodegradability, which means they are able to be replaced by repaired tissue without leaving toxic debris (Blanco et al. 2008; Kesharwani et al. 2021; Caliari and Burdick 2016). See Fig. 9.2, where the most used biopolymers (which will be further discussed in this chapter) for these applications appear with their chemical structure and functional groups, susceptible to chemical modification to enhance the potential of these natural polymers for regenerative medicine.

Given the importance of the ECM for a functional tissue regeneration, there is a special interest in selecting hydrogels that will mimic the natural ECM properties appropriately (Tibbitt and Anseth 2009). One advantage of natural hydrogels is that their aqueous, soft, three-dimensional structure mimics the in vivo ECM, facilitating their interaction with the host tissue (Kesharwani et al. 2021; Caliari and Burdick

2016; Tibbitt and Anseth 2009). Hydrogels can be cast into different shapes, such as films, particles, and fibers, or used to fill empty spaces, see Fig. 9.1.

Drug delivery system (DDS) is a conventional technique that involves the repeated administration of an active compound aiming to deliver adequate levels of therapeutic in the body to maintain the efficacy without side effects that is common in case of high doses. In view of achieving controlled and local drug delivery, it is necessary to use nanostructured biomaterials systems, where the active compound is incorporated (Dreiss 2020).

With the advancement of knowledge in the area of DDS, this technology began to be explored associated with other techniques in the health area, as in the case of Tissue Engineering (TE). TE emerged from the technological advance in the field of biomaterials and refers to the association of scaffolds, cells, and biologically active molecules, aiming at damaged tissues or organs regeneration and thus restoring lost function. Thus, TE uses three-dimensionally structured biomaterials to create a suitable microenvironment for the recruitment, adhesion, proliferation and differentiation of cells, and the scaffold (Caddeo et al. 2017).

Mimicry of the biological ambient can be improved by the addition of molecules – therapeutic drugs, growth factor, cellular signaling, or binding molecules – that will facilitate cellular function and tissue regeneration (Rambhia and Ma 2015). Moreover, as the field evolves toward clinical translation and the development of implantable devices, the necessity for additional features needs to be addressed. For example, the use of biomaterials for bone regeneration can be accompanied by infections that will decrease blood flow, impairing new bone formation. The addition of antibiotics within the scaffold can address these issues, but there will still be the need to develop controlled release from the scaffold to make sure the release itself does not hinder tissue formation (Ekenseair et al. 2013).

Taking this into account, researchers have been exploring methods to have a better control over the release profile of drugs, in order to enhance the speed, quantity, and quality of tissue regeneration (Li and Mooney 2016). The addition of therapeutics in the scaffold can accelerate tissue regeneration process and offer the drug directly to the injured site, avoiding systemic effects. Drug delivery systems (DDSs) also reduce the necessary dose to obtain the desired effect, and amount of drugs needed to counter-effect adverse reactions. Hydrogels are among the most used materials for DDSs. Hydrogels' DDSs main issue is that they usually display a burst release in the beginning, and the drug load hardly extends beyond a week. However, it is possible to promote a better control over drug release profile through the method of drug loading, hydrogel cross-linking, and even chemical or physical modifications to the surface of the material (Blanco et al. 2008; Martins et al. 2021).

9.2 Functionalization of Hydrogels for Drug Delivery Systems

Polysaccharides, which are the most abundant biopolymers on earth, consist of monosaccharides (sugars) linked together by O-glycosidic bonds. Differences in monosaccharide composition, binding types and patterns, chain shape, and molecular weight determine its physical properties, such as solubility, viscosity, gelation potential, and surface and interfacial properties (Khan and Ahmad 2013). Additionally, many scientists have been investigating their obtaining and applications since it was discovered that polysaccharides have bioactivities that were applied to clinical practice (Li et al. 2016). Proving that they have, mostly, properties such as biocompatibility and biodegradability, and for some of them, it has been reported that they have various activities as antioxidants (Yen et al. 2008), immunoregulators (Meng et al. 2016), antitumor agents (Zhang et al. 2023), antimicrobial (Chien et al. 2016), anti-inflammatory (Chang et al. 2019), anti-HIV (Mobarakeh et al. 2019), antimutagenic (Rani et al. 2021), and anticoagulants (Zakerikhoob et al. 2021). That is why polysaccharides play different physiological functions and, consequently, have great potential applications in biomedical fields. However, the use of natural polymers can be compromised by their rapid degradation process, which leads to the possible loss of biological properties during formulation and storage (Khan and Ahmad 2013). Also, it has been detected that some of the bioactivities of the polysaccharides are very weak.

However, some polysaccharides have other reactive functional groups, susceptible to modifications. Such is the case of chitosan, alginate, and hyaluronic acid. Both alginate and hyaluronic acid have carbonyl groups (-COOH) throughout their skeleton (Martins et al. 2021). Chitosan and hyaluronic acid, in turn, have N-acetylated units (-NH-CO-CH_3) where the former also has amino groups (-NH_2), depending on the degree of deacetylation (DD) (Khan and Ahmad 2013; Juncan et al. 2021). Figure 9.2 shows the structures and functional groups characteristic of alginate, hyaluronic acid, and chitosan, polysaccharides whose modifications are possible for opening a window of applications in biomedical and pharmacological areas, and whose main properties will be briefly mentioned below.

Chitosan

Among the most abundant polysaccharides in nature, special attention has been given to chitosan, polyamino-glucoside obtained by partial deacetylation of chitin, a structural part of the exoskeleton of crustaceans and insects, but also found in some classes of fungi. Chemically, chitin is a linear polymer composed of units of D-glucosamine (deacetylated unit) randomly linked by β ($1 \rightarrow 4$) glycosidic bonds to N-acetyl-D-glucosamine (acetylated unit), as shown in Fig. 9.2 (Wang et al. 2020.

When the number of deacetylates unites in chitin is $\geq$50%, it goes from being entirely insoluble to soluble in aqueous acidic media, being then considered chitosan. The poor solubility of chitosan when pH > 6.5 is a severe drawback in many of its potential applications (Pillai et al. 2009). Therefore, many studies have focused on modifying this polysaccharide toward water-soluble derivatives. The structural modification of a compound by appropriate chemical, physical, or biological methods allows obtaining numerous derivatives of the starting product, which helps to expand its applications (An et al. 2009).

In the case of chitosan chains, the modification site is dictated by the desired use of the final chitosan derivative. It is of great importance because the basic skeleton of chitosan and its original exceptional physicochemical and biochemical properties are preserved after modification while achieving desirable structures and functions particularly suitable for tissue repair and regeneration applications (Liu et al. 2011). There is a wide variety of chitosan derivatives, and reaction selection highly dependable on the final hydrogel structure (particle, film, three-dimensional structure, gel, etc.) and its desired application (Xu et al. 2021). The most frequent chemical modifications include quaternization, acylation, tosylation, Schiff base formation, O-carboxymethylation, N-carboxyalkylation, N-succinylation, and graft copolymerization (An et al. 2009; Liu et al. 2011; Jain et al. 2013).

As already mentioned, chitosan is made up of D-glucosamine units, which can be acetylated. Depending on the degree of deacetylation (DD), the chitosan can have different physical-chemical properties concerning parameters such as solubility, viscosity, and degradability. The solubility of chitosan in acid medium is given by the protonation of the amino groups present in its skeleton, thus conferring polycationic behavior (Wang et al. 2020). Its rigid semi-crystalline structure is due to inter, and intramolecular hydrogen bonds (Wang et al. 2020) and its average molecular weight (Mw) can range between 300 and 1000 kDa, depending on the source and method of preparation. Mw and DD are the predominant parameters that influence the solubility, mechanical resistance, and degradation properties of chitosan and its biomaterials (Caliari and Burdick 2016; Xu et al. 2021).

Alginate

Alginate is a linear polysaccharide, soluble in water, extracted from the cell wall of brown seaweeds. It has in its conformation α-(1 → 4)-L-guluronic acid (G) and β-(1 → 4)-D-mannuronic acid (M) junctions. It is composed of homopolymeric blocks M-M or G-G and blocks with an alternating sequence of the M-G blocks (Aarstad et al. 2012). The composition of the block sequence M and G and the molecular weight of the alginate depend on the variability of existing algal species. This diversity is used as a tool to control the properties of a series of natural functions in this polysaccharide. For example, the stem of the algae Laminaria Hyperborea requires high resistance and rigidity, therefore, the alginate that is extracted from them has a higher fraction of residues G. On the contrary, the leaves of the plants that require flexibility to float in fluid waters have a higher fraction of M residues (Brus et al. 2017).

While alginate is biocompatible, it lacks specific cell signals for attachment and therefore its biological performance is often lacking when compared to other biopolymers, such as chitosan and hyaluronic acid. For that reason, alginate is often used combined with positively charged polymers to improve overall cell attachment and viability. Moreover, the molecular weight of commercial sodium alginate varies in the range of 30–400 kDa. The viscosity of the alginate solutions increases when the pH decreases and reaches a maximum of around pH 3.0–3.5 when the carboxyl groups in the structure are protonated and form hydrogen bonds. By increasing the molecular weight of the alginate, the physical properties of the resulting gel can be improved. However, an alginate solution generated by a high molecular weight polymer is extremely viscous and becomes a challenge when it is desired to process (injection, rheology analysis, transport, etc.) (Fu et al. 2011).

Hyaluronic Acid

Hyaluronic acid (HA) is a linear glycosaminoglycan of natural origin that consists of repetitive units of D-glucuronic acid and N-acetyl-D-glucosamine linked by glycosidic links β-(1 → 4) and β-(1 → 3) (see Fig. 9.2); it was discovered in 1934, in the vitreous humor of the eyes of the bovine species (Amagai et al. 2009). It can be found in the extracellular matrix (ECM) of all tissues, and it is present in higher concentration in the connective, epithelial, and neural tissue; because it is the only glycosaminoglycan that is not sulfated and is synthesized in the internal plasma membrane by a family of transmembrane proteins called hyaluronan synthase (Antich et al. 2020). It is expelled in the MEC, where it resides for 3 or 5 days before being degraded by a family of enzymes called hyaluronidases (Amagai et al. 2009). As it is one of the main components of the extracellular matrix, it contributes significantly to cell migration and proliferation (Antich et al. 2020).

Under physiological conditions, HA is found in the sodium salt form, which presents negatively charged and it is known as sodium hyaluronate. Under these conditions, it is highly hydrophilic and is surrounded by water molecules linked by hydrogen bonding, for these reasons, its physicochemical properties have been extensively investigated and reviewed (Fagien et al. 2019). This polysaccharide can possess 25,000 repetitive units of disaccharides in length, and it has been reported that its molecular weight can reach $8{,}10^6$ Da, 10^7 Da, or 10^8 Da, according to different authors and depending on the enzyme that catalyzed its synthesis (Amagai et al. 2009). Due to its high molecular weight and strong intermolecular interactions, the aqueous HA solution is very viscous. Under alkaline or acidic conditions, the hydrolysis of the HA chains is induced (Antich et al. 2020; Fagien et al. 2019). Hyaluronic acid is a biocompatible material, but when it is injected into a physiological environment, a rapid degradation process often occurs (it is degraded in vivo by the enzyme hyaluronidase). For this reason, the hyaluronic acid must be chemically modified to obtain a more stable material and at the same time maintain its fundamental properties (Ding et al. 2022).

As already mentioned, the bioactivities of polysaccharides and their physicochemical properties are restricted directly or indirectly by their structure, so some methods to modify the structure of polysaccharides are the fundamental way to solve the problems described above (Li et al. 2016). These modifications can be made by physical, chemical, or biological procedures (Khan and Ahmad 2013), to obtain numerous derivatives that allow a better study of the structure-activity relationship, as well as enrich the variety of polysaccharides and expand or enhance their properties (Yen et al. 2008; Khan and Ahmad 2013; Li et al. 2016).

Chemical modification is a method that allows changing the structure of a polymer by introducing substituent groups, and in the case of polysaccharides that have bioactivities, these can be strengthened, or generated due to functionalization (Li et al. 2016). This type of modification also allows for modifying the hydrophobic character of the materials under study. The methods of chemical modification include sulfation, amidation, alkylation, carboxymethylation, phosphorylation, and acetylation, among others and their execution on the polysaccharide will depend on the reactivity of the functional groups present in the natural state of these biopolymers (Katime et al. 1996; Li et al. 2016).

Since chemical modification significantly improves the biocompatibility of polysaccharides, this characteristic could allow the use of these in 3D bioprinting techniques and thus expand their application in the biomedical area. Currently, there is great interest in the use of the 3D printing technique, also known as additive manufacturing, through which, by depositing a series of layers, prototypes of various types of structures can be built. The 3D printing technique in comparison with the processes of traditional use has the advantage of producing complex devices in addition to a personalized way, in this sense, it allows saving time, easy manufacturing, use of multiple materials, and controlling cell growth (Derakhshanfar et al. 2018; Melocchi et al. 2020).

3D printing has made various areas of science, such as tissue engineering and biomedicine, integrate the development of 3D biofabrication techniques which allows the development of cell-loaded scaffolds to improve cell adhesion and promote tissue growth (Akilbekova and Turlybekuly 2023). Some treatments have been implemented that involve the construction of the pinna, wound dressings, bone repair, and craniofacial regeneration. To carry out 3D bioprinting, the use of a suspension of a biomaterial-like gel known as bioink is required, which will be deposited layer by layer to generate the scaffold to be manufactured. The bioink must be designed from a biocompatible material that does not impede cell growth. Therefore, the invention of inks of biological origin constitutes a challenge since it must comply with the qualities that are required where both the biological needs and the rheological properties of the material are balanced (Montero et al. 2019).

Improving the bioactivity of polymers through chemical modification also increases the probability that it will promote tissue growth in a scaffold constructed by 3D printing. Bioactivity can promote an increase in the biocompatibility of scaffolds used in tissue engineering, increasing the possibilities that this material is recognized by cells and favoring the processes of cell adhesion and proliferation. Various materials of natural origin have been used as a gel to manufacture bioink (Li and Mooney 2016; Montero et al. 2019), among them we can find the polysaccharides: chitosan, alginate, and hyaluronic acid, as well as mixtures between them or with other biopolymers.

In this way, the functionalization of hydrogels can be improved by the addition of molecules – therapeutic drugs, growth factors, cellular signaling, or binding molecules – that will facilitate cellular function and tissue regeneration. Drugs can be adsorbed into the hydrogel scaffolds by permeation, entrapment, or covalent binding, as illustrated by Fig. 9.3.

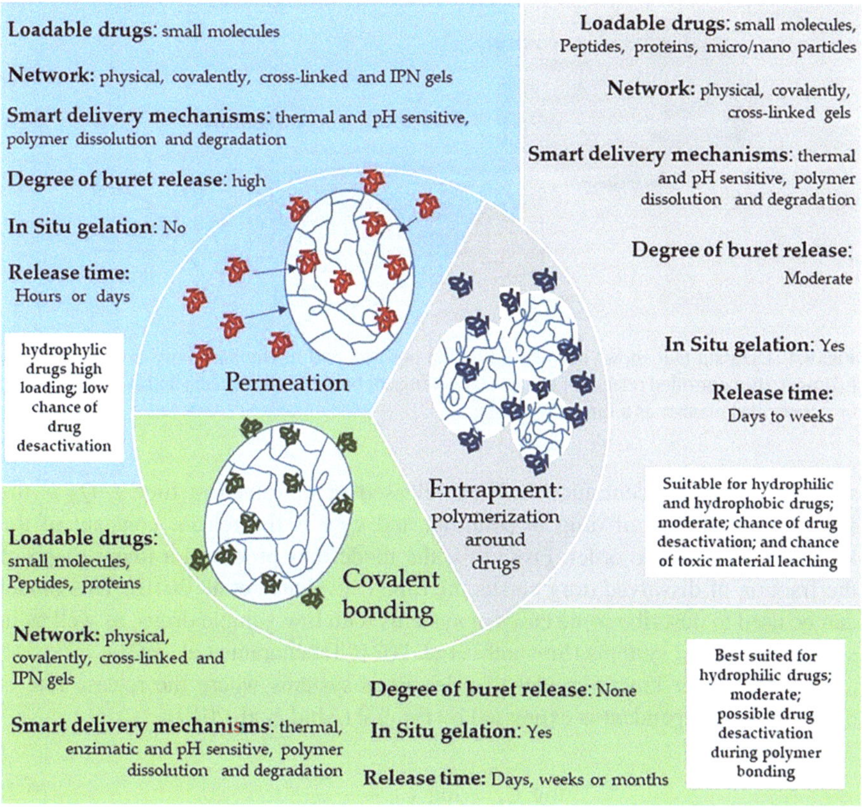

Fig. 9.3 Drug loading methods for hydrogel DDSs. Method of drug loading is an important decision since it affects the mechanism of drug release after implantation

9.3 Kinetic Models of Drug Release

The evaluation of kinetic parameters of drug release is essential because it provides information about the feasibility and efficiency of drug release in the matrix. Several kinetic models have been reported over the years, which allowed the correct interpretation of the release mechanisms. For this, a good reproducibility of the data is required and statistical analysis needs to be performed using 95% confidence limits.

A diagram that allows visualizing the behavior of the diffusion of a molecule that is being released from a hydrogel is represented in Fig. 9.4, and it is what can be estimated through the equations that are going to be studied below.

The linear expression for the zero-order kinetic model is represented by Eq. 9.1:

$$Q = Q_0 + k_0 t \quad (9.1)$$

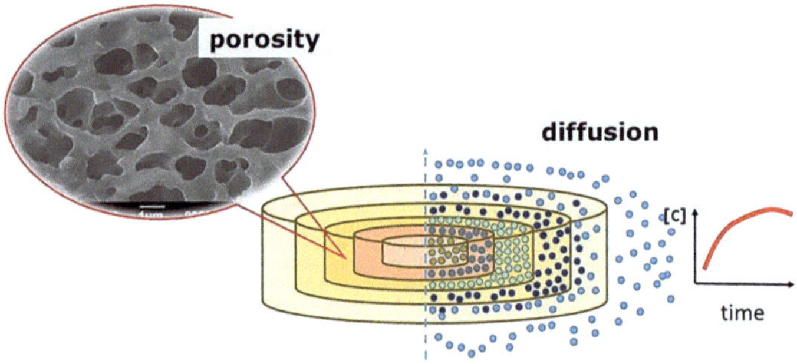

Fig. 9.4 Diagram that shows how, based on the porosity and interconnectivity of the pores of a hydrogel, the controlled release of encapsulated drugs or biomolecules occurs and they are released in a controlled manner as a function of time

where Q is the concentration of drug released or dissolved in time t, Q_0 is the initial concentration of drug in solution, and k_0 é is the release constant of the kinetic model of zero-order. From this, the model has produced a linear graph of the fraction of dissolved drug versus the time (Suvakanta et al. 2010). This model can be used to describe some cases of systems with low-soluble drugs, as well as in some transdermal systems (Sravanthi et al. 2021; Tebcharani et al. 2021).

The first-order kinetic model that describes systems where the release rate is concentration-dependent is expressed by Eq. 9.2 (Abid et al. 2019):

$$\log\ C_t = \log\ C_0 - \frac{k_t}{2,303} \tag{9.2}$$

where C_t is the concentration of the drug in time t, C_0 is the initial concentration of the incorporated drug, and k_t is the constant of the first-order kinetic model. The k_t value can be determined from the slope of the graph between cumulative log of the remaining drug and time. This model can be used to describe controlled release systems with hydrophilic drugs in porous matrices.

Initially designed for one-dimensional matrices, Higuchi model was proposed following some hypotheses that (i) the concentration of the drug in the system is greater than its solubility; (ii) diffusion of the unidirectional drug; (iii) the dimensions of the drug are smaller than the thickness in the matrix; (iv) swelling and dissolution of the matrix are negligible; (v) constant drug diffusion; and (vi) perfect sinking condition (Gohel et al. 2000). Linear form of the Higuchi model is given by Eq. 9.3:

$$Q_t = k_H\ \sqrt{t} \tag{9.3}$$

where Q_t is the concentration of the released drug and k_H is the constant rate of the Higuchi model. The k_H value is evaluated according to the linear graph between cumulative % of drug release versus time. This model can be extended to other types of matrix DDS, such as for transdermal patches or even for oral administration systems (Okonogi et al. 2021).

To identify whether there is a change in the surface area of the matrix at the time of release, the Hixson and Crowell model governs kinetics. The linearized form of the Hixson and Crowell model is observed in Eq. 9.4 (Friuli et al. 2022):

$$Q_t^{\frac{1}{3}} = Q_0^{\frac{1}{3}} + k_{HC}t \quad (9.4)$$

where Q_t is the concentration of the drug released in time t and k_{HC} is the release constant of the kinetic model of Hixson and Crowell. To study the release kinetics of this model, cubic root plots of the percentage of the remaining drug versus time are plotted.

To validate the model suggested by Hixson and Crowell some assumptions are made, (i) the model considers spheroidal materials and monodisperse materials; (ii) dissolution occurs on the surface; and (iii) the medium needs to be constantly shaken, so that it promotes a slow diffusion (Friuli et al. 2022). The Hixson and Crowell model applies to therapeutic dosage forms such as tablets, where there is a change in surface area and diameter.

Korsmeyer–Peppas model describes a release speed directly proportional to time from a polymeric system. Equation 9.5 considers the diffusion of the drug in the means of dissolution and the relaxation of the matrix (Wu et al. 2019).

$$\frac{m_t}{m_\infty} = k_{PT}t^n \quad (9.5)$$

where $\frac{m_t}{m_\infty}$ represents the fraction of the drug released as a function of time t, k_{PT} is constant, and n is the exponent of release. Release exponent n indicates the drug transport mechanisms that release systems based on cylindrical structures play from this value (Olejnik et al. 2017). To discover exponent n, only the part of the release curve where the release curve $\frac{m_t}{m_\infty} < 0,6$ is used. The model is plotted as a log of the cumulative percentage of drug release versus time log.

This model can be extended to other types of matrix drug delivery systems, such as for liposome systems and even for carbon points loaded with antibiotics (Saravanan et al. 2020; Wu et al. 2019).

The release of drugs in spherical matrices was described by Baker and Lonsdale from Higuchi's model (Baker and Lonsdale, 1974), according to Eq. 9.6:

$$f = \frac{3}{2}\left[1 - \left(1 - \frac{M_t}{M_\infty}\right)^{\frac{1}{3}}\right] - \frac{M_t}{M_\infty} = k_t \quad (9.6)$$

where $\frac{m_t}{m_\infty}$ is the fraction of the drug released in time t and can be simplified by the letter Q. The model proposed by Baker and Lonsdale is plotted as $\frac{3}{2}\left[1 - (1 - Q)^{\frac{1}{3}}\right] - Q$ versus the time, to which the k_t corresponds to the slope.

This model is applied in the linearization of data regarding the release of drugs from microspheres or microcapsules.

Weibull's model describes different dissolution processes applied to controlled drug release systems (Langenbucher 1972), described through Eq. 9.7.

$$M = M_0 \left[1 - e^{-\frac{(t-T)^b}{a}} \right] \tag{9.7}$$

where M is the amount of drug dissolved as a function of time and M_0 is the total amount of drug that is being released. T represents the delay time measured as a result of the dissolution process. Parameter a has dependency over time, and parameter b describes the shape of the dissolution curve progression. When $b = 1$, the curve corresponds to an exponential profile, if $b > 1$, the shape of the curve becomes sigmoidal with an inflection point, while the shape of the curve with $b < 1$ would show a sharper increase than that with $b = 1$. The Weibull's model is particularly useful for the comparison of different drug release profiles in a given matrix.

Hopfenberg developed a model that correlates drug release with surface erosion of the polymeric system and provided that the surface area remains constant during the degradation process. The commutative fraction of the drug as a function of time is given according to Eq. 9.8 (Hopfenberg and Hsu 1978):

$$\frac{M_t}{M_\infty} = 1 - \left[1 - \frac{k_0 t}{C_L a} \right]^n \tag{9.8}$$

where C_L is the initial load of the drug through the system, k_0 is the constant that describes the process of surface erosion of the polymer, a is the half thickness of the matrix system, and n is an exponent dependent on geometry ($n = 1$ flat geometry, $n = 2$ cylindrical geometry, and $n = 3$ spherical geometry).

The Hopfenberg model can be applied to identify the mechanism of release of oil spheres from hydrogels, with good solubility and intermediate release rate. The in vitro release profile is described by an exponential model described by Gompertz, as shown in Eq. 9.9 (Permanadewi et al. 2019).

$$X(t) = X_{\max} \exp \left[-\alpha \, e^{\beta \, logt} \right] \tag{9.9}$$

where $X(t)$ is the percentage of the drug dissolved in time t divided by 100; $X_{\max}$ is the maximum dissolution; α is the location or scale parameter, which determines the undissolved part in $t = 1$; and β is the parameter that describes the dissolution rate per unit of time. This model is useful in cases where $b = 1$.

The model proposed by Gompertz is empirical, so it becomes useful to compare the profiles of the Weibull model (Li et al. 2005). Drug release mechanisms from biodegradable polymers, with combinations between degradation and diffusion, can follow the model proposed by Gallagher and Corrigan, Eq. 9.10 (García-González et al. 2015).

$$\frac{M_t}{M_\infty} = \left(1 - e^{-k_1 t}\right) + \left(1 - \frac{M_b}{M_\infty}\right) \frac{e^{k_2(t-t_{\max})}}{1 + e^{k_2(t-t_{\max})}} \tag{9.10}$$

where M_b is the fraction of the drug released during the first stage (the explosion effect), k_1 (min^{-1}) and k_2 (min^{-1}) are the first-order kinetic constants (1st stage of release) and the kinetic constant for the 2nd stage of the release phase (polymer degradation), respectively, and $t_{\max}$ (min) is the time for maximum release of the drug. The model describes well systems consisting of two-stage, first stage diffusion of the drug under an explosion effect followed by a release controlled by the slow erosion of the mass of the polymer matrix.

Regarding controlled release systems to which the assumption is made that there is only the kinetic process of zero order, we have the Cooney model. This model provides details of the surface erosion of spheres and cylinders, considering that the release is proportional to the surface area of the device, which is time dependent. For a cylinder with the initial length L_0 and initial diameter D_0, we have Eq. 9.11 which was derived by quantifying the release rate of drug f as a function of time t (Cooney 1972).

$$f = \frac{(D_0 - 2K_t)^2 + 2(D_0 - 2K_t)(L_0 - 2K_t)}{D^2 + 2D_0L_0} \tag{9.11}$$

where K_t is a model constant. The considerations of these models are made based on the ratio L_0/D_0, when it is approaching zero (the curves turn into a horizontal line with a constant relative drug release rate of 1. For disc-shaped cylinders ($L_0/D_0 < 1$ ratios), the relative rate of drug release remains finite until it completes drug release (Siepmann and Siepmann 2008).

9.4 Examples

In continuation, an example of these hydrogels applications are considered and some results related to an investigation will be presented. In a study entitled *Ibuprofen loaded Chitosan Films:* In Vitro *Assessment of Drug Release Profile and Cell Viability on Primary Neurons Culture* (Juraski et al. 2021), the authors prepared ibuprofen-loaded chitosan films (ICH) and evaluated its release profile and cell viability regarding primary neurons from rats spinal cord.

The release of ibuprofen from ICH films was evaluated in PBS at 37 °C in sink conditions. Under these experimental settings, over 45% of the drug was released within the first 24 h of incubation in PBS. Figure 9.5 (close up) shows that ibuprofen is continuously released into the PBS media, with a burst effect within the first 24 h. Figure 9.5 shows that ibuprofen continued to be released at this rate up to 30 days, where another 35% was slowly released into the PBS media. The release

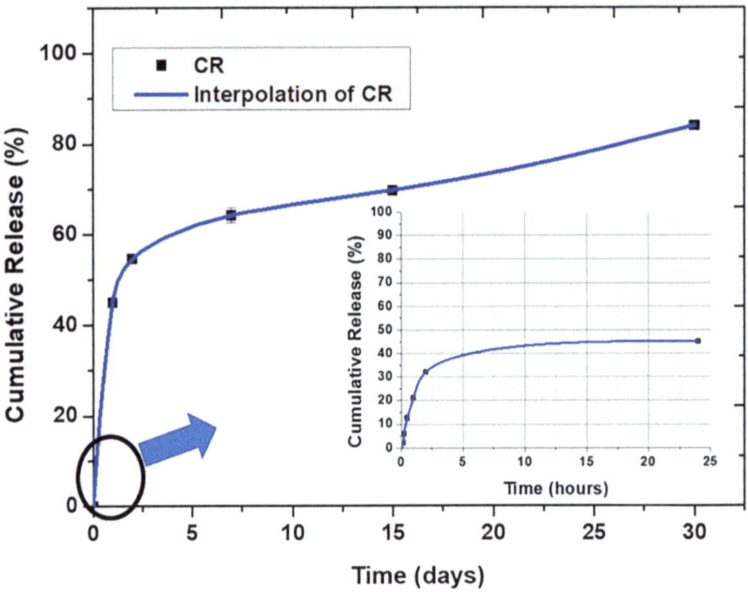

Fig. 9.5 Cumulative ibuprofen release from ICH films. Close-up shows the release profile in the first 24 h of immersion in PBS (pH = 7.4). Data represents the absorbance and concentration for each sample ± standard error

Table 9.1 Equations for the different models and the regression coefficient (R^2) of ICH films in vitro in the first 24 h and after 30 days of immersion in PBS (pH = 7.4)

Model	Regression coefficients (R^2)	
	24 h	30 days
Zero-order	0.81	0.55
First-order	0.86	0.82
Higuchi	0.95	0.78
Kosmeyer–Peppas	0.70	0.98
Hixon	0.84	0.51

profile changed due to the origin of particles released; initially, it releases the loosely entrapped surface drug particles and drug release shows a burst release effect. As the film was degraded, particles entrapped inside the polymeric matrix were also released into the media, and cumulative drug release reached 80% after 30 days, indicating that there were still some drug particles yet to be released.

The mechanism of drug release was determined by fitting the curves on distinct models and comparing the regression coefficients (R^2). Table 9.1 shows the R^2 values for the curves presented in Fig. 9.5.

Table 9.1 shows that the model that best fitted the in vitro release profile in the first 24 h was Higuchi, but for the entire 30 days of immersion, drug release profile is better described by the Kosmeyer–Peppas model. That's because the in vitro drug release profile showed a fast release within the first hours, with stabilization within 5 h and that lasted up to 24 h of drug release (Fig. 9.5, close up). Over 45% of the drug loaded in the samples was removed in the first 24 h. This is very important when considering the local release of anti-inflammatory drugs. The inflammatory processes usually have peaks within the first 24 h after injury, therefore, any drug applied with the intention to lessen the damaging effects of the inflammatory response needs to act within this time frame (Juraski et al. 2021). The driving force behind the burst effect observed in the first 24 h of IBU release was the drug concentration gradient. The release was mostly slow and steady, indicating that IBU was well distributed in the polymeric matrix, both in the film surface and inside it.

The sustained release is also explained by its drug-loading content. Higher drug-loading content causes a sustained release profile because after all the surface particles were removed there were still the ones entrapped inside the matrix, which started to diffuse through it (Fig. 9.5), causing another increase in the release rate from 15 to 30 days. The drug release profile is crucial for antiinflammatory drug release because most inflammatory peaks occur in the first 24 h after injury. Therefore, drugs that aim to lessen the damaging effects of the inflammatory response must act within this period. Furthermore, IBU prolonged administration at the lesion site has been associated with controlled inflammation and tissue protection.

9.5 Conclusions

This chapter aimed to present the characteristics of nature-sourced hydrogels and their advantageous use as drug delivery devices. There are several methods for fabricating drug-releasing hydrogels, and both material choice and device format have a significant impact on the release profile. One of the most notorious aspects of natural hydrogels is their unpredictable degradation rate, which can interfere with the drug release profile. In this sense, the next generation of natural hydrogels for drug delivery studies would benefit from focusing on the mathematical models for describing drug release profiles. These mathematical tools could help address the need for standardization of natural hydrogels in vitro testing, and therefore speed the process into in vivo and eventually clinical studies.

Funding Declaration/Acknowledgments M.A. Sabino wants to thank FAPESP for the grant (# 2021/13949-5) as a visiting researcher at the CTI Renato Archer and the financial support of the CNPq/PCI program through the K. F. Nascimento fellowship. Amanda C Juraski would like to thank CNPq for grant #140574/2019-0.

References

Aarstad, OA, Tøndervik, A, Sletta, H, & Skjåk-Bræk, G (2012). Alginate sequencing: an analysis of block distribution in alginates using specific alginate degrading enzymes. Biomacromolecules, 13(1), 106–116. https://doi.org/10.1021/bm2013026

Abid, S, Hussain, T, Nazir, A, Zahir, A, & Khenoussi, N (2019). A novel double-layered polymeric nanofiber-based dressing with controlled drug delivery for pain management in burn wounds. Polymer Bulletin, 76(12), 6387–6411. https://doi.org/10.1007/s00289-019-02727-w

Akilbekova, D, & Turlybekuly, A (2023). Patient-specific 3D bioprinting for in situ tissue engineering and regenerative medicine. In *3D* Printing in Medicine (pp. 149–178). Woodhead Publishing. https://doi.org/10.1016/B978-0-323-89831-7.00003-1

Amagai, I, Tashiro, Y, & Ogawa, H (2009). Improvement of the extraction procedure for hyaluronan from fish eyeball and the molecular characterization. Fisheries Science, 75, 805–810. https://doi.org/10.1007/s12562-009-0092-2

An, NT, Dong, NT, & Le Dung, P (2009). Water-soluble N-carboxymethylchitosan derivatives: Preparation, characteristics and its application. Carbohydrate Polymers, 75(3), 489–497. https://doi.org/10.1016/j.carbpol.2008.08.017

Antich, C, de Vicente, J, Jiménez, G, Chocarro, C, Carrillo, E, Montañez, E, ... & Marchal, JA (2020). Bio-inspired hydrogel composed of hyaluronic acid and alginate as a potential bioink for 3D bioprinting of articular cartilage engineering constructs. Acta biomaterialia, 106, 114–123. https://doi.org/10.1016/j.actbio.2020.01.046

Baker, RW, & Lonsdale, HS (1974). Controlled relaese of biologically active agents. In Synthetic Membranes: Science, Engineering and Applications. Springer Netherlands. https://doi.org/10.1007/978-94-009-4712-2

Blanco, MD, Guerrero, S, Teijon, C, Olmo, R, Pastrana, L, Katime, I, & Teijón, JM (2008). Preparation and characterization of nanoparticulate poly (N-isopropylacryl-amide) hydrogel for the controlled release of anti-tumour drugs. Polymer International, 57(11), 1215–1225. https://doi.org/10.1002/pi.2457

Brus, J, Urbanova, M, Czernek, J, Pavelkova, M, Kubova, K, Vyslouzil, J., ... & Kulich, P (2017). Structure and dynamics of alginate gels cross-linked by polyvalent ions probed via solid state NMR spectroscopy. Biomacromolecules, 18(8), 2478–2488. https://doi.org/10.1021/acs.biomac.7b00627

Caddeo, S, Boffito, M and Sartori, S (2017). Tissue Engineering Approaches in the Design of Healthy and Pathological In Vitro Tissue Models. Front. Bioeng. Biotechnol, 5:40. https://doi.org/10.3389/fbioe.2017.00040

Caliari, SR, & Burdick, JA (2016). A practical guide to hydrogels for cell culture. Nature methods, 13(5), 405–414. https://doi.org/10.1038/nmeth.3839

Chang, SH, Lin, YY, Wu, GJ, Huang, CH, & Tsai, GJ (2019). Effect of chitosan molecular weight on anti-inflammatory activity in the RAW 264.7 macrophage model. International journal of biological macromolecules, 131, 167–175. https://doi.org/10.1016/j.ijbiomac.2019.02.066

Chien, RC, Yen, MT, & Mau, JL (2016). Antimicrobial and antitumor activities of chitosan from shiitake stipes, compared to commercial chitosan from crab shells. Carbohydrate polymers, 138, 259–264. https://doi.org/10.1016/j.carbpol.2015.11.061

Cooney, DO (1972). Effect of Geometry on the Dissolution of Pharmaceutical Tablets and Other Solids: Surface Detachment Kinetics Controlling. AIChE Journal, 18(2). https://doi.org/10.1002/aic.690180234

Derakhshanfar, S, Mbeleck, R, Xu, K, Zhang, X, Zhong, W, & Xing, M (2018). 3D bioprinting for biomedical devices and tissue engineering: A review of recent trends and advances. Bioactive materials, 3(2), 144–156. https://doi.org/10.1016/j.bioactmat.2017.11.008

Ding, YW, Wang, ZY, Ren, ZW, Zhang, XW, & Wei, DX (2022). Advances in modified hyaluronic acid-based hydrogels for skin wound healing. Biomaterials Science. https://doi.org/10.1039/D2BM00397J

Dreiss, C (2020). Hydrogel design strategy for drug delivery. Current Opinion in Colloid & Interface Science. 48, 1–17. https://doi.org/10.1016/j.cocis.2020.02.001

Ekenseair, AK, Kasper FK., Mikos, AG (2013). Perspectives on the interface of drug delivery and tissue engineering. Adv Drug Deliv Rev, 65(1), 89–92. https://doi.org/10.1016/j.addr.2012.08.017.

Fagien, S, Bertucci, V, von Grote, E, & Mashburn, JH (2019). Rheologic and physicochemical properties used to differentiate injectable hyaluronic acid filler products. Plastic and reconstructive surgery, 143(4), 707. https://doi.org/10.1097/2FPRS.0000000000005429

Friuli, V, Pisani, S, Conti, B, Bruni, G, & Maggi, L (2022). Tablet Formulations of Polymeric Electrospun Fibers for the Controlled Release of Drugs with pH-Dependent Solubility. Polymers, 14(10). https://doi.org/10.3390/polym14102127

Fu, S, Thacker, A, Sperger, DM, Boni, RL, Buckner, IS, Velankar, S, ... & Block, LH (2011). Relevance of rheological properties of sodium alginate in solution to calcium alginate gel properties. Aaps Pharmscitech, 12, 453–460. https://doi.org/10.1208/s12249-011-9587-0

García-González, CA, Jin, M, Gerth, J, Alvarez-Lorenzo, C, & Smirnova, I (2015). Polysaccharide-based aerogel microspheres for oral drug delivery. Carbohydrate Polymers, 117, 797–806. https://doi.org/10.1016/j.carbpol.2014.10.045

Gohel, MC, Panchal, MK, & Jogani, VV (2000). Novel Mathematical Method for Quantitative Expression of Deviation from the Higuchi Model. In AAPS PharmSciTech (Vol. 1, Issue 4). https://doi.org/10.1208/pt010431

Hopfenberg, HB, & Hsu, KC (1978). Swelling-Controlled, Constant Rate Delivery Systems. Polymer Engineering and Science, 18(15). https://doi.org/10.1002/pen.760181511

Jain, A, Gulbake, A, Shilpi, S, Jain, A, Hurkat, P, & Jain, SK (2013). A new horizon in modifications of chitosan: syntheses and applications. Critical Reviews™ in Therapeutic Drug Carrier Systems, 30(2). https://doi.org/10.1615/CritRevTherDrugCarrierSyst.2013005678

Juncan, AM, Moisă, DG, Santini, A, Morgovan, C, Rus, LL, Vonica-Ţincu, AL, & Loghin, F (2021). Advantages of hyaluronic acid and its combination with other bioactive ingredients in cosmetics. Molecules, 26(15), 4429. https://doi.org/10.3390/molecules26154429

Juraski, AC, Simbara, MO, Paschon, V, Malmonge, SM, Daguano, JKMB (2021). Ibuprofen loaded Chitosan Films: in Vitro Assessment of Drug Release Profile and Cell Viability on Primary Neurons Culture. Iranian Journal of Materials Science and Engineering, 19 (2). https://doi.org/10.22068/ijmse.2311

Katime, I, Velada, JL, Novoa, R, de Apodaca, ED, Puig, J, & Mendizabal, E (1996). Swelling kinetics of poly (acrylamide)/poly (mono-n-alkyl itaconates) hydrogels. Polymer international, 40(4), 281–286. https://doi.org/10.1002/(SICI)1097-0126(199608)40:4%3C281::AID-PI555%3E3.0.CO;2-H

Kesharwani, P, Bisht, A, Alexander, A, Dave, V, & Sharma, S (2021). Biomedical applications of hydrogels in drug delivery system: An update. Journal of Drug Delivery Science and Technology, 66, 102914. https://doi.org/10.1016/j.jddst.2021.102914

Khan, F, & Ahmad, SR (2013). Polysaccharides and their derivatives for versatile tissue engineering application. Macromolecular Bioscience, 13(4), 395–421. https://doi.org/10.1002/mabi.201200409

Langenbucher, F (1972). Linearization of dissolution rate curves by the Weibull distribution. J. Pharm. PharmaC., *24*(979). https://doi.org/10.1111/j.2042-7158.1972.tb08930.x

Li, J, & Mooney, DJ (2016). Designing hydrogels for controlled drug delivery. Nature Reviews Materials, 1(12), 1–17. https://doi.org/10.1038/natrevmats.2016.71

Li, J, Gu, JD, & Pan, L (2005). Transformation of dimethyl phthalate, dimethyl isophthalate and dimethyl terephthalate by Rhodococcus rubber Sa and modeling the processes using the modified Gompertz model. International Biodeterioration and Biodegradation, 55(3), 223–232. https://doi.org/10.1016/j.ibiod.2004.12.003

Li, S, Xiong, Q, Lai, X, Li, X, Wan, M, Zhang, J, ... & Lin, Y (2016). Molecular modification of polysaccharides and resulting bioactivities. Comprehensive Reviews in Food Science and Food Safety, 15(2), 237–250. https://doi.org/10.1111/1541-4337.12161

Liu, X, Ma, L, Mao, Z, & Gao, C. (2011). Chitosan-based biomaterials for tissue repair and regeneration. Chitosan for biomaterials II, 81–127. https://doi.org/10.1007/12_2011_118

Martins, M, Sato, A CK, Ogino, K, & Goldbeck, R (2021). Evaluating the addition of xylooligosaccharides into alginate-gelatin *hydrogels. Food Research International, 147,* 110516. https://doi.org/10.1016/j.foodres.2021.110516

Melocchi, A, Uboldi, M, Maroni, A, Foppoli, A, Palugan, L, Zema, L, & Gazzaniga, A (2020). 3D printing by fused deposition modeling of single-and multi-compartment hollow systems for oral delivery–A review. International journal of pharmaceutics, 579, 119155. https://doi.org/10.1016/j.ijpharm.2020.119155

Meng, X, Liang, H, & Luo, L (2016). Antitumor polysaccharides from mushrooms: a review on the structural characteristics, antitumor mechanisms and immunomodulating activities. *Carbohydrate research, 424*, 30–41. https://doi.org/10.1016/j.carres.2016.02.008

Mobarakeh, VI, Modarressi, MH, Rahimi, P, Bolhassani, A, Arefian, E, Atyabi, F, & Vahabpour, R (2019). Optimization of chitosan nanoparticles as an anti-HIV siRNA delivery vehicle. International journal of biological macromolecules, 129, 305–315. https://doi.org/10.1016/j.ijbiomac.2019.02.036

Montero, FE, Rezende, RA, Da Silva, JV, & Sabino, MA (2019). Development of a smart bioink for bioprinting applications. *Frontiers in Mechanical Engineering, 5,* 56. https://doi.org/10.3389/fmech.2019.00056

Okonogi, S, Phumat, P, Khongkhunthian, S, Chaijareenont, P, Rades, T, & Müllertz, A (2021). Development of self-nanoemulsifying drug delivery systems containing 4-allylpyrocatechol for treatment of oral infections caused by candida albicans. Pharmaceutics, 13(2), 1–16. https://doi.org/10.3390/pharmaceutics13020167

Olejnik, A, Kapuscinska, A, Schroeder, G, & Nowak, I (2017). Physico-chemical characterization of formulations containing endomorphin-2 derivatives. Amino Acids, 49(10), 1719–1731. https://doi.org/10.1007/s00726-017-2470-x

Permanadewi, I, Kumoro, AC, Wardhani, DH, & Aryanti, N (2019). Modelling of controlled drug release in gastrointestinal tract simulation. Journal of Physics: Conference Series, 1295(1). https://doi.org/10.1088/1742-6596/1295/1/012063

Pillai, CK, Paul, W, & Sharma, CP (2009). Chitin and chitosan polymers: Chemistry, solubility and fiber formation. Progress in polymer science, 34(7), 641–678. https://doi.org/10.1016/j.progpolymsci.2009.04.001

Rambhia, KJ, & Ma, PX (2015). Controlled drug release for tissue engineering. Journal of Controlled Release, 219, 119–128. https://doi.org/10.1016/j.jconrel.2015.08.049

Rani, A, Arfat, Y, Aziz, RS, Ali, L, Ahmed, H, Asim, S, ... & Hocart, CH (2021). Enzymatically assisted extraction of antioxidant and anti-mutagenic compounds from radish (Raphanus sativus). Environmental Technology & Innovation, 23, 101620. https://doi.org/10.1016/j.eti.2021.101620

Saravanan, A, Maruthapandi, M, Das, P, Ganguly, S, Margel, S, Luong, JHT, & Gedanken, A (2020). Applications of N-Doped Carbon Dots as Antimicrobial Agents, Antibiotic Carriers, and Selective Fluorescent Probes for Nitro Explosives. ACS Applied Bio Materials, 3(11), 8023–8031. https://doi.org/10.1021/acsabm.0c01104

Siepmann, J, & Siepmann, F (2008). Mathematical modeling of drug delivery. In International Journal of Pharmaceutics (Vol. 364, Issue 2, pp. 328–343). https://doi.org/10.1016/j.ijpharm.2008.09.004

Sravanthi, A, Reddy M, Sunitha, & Jaswanth, A (2021). In vivo evaluation of a novel zero order drug releasing transdermal system of rotigotine. Asian Journal of Pharmacy and Pharmacology, 7(3), 126–130. https://doi.org/10.31024/ajpp.2021.7.3.3

Suvakanta, D, Murthy, PN, Lilakanta, N & Prasanta, C (2010). Kinetic Modeling on drug release from controlled drug delivery systems. Acta Poloniae Pharmaceutica, 67(3), 217–223. PMID: 20524422

Tebcharani, L, Wanzke, C, Lutz, TM, Rodon-Fores, J, Lieleg, O, & Boekhoven, J (2021). Emulsions of hydrolyzable oils for the zero-order release of hydrophobic drugs. Journal of Controlled Release, 339, 498–505. https://doi.org/10.1016/j.jconrel.2021.10.014

Tibbitt, MW, & Anseth, KS (2009). Hydrogels as extracellular matrix mimics for 3D cell culture. Biotechnology and bioengineering, 103(4), 655–663. https://doi.org/10.1002/bit.22361

Wang, W, Xue, C, & Mao, X (2020). Chitosan: Structural modification, biological activity and application. International Journal of Biological Macromolecules, 164, 4532–4546. https://doi.org/10.1016/j.ijbiomac.2020.09.042

Wu, IY, Bala, S, Škalko-Basnet, N, & di Cagno, MP (2019). Interpreting non-linear drug diffusion data: Utilizing Korsmeyer-Peppas model to study drug release from liposomes. European Journal of Pharmaceutical Sciences, 138(July), 105026. https://doi.org/10.1016/j.ejps.2019.105026

Xu, W, Zhu, Y, Ravichandran, D, Jambhulkar, S, Kakarla, M, Bawareth, M, ... & Song, K (2021). Review of fiber-based three-dimensional printing for applications ranging from nanoscale nanoparticle alignment to macroscale patterning. ACS Applied Nano Materials, 4(8), 7538–7562. https://doi.org/10.1021/acsanm.1c01408

Yen, MT, Yang, JH, & Mau, JL (2008). Antioxidant properties of chitosan from crab shells. Carbohydrate polymers, 74(4), 840–844. https://doi.org/10.1016/j.carbpol.2008.05.003

Zakerikhoob, M, Abbasi, S, Yousefi, G, Mokhtari, M & Noorbakhsh, MS (2021). Curcumin-incorporated crosslinked sodium alginate-g-poly (N-isopropyl acrylamide) thermo-responsive hydrogel as an in-situ forming injectable dressing for wound healing: In vitro characterization and in vivo evaluation. Carbohydrate Polymers, 271, 118434. https://doi.org/10.1016/j.carbpol.2021.118434

Zhang, H, Dong, X, Ji, H, Yu, J, & Liu, A (2023). Preparation and structural characterization of acid-extracted polysaccharide from Grifola frondosa and antitumor activity on S180 tumor-bearing mice. International Journal of Biological Macromolecules, 123302. https://doi.org/10.1016/j.ijbiomac.2023.123302

Chapter 10
Techniques for Estimating Parameters of Sampled Sinusoidal Signals in Electrical Impedance Tomography

Erick Dario León Bueno de Camargo, Fernando Silva de Moura, Olavo Luppi Silva, Edson Rodrigues, and Mirela Oliveira Tomazini

Abstract EIT is a non-invasive imaging technique that estimates the distribution of electrical resistivity within the body. This is done by using electrical potential measurements taken on the skin which are induced by a sinusoidal current source. One of the preprocessing steps of the sampled sinusoid—necessary to feed the image estimation algorithm—is the determination of the parameters of the sampled sinusoidal signal, i.e., its amplitude, phase, frequency, and offset. In this chapter, we will present different techniques used to estimate these parameters, whether in real time or offline. We will also present numerical implementation examples and results with simulated noisy signals.

Keywords Parameter estimation · Least squares · Sinusoid · DFT · Electrical impedance tomography

10.1 Introduction

Electrical Impedance Tomography (EIT) is an imaging technique that estimates the distribution of tissue resistivity within the body. Since each tissue has a different average resistivity and this resistivity can also change over time, EIT is a valuable tool for making non-invasive images of the inside of the body. Additionally, these images can be shown in real time, reflecting both anatomical and physiological

E. D. L. B. de Camargo · O. L. Silva · E. Rodrigues · M. O. Tomazini
Graduate Program in Biomedical Engineering, Federal University of ABC—UFABC, São Bernardo do Campo, SP, Brazil

F. S. de Moura (✉)
Graduate Program in Biomedical Engineering, Federal University of ABC—UFABC, São Bernardo do Campo, SP, Brazil

Graduate Program in Biomedical Engineering, Federal University of ABC—UFABC Al. da Universidade, São Bernardo do Campo, SP, Brazil
e-mail: fernando.moura@ufabc.edu.br

C. B. Lombello, P. A. da Ana (eds.), *Current Trends in Biomedical Engineering*,
https://doi.org/10.1007/978-3-031-38743-2_10

conditions. The most successful biomedical application of EIT is in lung monitoring under mechanical ventilation (Bachmann et al. 2018; Castro et al. 2019), including COVID-19 patients (Perier et al. 2020; Nascimento et al. 2021).

To produce a resistivity image, the EIT system needs a measurement hardware with J electrodes attached to the skin and an image estimation software. The hardware has a voltage-controlled current source that is used to apply a sinusoidal current at the patient's body and several high input impedance voltage measurement channels whose analog to digital converters (ADCs) will sample the electric potential evoked by the application of electrical current into the body. To produce one image, J^2 voltage measurements and J current measurements are needed, in different current and voltage patterns as explained in Silva et al. (2017).

On the other hand, the software receives from the hardware module (i) the amplitude of J^2 measured voltages, (ii) the amplitude of J measured currents, and (iii) the J^2 phases between the respective voltage and current. It also considers as input a geometrical description of the domain to be imaged as a finite element mesh in order to produce one resistivity distribution map by solving an ill-posed nonlinear inverse problem.

However, the ADC cannot directly measure the amplitude and phase of the sinusoidal voltage or current. According to the Nyquist theorem, to correctly sample a signal, ADC's sample frequency f_s should be greater than twice the frequency f of the sinusoidal voltage. In practice, $3f \leq f_s \leq 20f$. Then, the number of sampled voltages per period is $N_p = f_s/f$ and, if Δt is the amount of time the ADC is set to sample the sinusoid signal of one single electrode, it will produce a vector with a total of $N = f_s \Delta t$ elements.

That is why we need algorithms to estimate the parameters of a sinusoid (namely the amplitude, the phase, the offset, and the frequency), from a vector of N measured possibly noisy voltages. This chapter presents the theory and example codes written in Python 3, to estimate those parameters from two different approaches: one based on the Least-Squares (LSQ) method and the other based on the Discrete Fourier Transform (DFT). We also consider a variation of the second method named Windowed Discrete Fourier Transform (WDFT). In all methods, the case where sinusoid frequency is known or not known is discussed, as well as the advantages and drawbacks to implement the methods in microcontrollers operating in real time. At the end, the results of a set of simulations are presented, discussing the error in the estimated parameters caused by a frequency drift of the current source signal generator, as well as the influence of sampling frequency and data length.

10.2 Problem Statement

Consider the following parameterized model of a sinusoidal signal:

$$s(t) = A \sin(2\pi f t + \phi) + c, \tag{10.1}$$

where A is the amplitude of the signal, f is the frequency, ϕ is the phase, and c is a constant offset. It is convenient to rewrite this equation using the trigonometric identity $\sin(a+b) = \sin a \cos b + \cos a \sin b$

$$\begin{aligned} s(t) &= A\sin(2\pi f t + \phi) + c \\ &= A\cos(\phi)\sin(2\pi f t) + A\sin(\phi)\cos(2\pi f t) + c. \end{aligned} \tag{10.2}$$

Assume that $s[n]$ are samples of this signal $s(t)$ at time t_n, corrupted by zero-mean additive noise $w(t)$

$$s[n] = s(t_n) + w(t_n), \qquad \text{for } n = 0, \cdots, N-1, \tag{10.3}$$

forming a set of N noisy measurement pairs $(t_n, s[n])$

$$M = \{\, (t_0, s[0]), (t_1, s[1]), \cdots, (t_{N-1}, s[N-1]) \,\}. \tag{10.4}$$

The demodulation problem is stated as follows: **find the set of parameters $\{A, f, \phi, c\}$ of the sinusoidal signal $s(t)$ that best fits the set of noisy measurements (10.4), assuming that noise is present in the dependent variable $s[n]$ only.**

There are two main demodulation cases of interest, depending on the parameters of the sinusoidal function to be estimated: (i) if f is known and we need to adjust $\{A, \phi, c\}$, (ii) if f is not known and the full set of parameters $\{A, f, \phi, c\}$ must be adjusted. The first method is used when f is sufficiently precise, accurate, and stable over time; otherwise, the second method is recommended.

10.3 Demodulation When the Frequency f Is Known

The next subsections present two demodulation methods that can be used when the frequency f of the sinusoidal function is known: one using the least-squares method (Sect. 10.3.1) and the other using the Discrete Fourier Transform method (Sect. 10.3.2).

10.3.1 Least-Squares Method (Known f)

To estimate the unknown parameters $\{A, \phi, c\}$ of a sinusoidal signal (10.2) with known frequency f and additive noise, one can use the least-squares (LSQ) method. This involves taking the mean of the squared error between the measurements (10.4) and the estimated sinusoidal signal.

If the frequency f is known and fixed, we can modify (10.2) defining two new variables P and Q such that

$$s(t) = P\sin(2\pi f t) + Q\cos(2\pi f t) + c \tag{10.5}$$

$$P = A\cos\phi \tag{10.6}$$

$$Q = A\sin\phi. \tag{10.7}$$

The unknown parameters are now $\{P, Q, c\}$ in this new formulation. Note that (10.5) is linear with respect to the unknowns and leads to a linear regression problem. Once P and Q are found, the amplitude A and the phase ϕ of the signal can be determined by

$$A = \sqrt{P^2 + Q^2} \tag{10.8}$$

$$\phi = \arctan(Q/P). \tag{10.9}$$

Care must be taken when finding the correct quadrant of ϕ from the arctan. Usually, the function `math.atan2` is used for that.

The unknown parameters can be adjusted to the measurement set (10.4) using the least-squares method to fit the data to the noiseless model (10.5). For that, first we define the residual at t_n, that is, the difference between the measured value $s[n]$ and fitted value $s(t_n)$ provided by the model.

$$\begin{aligned} r(t_n) &= s[n] - s(t_n) \\ &= s[n] - P\sin(2\pi f t_n) + Q\cos(2\pi f t_n) + c \\ &= s[n] - \begin{bmatrix}\sin(2\pi f t_n) & \cos(2\pi f t_n) & 1\end{bmatrix} \begin{bmatrix} P \\ Q \\ c \end{bmatrix} \\ &= s[n] - \boldsymbol{h}^{\mathrm{T}}(t_n) \cdot \boldsymbol{x}, \end{aligned} \tag{10.10}$$

where two vectors were defined: $\boldsymbol{h}(t) = [\ \sin(2\pi f t) \quad \cos(2\pi f t) \quad 1\ ]^{\mathrm{T}}$ is a function of time only and $\boldsymbol{x} = [\ P \quad Q \quad c\ ]^{\mathrm{T}}$ is a vector with the parameters to be determined.

For N measurements, we have N residual equations that can be computed

$$\begin{aligned} r(t_0) &= s[0] - \boldsymbol{h}^{\mathrm{T}}(t_0) \cdot \boldsymbol{x} \\ r(t_1) &= s[1] - \boldsymbol{h}^{\mathrm{T}}(t_1) \cdot \boldsymbol{x} \\ &\vdots \\ r(t_{N-1}) &= s[N-1] - \boldsymbol{h}^{\mathrm{T}}(t_{N-1}) \cdot \boldsymbol{x}, \end{aligned} \tag{10.11}$$

where $\boldsymbol{x}$ must be the same for all equations. For short, (10.11) can be written in matrix form

$$\boldsymbol{r} = \boldsymbol{s} - \boldsymbol{H}\boldsymbol{x} \tag{10.12}$$

$$\boldsymbol{r} = \begin{bmatrix} r(t_0) \\ r(t_1) \\ \vdots \\ r(t_{N-1}) \end{bmatrix}, \qquad \boldsymbol{s} = \begin{bmatrix} s[0] \\ s[1] \\ \vdots \\ s[N-1] \end{bmatrix}, \qquad \boldsymbol{H} = \begin{bmatrix} \boldsymbol{h}^{\mathrm{T}}(t_0) \\ \boldsymbol{h}^{\mathrm{T}}(t_1) \\ \vdots \\ \boldsymbol{h}^{\mathrm{T}}(t_{N-1}) \end{bmatrix},$$

where $\boldsymbol{r} \in \mathbb{R}^N$ is the residual vector, $\boldsymbol{s} \in \mathbb{R}^N$ is the measurement vector, and $\boldsymbol{H} \in \mathbb{R}^{N \times 3}$ is a matrix. Note that $\boldsymbol{H}$ depends only on the time t_n when the samples were measured. All the unknowns are in vector $\boldsymbol{x}$.

The goal of this method is to find $\boldsymbol{x}$ that minimizes the sum of the squared residuals. This is equivalent to finding $\boldsymbol{x}$ that minimizes the squared Euclidean norm of $\boldsymbol{r}$

$$\boldsymbol{x}_{LSQ} = \arg\min_{\boldsymbol{x}} \left\{ \sum_{i=1}^{N} r^2(t_n) \right\} = \arg\min_{\boldsymbol{x}} \|\boldsymbol{r}\|^2. \tag{10.13}$$

The Euclidean norm of $\boldsymbol{r}$ is

$$\|\boldsymbol{r}\|^2 = \boldsymbol{r}^{\mathrm{T}}\boldsymbol{r} = (\boldsymbol{s} - \boldsymbol{H}\boldsymbol{x})^{\mathrm{T}}(\boldsymbol{s} - \boldsymbol{H}\boldsymbol{x}), \tag{10.14}$$

and $\boldsymbol{x}_{LSQ}$ can be calculated by taking the partial derivative of (10.14) with respect to $\boldsymbol{x}$, setting it to zero, and solving for $\boldsymbol{x}_{LSQ}$

$$\begin{aligned} \left.\frac{\partial \|\boldsymbol{r}\|^2}{\partial \boldsymbol{x}}\right|_{\boldsymbol{x}_{LSQ}} &= 0 \\ -\frac{1}{2}\boldsymbol{H}^{\mathrm{T}}(\boldsymbol{s} - \boldsymbol{H}\boldsymbol{x}_{LSQ}) &= 0 \\ \boldsymbol{H}^{\mathrm{T}}\boldsymbol{H}\boldsymbol{x}_{LSQ} &= \boldsymbol{H}^{\mathrm{T}}\boldsymbol{s}. \end{aligned} \tag{10.15}$$

Finally, we can isolate the solution $\boldsymbol{x}_{LSQ}$

$$\begin{aligned} \boldsymbol{x}_{LSQ} &= (\boldsymbol{H}^{\mathrm{T}}\boldsymbol{H})^{-1}\boldsymbol{H}^{\mathrm{T}}\boldsymbol{s} \\ &= \boldsymbol{H}^{\dagger}\boldsymbol{s}, \end{aligned} \tag{10.16}$$

where $\boldsymbol{H}^{\dagger} = (\boldsymbol{H}^{\mathrm{T}}\boldsymbol{H})^{-1}\boldsymbol{H}^{\mathrm{T}}$ is called pseudo-inverse of $\boldsymbol{H}$. Therefore, the least-squares algorithm boils down to multiplying the $\boldsymbol{H}^{\dagger}$ matrix by the measurement vector $\boldsymbol{s}$.

Note that $\boldsymbol{H}^{\dagger}$ does not depend on the sampled sinusoid $s[n]$. As long as t_n and the frequency f do not change, the same matrix can be used to demodulate multiple signals. For real-time implementation in microcontrollers, it is convenient to pre-compute $\boldsymbol{H}^{\dagger}$ and store it in memory and then execute matrix–vector multiplication to demodulate a signal.

In order to complete the presentation of the method, it is also necessary to check that $\boldsymbol{x}_{LSQ}$ is indeed a minimum. For that, we need to compute the second partial derivative of the residual

$$\frac{\partial^2 \|\boldsymbol{r}\|^2}{\partial \boldsymbol{x}^2} = \frac{1}{2}\boldsymbol{H}^{\mathrm{T}}\boldsymbol{H}. \tag{10.17}$$

Note that if $\boldsymbol{H}$ is real and full rank, then $\boldsymbol{H}^{\mathrm{T}}\boldsymbol{H}$ must be symmetrical, and all its eigenvalues are real and positive. Therefore, $\boldsymbol{H}^{\mathrm{T}}\boldsymbol{H}$ is a positive definite matrix, and the quadratic equation (10.14) describes a convex function with a minimum at $\boldsymbol{x}_{LSQ}$. Matrix $\boldsymbol{H}$ is full rank when it has three linearly independent rows. Observing the structure of matrix $\boldsymbol{H}$ in (10.12), we can easily find a set of times $\{t_n\}$ so that this is true. As one example, one can show that if the Nyquist–Shannon sampling theorem holds and $N \geq 3$, this will be true. Often, we choose $N \gg 3$.

Computational Implementation

Sample Code 10.1: Implements the Least Squares Method when f is known using the pseudo-inverse $\boldsymbol{H}^{\dagger}$

```
import numpy as np

# First we define the parameters of the sinusoidal signal s(t)
f = 10                                              # Signal frequency in Hz
A = 2                                               # Signal amplitude
phi = np.pi/4                                       # Signal phase

# Now we simulate the measurement process with additive noise
N = 1000                                            # Number of measurements
sampling_freq = 1000                                # Sampling frequency, in Hz
t_n = np.array(range(N))/sampling_freq              # Time array, in seconds

w = np.random.normal(scale=0.1, size=len(t_n))# Random noise
s_n = A * np.sin(2 * np.pi * f * t_n + phi) + w     # Sinusoid array with the samples s_n

# Next we need to compute matrix H and its pseudo-inverse
# Compute H
col1 = np.sin(2 * np.pi * f * t_n)
col2 = np.cos(2 * np.pi * f * t_n)
col3 = np.ones(t_n.shape)
H = np.column_stack((col1, col2, col3))

# Compute the pseudo-inverse of H
pinvH = np.linalg.pinv(H)

# Find x_lsq = pinvH * s_n
x_lsq = np.matmul(pinvH, s_n)

P_lsq = x_lsq[0]
Q_lsq = x_lsq[1]
c_lsq = x_lsq[2]

#computes the amplitude and phase from estimated P and Q
A_est = np.sqrt(P_lsq**2 + Q_lsq**2)  # Estimated amplitude
phi_est = np.arctan2(Q_lsq,P_lsq)  # Estimated phase

# Print the estimated parameters
print(f'Real amplitude: {A:.4f}; Estimated amplitude: {A_est:.4f}; Error: {100*(A_est-A)/A:.2f}%')
print(f'Real phase: {phi:.4f}; Estimated phase: {phi_est:.4f}; Error: {100*(phi_est-phi)/phi:.2f}%')
```

Output of Sample Code 10.1

```
Real amplitude: 2.0000; Estimated amplitude: 2.0011; Error: 0.06%
Real phase: 0.7854; Estimated phase: 0.7888; Error: 0.43%
```

10.3.2 Discrete Fourier Transform Method (Known f)

It is also possible to use the Discrete Fourier Transform (DFT) to estimate the amplitude and phase of a sinusoidal signal from noisy measurements. The DFT is widely used in digital signal processing, communications, audio processing, image processing, and many other fields. The DFT was first introduced by J.W. Cooley and J.W. Tukey in (1965), and it is a mathematical tool that allows analyzing the frequency content of a discrete-time signal.

The DFT transforms a sequence $\{x[0], \ldots x[N-1]\}$, sampled from a signal $x(t)$, at constant sampling period $T_s = 1/f_s$, that is, $x[n] = x(nT_s)$ for $n = 0, \cdots, N-1$ into a sequence of complex coefficients $X[k]$ in the (discrete) frequency domain. The transform is defined as

$$X[k] = \sum_{n=0}^{N-1} x[n] \exp(-ik\Omega_0 n), \quad \text{for } k = 0, \cdots, N-1 \tag{10.18}$$

$$\Omega_0 = \frac{2\pi}{N}, \tag{10.19}$$

where i is the imaginary unit and Ω_0 is the fundamental normalized angular frequency, in rad/sample. $X[k]$ is the kth component of the DFT and is associated with the normalized angular frequency Ω_k, given by

$$\Omega_k = k\Omega_0, \tag{10.20}$$

where Ω_k is a multiple of the fundamental frequency Ω_0. Note that Ω_0 defines the frequency resolution of $S[k]$. The normalization of the angular frequency Ω_k is such that the frequency range $0 \le f_k \le \frac{f_s}{2}$ (e.g., in Hertz) is linearly interpolated into the interval $[0, \pi]$. Note that the Nyquist frequency $f_{\text{nyquist}} = \frac{f_s}{2}$ is mapped into $\Omega_{\text{nyquist}} = \pi$ rad/sample. Therefore, the conversion is given by

$$f_k = \frac{f_s/2}{\pi} \Omega_k. \tag{10.21}$$

Replacing (10.19) and (10.20) in (10.21), we get a relation between the index k and f_k

$$k = \frac{N f_k}{f_s} \qquad \text{for } 0 \leq f_k \leq \frac{f_s}{2}. \tag{10.22}$$

To bring back the complex coefficients $X[k]$ to the time domain, the Inverse Discrete Fourier Transform (IDFT) is defined as

$$x[n] = \frac{1}{N} \sum_{k=0}^{N-1} X[k] \exp(ik\Omega_0 n), \qquad \text{for } n = 0, \cdots, N-1. \tag{10.23}$$

Let us now apply the DFT to the sampled sinusoidal signal $s(t) = A\sin(2\pi f t + \phi)$ with period $T = 1/f$ and sampling period $T_s = 1/f_s$. For now, let us ensure that the N samples $s[n]$ contain an integer number $m > 0$ of sinusoidal cycles.

With this constraint, note that the elapsed time after m sinusoidal periods T must be the same time necessary to collect N samples, that is, $m\,T = N\,T_s$ or, in terms of f and f_s,

$$f_s = \frac{N}{m} f, \quad m \in \mathbb{N}. \tag{10.24}$$

This equation can be used to set the sampling frequency such that the samples $s[n]$ encompass $m \in \mathbb{N}$ periods. Let us also ensure that the Nyquist–Shannon sampling theorem holds. This theorem states that $f_s > 2f$ to be possible to recover the signal $s(t)$ from the samples $s[n]$. If (10.24) is inserted into this inequality, we get $m < N/2$. Therefore, we conclude that m must be further restricted to

$$f_s = \frac{N}{m} f, \quad 0 < m < N/2, \quad m \in \mathbb{N}. \tag{10.25}$$

Considering that the samples $s[n]$ are taken at $t_n = nT_s$ following (10.25), then

$$\begin{aligned} s[n] &= A\sin(2\pi f t_n + \phi) \\ &= A\sin(2\pi f n T_s + \phi) \\ &= A\sin\left(2\pi \frac{m}{N} f_s n T_s + \phi\right) \\ &= A\sin\left(2\pi \frac{m}{N} n + \phi\right). \end{aligned} \tag{10.26}$$

We can express the previous sine function in complex exponential form $\sin(x) = \frac{1}{2i}[\exp(ix) - \exp(-ix)]$

$$s[n] = \frac{A}{2i}\left[\exp\left(i2\pi\frac{m}{N}n + i\phi\right) - \exp\left(-i2\pi\frac{m}{N}n - i\phi\right)\right]. \tag{10.27}$$

Inserting (10.27) into (10.18), we can compute the DFT for the desired sinusoidal function

$$\begin{aligned}
S[k] &= \sum_{n=0}^{N-1} \frac{A}{2i}\left[\exp\left(i2\pi\frac{m}{N}n\right)\exp(i\phi) - \exp\left(-i2\pi\frac{m}{N}n\right)\exp(-i\phi)\right] \\
&\quad \exp\left(-ik\Omega_0 n\right) \\
&= \frac{A\exp(i\phi)}{2i}\sum_{n=0}^{N-1}\exp\left(i2\pi\frac{m}{N}n\right)\exp\left(-ik\frac{2\pi}{N}n\right) \\
&\quad - \frac{A\exp(-i\phi)}{2i}\sum_{n=0}^{N-1}\exp\left(-i2\pi\frac{m}{N}n\right)\exp\left(-ik\frac{2\pi}{N}n\right) \\
&= \frac{A\exp(i\phi)}{2i}\sum_{n=0}^{N-1}\exp\left(-i2\pi\frac{k-m}{N}n\right) \\
&\quad - \frac{A\exp(-i\phi)}{2i}\sum_{n=0}^{N-1}\exp\left(-i2\pi\frac{k+m}{N}n\right).
\end{aligned} \tag{10.28}$$

Isolating the first sum on the left and defining a new variable $p = -i2\pi\frac{k-m}{N}$, we get

$$\begin{aligned}
S_1 &= \sum_{n=0}^{N-1}\exp\left(-i2\pi\frac{k-m}{N}n\right) \\
&= \sum_{n=0}^{N-1}\exp(pn).
\end{aligned} \tag{10.29}$$

If $m = k$, then $p = 0$ and

$$S_1 = \sum_{n=0}^{N-1}\exp(0) = N \qquad \text{if } k = m. \tag{10.30}$$

In the case that $k \neq m$, Eq. (10.29) becomes the sum of the first N terms of a geometric series with common ratio $r = \exp(p)$.

$$S_1 = 1 + \exp(p) + \exp(2p) + \cdots \exp[(N-1)p]. \tag{10.31}$$

From the formula for the sum of a geometric series, we get

$$\begin{aligned} S_1 &= \frac{1 - \exp(Np)}{1 - \exp(p)} \\ &= \frac{1 - \exp(-i2\pi(k-m))}{1 - \exp(-i2\pi\frac{k-m}{N})} \\ &= \frac{1 - 1}{1 - \exp(-i2\pi\frac{k-m}{N})} \\ &= 0, \qquad \text{if } k \neq m, \end{aligned} \tag{10.32}$$

where the numerator is zero because $2\pi(k - m)$ is an integer multiple of 2π. Therefore, the sum S_1 reduces to

$$S_1 = \begin{cases} N \text{ for } k = m \\ 0 \;\; \text{for } k \neq m. \end{cases} \tag{10.33}$$

Proceeding in the same way for the other sum on the right of (10.28), we can show that

$$S_2 = \begin{cases} N \text{ for } k = -m \\ 0 \;\; \text{for } k \neq -m. \end{cases} \tag{10.34}$$

In summary, considering what has been exposed in the previous paragraphs, we can conclude that (10.28) can be expressed as

$$S[k] = \begin{cases} \dfrac{NA\exp(i\phi)}{2i} & \text{for } k = m \\ -\dfrac{NA\exp(-i\phi)}{2i} & \text{for } k = N - m \\ 0 & \text{otherwise,} \end{cases} \tag{10.35}$$

where the periodicity property of S[k], that is, $S[k] = S[N + k]$, was used to keep $k > 0$.

In conclusion, the pure sinusoidal function has only two nonzero elements in $S[k]$, located at indices $k = m$ and $k = N - m$, where m is given by (10.25). These values are complex conjugate.

The amplitude A and the phase ϕ can be extracted from $S[k]$ in (10.35) for $k = m$,

$$S[m] = \frac{NA\exp(i\phi)}{2i}$$

$$= \frac{NA}{2} \frac{\exp(i\phi)}{\exp(i\pi/2)}$$

$$= \frac{NA}{2} \exp[i(\phi - \pi/2)]. \tag{10.36}$$

First, we write $S[m]$ in polar form as $S[m] = |S_m| \exp(i\theta_m)$, then we compare with (10.36) to extract expressions for the magnitude and phase

$$|S_m| = \frac{NA}{2} \tag{10.37}$$

$$\theta_m = \phi - \pi/2, \tag{10.38}$$

and then, isolating the amplitude A and the phase ϕ,

$$A = \frac{2|S_m|}{N} \tag{10.39}$$

$$\phi = \theta_m + \frac{\pi}{2}. \tag{10.40}$$

Note that we do not need to compute the DFT for all k since we are interested in a single component $k = m$. Therefore, $S[m]$ can be computed very quickly with (10.18).

Zero-Padding If we do not have an integer number of cycles in the N samples, then (10.25) does not hold and no $S[k]$ will be nonzero. It is still possible to use a technique called zero-padding to demodulate the signal. Zero-padding is a technique used in DFT to increase the frequency resolution of $S[k]$ that involves adding zeros to the end of $s[n]$ before computing its DFT (Smith 2008).

The new signal $s_{zp}[n]$ with zero-padding will have length $N_{zp} > N$ and can be defined as

$$s_{zp}[n] = \begin{cases} s[n] & \text{for } 0 \le n \le N-1 \\ 0 & \text{for } N \le n \le N_{zp} - 1. \end{cases} \tag{10.41}$$

This procedure reduces the fundamental angular frequency to $\Omega_0^{zp} = 2\pi/N_{zp}$ and, therefore, reduces the distance between adjacent frequency bins Ω_k. To find a suitable N_{zp} for the zero-padding, we go back to (10.25) and find a new length N_{zp} such that m becomes integer

$$m_{zp} = \frac{N_{zp}}{f_s} f, \quad 0 < m_{zp} < N_{zp}/2, \quad m_{zp} \in \mathbb{N}. \tag{10.42}$$

With the new value m_{zp}, one can extract the parameters A and ϕ from $S[m_{zp}]$ following the same procedure. Note that the zeros added to the end of the signal will not affect the sum in (10.18), so only the first N values need to be considered in the sum.

One can also show that this procedure is equivalent to interpolating the original grid of $S[k]$ into a finer grid $\Omega_k^{zp} = k\Omega_0^{zp}$ using a sinc function.

Computational Implementation

Sample Code 10.2: Implements the Discrete Fourier Transform method when frequency is known

```
import numpy as np

# First we define the parameters of the sinusoidal signal s(t)
# this are the values we want to estimate!
f = 10                                          # Signal frequency in Hz
A = 2                                           # Signal amplitude
phi = np.pi/4                                   # Signal phase

# Now we simulate the measurement process with additive noise
N = 1000                                        # Number of measurements
m = 10                                          # number of cycles (integer value)
fs = N/m*f                                      # Sampling frequency, in Hz
Ts = 1/fs                                       # Sampling period
t_n = np.array(range(N)) * Ts                   # Time array in seconds, from 0 to (N-1)*Ts
w = np.random.normal(scale=0.1, size=len(t_n))  # Random noise
s_n = A * np.sin(2 * np.pi * f * t_n + phi) + w # Sampled signal s[n]

# DFT: Compute S[k] only for the specific k=m
Omega_0 = 2 * np.pi / N                         # fundamental angular frequency
k=m
S_m = 0
for n in range(N):
  S_m += s_n[n] * np.exp(-1j * k * Omega_0 * n)

# Extract A and phi from S[k]
A_est = 2 * np.abs(S_m) / N                    # Estimated amplitude
phi_est = np.angle(S_m) + np.pi / 2            # Estimated phase

print(f'Real amplitude: {A:.4f}; Estimated amplitude: {A_est:.4f}; Error: {100*(A_est-A)/A:.2f}%')
print(f'Real phase: {phi:.4f}; Estimated phase: {phi_est:.4f}; Error: {100*(phi_est-phi)/phi:.2f}%')
```

Output of Sample Code 10.2

```
Real amplitude: 2.0000; Estimated amplitude: 1.9976; Error: -0.12%
Real phase: 0.7854; Estimated phase: 0.7843; Error: -0.14%
```

Sample Code 10.3: Modification of Sample Code 2 to include zero padding technique

```
import numpy as np

# Generate a discrete sinusoidal signal with additive noise
N = 2000                                        # Number of measurements (21 cycles)
sampling_freq = 1000                            # Sampling frequency, in Hz
t = np.array(range(N))/sampling_freq            # Time array, in seconds
f = 10.5                                        # Signal frequency in Hz
A = 2                                           # Signal amplitude
phi = np.pi/4                                   # Signal phase
w = np.random.normal(scale=0.1, size=len(t))    # Random noise
y = A * np.cos(2 * np.pi * f * t + phi) + w     # Sinusoid array

# Compute DFT component
N_zp = N*2              # considering N zeros added to signal
df = sampling_freq/N_zp
freq = f;               # expected frequency in Hz
k = np.round(freq/df)
X = 0;
for n in range(N):   # zeros do not need to be added
  X += y[n] * np.exp(-2j * np.pi * k * n / N_zp)

# Estimate amplitude and phase at peak frequency
A_est = 2 * np.abs(X) / N                # Estimated amplitude
phi_est = np.arctan2(X.imag, X.real)     # Estimated phase

print(f'Real amplitude: {A:.4f}; Estimated amplitude: {A_est:.4f}; Error: {100*(A_est-A)/A:.2f}%')
print(f'Real phase: {phi:.4f}; Estimated phase: {phi_est:.4f}; Error: {100*(phi_est-phi)/phi:.2f}%')

# Second approach, at any given frequency
X_f = 0
for n in range(N):   # zeros do not need to be added
  X_f += y[n] * np.exp(-2j * np.pi * freq * n / sampling_freq)

# Estimate amplitude and phase at peak frequency
A_est_f = 2 * np.abs(X_f) / N                  # Estimated amplitude
phi_est_f = np.arctan2(X_f.imag,X_f.real)      # Estimated phase

print('Second approach:')
print(f'Real amplitude: {A:.4f}; Estimated amplitude: {A_est_f:.4f}; Error: {100*(A_est_f-A)/A:.2f}%')
print(f'Real phase: {phi:.4f}; Estimated phase: {phi_est_f:.4f}; Error: {100*(phi_est_f-phi)/phi:.2f}%')
```

Output of Sample Code 10.3

```
Real amplitude: 2.0000; Estimated amplitude: 1.9935; Error: -0.32%
Real phase: 0.7854; Estimated phase: 0.7848; Error: -0.08%
Second approach:
Real amplitude: 2.0000; Estimated amplitude: 1.9935; Error: -0.32%
Real phase: 0.7854; Estimated phase: 0.7848; Error: -0.08%
```

10.4 Demodulation When the Frequency f Is Not Known

10.4.1 Least-Squares Method (Unknown f)

If the signal's frequency f is not known, the problem becomes nonlinear due to the product of unknown parameters in the model (10.2). Thus, we will need an iterative method to approximate the solution. This is done by linearization of the model around the latest estimate of the unknown parameters. For this purpose, Eq. (10.2) can be regarded as a function of time t and a vector or parameters to be estimated $\boldsymbol{x} = [\ A \ \ f \ \ \phi \ \ c\ ]^{\mathrm{T}}$.

To linearize the model, we use the truncated Taylor series (10.43) around a point $\boldsymbol{x}_0 = [\ A_0 \quad f_0 \quad \phi_0 \quad c_0\ ]^{\mathrm{T}}$, which represents the latest estimates of unknown parameters,

$$
\begin{aligned}
s(t,\boldsymbol{x}) \approx s(t,\boldsymbol{x}_0) &+ \frac{\partial s(t,\boldsymbol{x}_0)}{\partial A}\cdot(A - A_0) + \frac{\partial s(t,\boldsymbol{x}_0)}{\partial f}\cdot(f - f_0) \\
&+ \frac{\partial s(t,\boldsymbol{x}_0)}{\partial \phi}\cdot(\phi - \phi_0) + \frac{\partial s(t,\boldsymbol{x}_0)}{\partial c}\cdot(c - c_0),
\end{aligned}
\tag{10.43}
$$

or in matrix notation,

$$
s(t,\boldsymbol{x}) \approx s(t,\boldsymbol{x}_0) + \boldsymbol{J}_s(t,\boldsymbol{x}_0)\cdot(\boldsymbol{x} - \boldsymbol{x}_0) \tag{10.44}
$$

$$
\boldsymbol{J}_s(t,\boldsymbol{x}) = \begin{bmatrix} \frac{\partial s}{\partial A} & \frac{\partial s}{\partial f} & \frac{\partial s}{\partial \phi} & \frac{\partial s}{\partial c} \end{bmatrix}, \tag{10.45}
$$

where $\boldsymbol{J}_s(t,\boldsymbol{x}_0) \in \mathbb{R}^4$ is the Jacobian (row) vector evaluated at the linearization point $\boldsymbol{x}_0$. The partial derivatives of (10.45) are

$$
\frac{\partial s}{\partial A} = \sin(2\pi f t)\cos\phi + \cos(2\pi f t)\sin\phi \tag{10.46}
$$

$$
\frac{\partial s}{\partial f} = 2A\pi t\cos(2\pi f t)\cos\phi - 2A\pi t\sin(2\pi f t)\sin\phi \tag{10.47}
$$

$$
\frac{\partial s}{\partial \phi} = -A\sin(2\pi f t)\sin\phi + A\cos(2\pi f t)\cos\phi \tag{10.48}
$$

$$
\frac{\partial s}{\partial c} = 1. \tag{10.49}
$$

From now on, the Jacobian $\boldsymbol{J}_s(t,\boldsymbol{x}_0)$ will be denoted by $\boldsymbol{J}_0(t)$ to simplify notation. Thus, the linearized noiseless model (10.43) can be written as

$$
s(t,\boldsymbol{x}) \approx s(t,\boldsymbol{x}_0) + \boldsymbol{J}_0(t)\cdot(\boldsymbol{x} - \boldsymbol{x}_0). \tag{10.50}
$$

Equation (10.50) can be used to find the least-squares solution $\boldsymbol{x}_{LSQ}$ from measurements vector $M = \{(t_0, s[0]), (t_1, s[1]), \cdots, (t_{N-1}, s[N-1])\}$. Proceeding in a similar way to what was done in Sect. 10.3.1, the residual at t_n is

$$\begin{aligned} r_0(t_n) &= s[n] - s(t_n, \boldsymbol{x}) \qquad (10.51)\\ &\approx s[n] - s(t_n, \boldsymbol{x}_0) - \boldsymbol{J}_0(t_n) \cdot (\boldsymbol{x} - \boldsymbol{x}_0) \\ &= s[n] - s(t_n, \boldsymbol{x}_0) + \boldsymbol{J}_0(t_n) \cdot \boldsymbol{x}_0 - \boldsymbol{J}_0(t_n) \cdot \boldsymbol{x} \\ &= s[n] - \bar{s}_0(t_n) - \boldsymbol{J}_0(t_n) \cdot \boldsymbol{x}. \qquad (10.52) \end{aligned}$$

In the previous equation, we define, for short, $\bar{s}_0(t_n) = s(t_n, \boldsymbol{x}_0) - \boldsymbol{J}_0(t_n) \cdot \boldsymbol{x}_0$ because it is completely specified by t_n and $\boldsymbol{x}_0$. Then, we can compute the residual for each measurement pair $(t_n, s[n])$

$$\begin{aligned} r_0(t_0) &= s[0] - \bar{s}_0(t_0) - \boldsymbol{J}_0(t_0) \cdot \boldsymbol{x} \\ r_0(t_1) &= s[1] - \bar{s}_0(t_1) - \boldsymbol{J}_0(t_1) \cdot \boldsymbol{x} \\ &\vdots \\ r_0(t_{N-1}) &= s[N-1] - \bar{s}_0(t_{N-1}) - \boldsymbol{J}_0(t_{N-1}) \cdot \boldsymbol{x} \end{aligned} \qquad (10.53)$$

or write all residuals in matrix form

$$\boldsymbol{r}_0 = \boldsymbol{s} - \bar{\boldsymbol{s}}_0 - \boldsymbol{H}_0 \boldsymbol{x} \qquad (10.54)$$

$$\boldsymbol{r}_0 = \begin{bmatrix} r_0(t_0) \\ r_0(t_1) \\ \vdots \\ r_0(t_{N-1}) \end{bmatrix}, \quad \boldsymbol{s} = \begin{bmatrix} s[0] \\ s[1] \\ \vdots \\ s[N-1] \end{bmatrix}, \quad \bar{\boldsymbol{s}}_0 = \begin{bmatrix} \bar{s}_0(t_0) \\ \bar{s}_0(t_1) \\ \vdots \\ \bar{s}_0(t_{N-1}) \end{bmatrix},$$

$$\boldsymbol{H}_0 = \begin{bmatrix} \boldsymbol{J}_0(t_0) \\ \boldsymbol{J}_0(t_1) \\ \vdots \\ \boldsymbol{J}_0(t_{N-1}) \end{bmatrix}, \qquad (10.55)$$

where $\boldsymbol{r}, \boldsymbol{s}, \bar{\boldsymbol{s}}_0 \in \mathbb{R}^N$ and $\boldsymbol{H}_0 \in \mathbb{R}^{N \times 4}$. Note that $\boldsymbol{H}_0$ and $\bar{\boldsymbol{s}}_0$ depend on the time t_n and linearization point $\boldsymbol{x}_0$ only.

The least-squares solution $\boldsymbol{x}_{0,LSQ}$ is given by the same procedure described in Sect. 10.3.1 with the modified residual Eq. (10.54)

$$\boldsymbol{x}_{0,LSQ} = \arg\min_{\boldsymbol{x}} \left\{ \sum r_0^2(t_n) \right\} = \arg\min_{\boldsymbol{x}} \|\boldsymbol{r}_0\|^2, \qquad (10.56)$$

where the Euclidean norm of $\boldsymbol{r}_0$ is

$$\|\boldsymbol{r}_0\|^2 = \boldsymbol{r}_0^{\mathrm{T}}\boldsymbol{r}_0 = (\boldsymbol{s} - \bar{\boldsymbol{s}}_0 - \boldsymbol{H}_0\boldsymbol{x})^{\mathrm{T}}(\boldsymbol{s} - \bar{\boldsymbol{s}}_0 - \boldsymbol{H}_0\boldsymbol{x}). \tag{10.57}$$

The solution $\boldsymbol{x}_{0,LSQ}$ can be calculated imposing zero partial derivative of $\|\boldsymbol{r}_0\|^2$ with respect to $\boldsymbol{x}$

$$\begin{aligned} \left.\frac{\partial \|\boldsymbol{r}\|^2}{\partial \boldsymbol{x}}\right|_{\boldsymbol{x}_{0,LSQ}} &= 0 \\ -\frac{1}{2}\boldsymbol{H}_0^{\mathrm{T}}(\boldsymbol{s} - \bar{\boldsymbol{s}}_0 - \boldsymbol{H}_0\boldsymbol{x}_{0,LSQ}) &= 0 \\ \boldsymbol{H}_0^{\mathrm{T}}\boldsymbol{H}_0\boldsymbol{x}_{0,LSQ} &= \boldsymbol{H}_0^{\mathrm{T}}(\boldsymbol{s} - \bar{\boldsymbol{s}}_0). \end{aligned} \tag{10.58}$$

Finally, we can isolate the solution $\boldsymbol{x}_{0,LSQ}$

$$\begin{aligned} \boldsymbol{x}_{0,LSQ} &= \left(\boldsymbol{H}_0^{\mathrm{T}}\boldsymbol{H}_0\right)^{-1}\boldsymbol{H}_0^{\mathrm{T}}(\boldsymbol{s} - \bar{\boldsymbol{s}}_0) \\ \boldsymbol{x}_{0,LSQ} &= \boldsymbol{H}_0^{\dagger}(\boldsymbol{s} - \bar{\boldsymbol{s}}_0). \end{aligned} \tag{10.59}$$

It is also possible to show that $\boldsymbol{x}_{0,LSQ}$ is the minimum to the problem by calculating the second derivative.

Since it is an iterative process, a new linearization point is obtained from (10.43) by making $\boldsymbol{x}_1 = \boldsymbol{x}_{0,LSQ}$. We often use a step size $0 < \lambda < 1$ to control the stability of the solution

$$\boldsymbol{x}_1 = \boldsymbol{x}_0 + \lambda(\boldsymbol{x}_{LSQ} - \boldsymbol{x}_0), \tag{10.60}$$

and however, it slows the convergence. The procedure is repeated from (10.43) to estimate $\boldsymbol{x}_{k+1}$ from the linearized point $\boldsymbol{x}_k$.

Often, we need to specify a stop criterion for the iterative process. One possibility is to observe the norm of the residuals between the measurements $\boldsymbol{s}$ and the predicted value of the nonlinear model (10.2) with the updated solution $\boldsymbol{x}_{k+1}$, that is,

$$\boldsymbol{r}_k^* = \begin{bmatrix} s[0] - s(t_0, \boldsymbol{x}_{k+1}) \\ s[1] - s(t_1, \boldsymbol{x}_{k+1}) \\ \vdots \\ s[N-1] - s(t_{N-1}, \boldsymbol{x}_{k+1}) \end{bmatrix}, \tag{10.61}$$

and stop when $\|\boldsymbol{r}_k^*\|$ becomes smaller than a small threshold $0 < \varepsilon$. Another possibility is to stop when $| \, \|\boldsymbol{r}_k^*\| - \|\boldsymbol{r}_{k-1}^*\| \, | \leq \epsilon$, where ϵ is a small value.

The following pseudocode implements the iterative process to find the parameters.

Algorithm 1 Iterative process

procedure ITERATIVE PROCEDURE($s, x_0, \lambda, N_{iter}^{max}, \varepsilon$)
  $x_k \leftarrow x_0$
  for $k \leftarrow 0, N_{iterMax}$ **do** ▷ We can set a maximum number of iterations
    Build the Jacobian $J_k(t)$ at x_k using (10.45)
    Compute $\bar{s}_k$ and H_k using (10.55)
    Compute $x_{k,LSQ}$ using (10.59)
    Compute x_{k+1} using (10.60)
    Compute the updated residual r_k^* using (10.61)
    if $| \, \|r_k^*\| - \|r_{k-1}^*\| \, | \leq \epsilon$ **then** ▷ Stop criteria computations
      return x_{k+1}
    end if
  end for
  return x_{k+1}
end procedure

Computational Implementation

Sample Code 10.4: Implements the iterative Least Squares Method when frequency is unknown.

```
import numpy as np

# First we define the parameters of the sinusoidal signal s(t)
f = 10.555                                    # Signal frequency in Hz
A = 2                                         # Signal amplitude
phi = np.pi/4                                 # Signal phase
c = 0.1                                       # Signal offset

# Now we simulate the measurement process with additive noise
N = 1000                                      # Number of measurements (aprox. 21 cycles)
sampling_freq = 1000                          # Sampling frequency, in Hz
t_n = np.array(range(N))/sampling_freq        # Time array, in seconds

w = np.random.normal(scale=0.1, size=len(t_n))# Random noise
s_n = A * np.sin(2 * np.pi * f * t_n + phi) + w       # Sinusoid array with the samples s_n

#  Initial guess
[Ao, fo, phio, co] = [2.5, 11, 0.0, 0.0]

# Convergence parameters
nIterMax = 300                                # Maximum number of iterations
lambda_step = 1                                 # This parameter can be reduced to improve stability
last_error = np.inf                           # Initialize error parameter

# Iterative process
for idx in range(nIterMax):
  if idx==0:
    A_k=Ao
    f_k=fo
    phi_k=phio
    c_k=co

  # Compute H_k and s_k
  #   partial derivatives
  angles = 2 * np.pi * f_k * t_n
  partialA = np.sin(angles) * np.cos(phi_k) + np.cos(angles) * np.sin(phi_k)
```

```
    partialf = ( 2 * np.pi * t_n * A_k * np.cos(angles) * np.cos(phi_k)
                -2 * np.pi * t_n * A_k * np.sin(angles) * np.sin(phi_k) )
    partialPhi = -A_k * np.sin(angles) * np.sin(phi_k) + A * np.cos(angles) * np.cos(phi_k)
    partialC = np.ones(t_n.shape)

    H_k = np.column_stack((partialA, partialf, partialPhi, partialC))
    s_bar_k =  A_k * np.sin(angles + phi_k) + c_k - np.matmul(H_k,[A_k, f_k, phi_k, c_k])

    # Compute the pseudo-inverse of H_k
    pinvH_k = np.linalg.pinv(H_k)

    # Find x_lsq = pinvH_k * (s_n - s_bar_k)
    x_lsq = np.matmul(pinvH_k,(s_n-s_bar_k))

    # Update current estimate x_k+1
    [A_k, f_k, phi_k, c_k] =[A_k, f_k, phi_k, c_k] + lambda_step*(x_lsq - [A_k, f_k, phi_k, c_k] )

    # Interrupt loop if converged
    r_k_star = s_n - (A_k * np.sin(2 * np.pi * f_k * t_n + phi_k) + c_k)
    error = np.linalg.norm(r_k_star)
    if np.abs(error - last_error)<1e-10:
      break
    last_error = error

# Print the estimated parameters
[A_est, freq, phi_est, c_est] = [A_k, f_k, phi_k, c_k]
print(f'Real frequency: {f:.4f}Hz; Estimated peak frequency = {freq:.4f}Hz; Error: {100*(freq-f)/f:.2f}%')
print(f'Real amplitude: {A:.4f}; Estimated amplitude: {A_est:.4f}; Error: {100*(A_est-A)/A:.2f}%')
print(f'Real phase: {phi:.4f}; Estimated phase: {phi_est:.4f}; Error: {100*(phi_est-phi)/phi:.2f}%')
```

Output of Sample Code 10.4

```
Real frequency: 10.5550Hz; Estimated peak frequency = 10.5539Hz; Error: -0.01%
Real amplitude: 2.0000; Estimated amplitude: 1.9908; Error: -0.46%
Real phase: 0.7854; Estimated phase: 0.7923; Error: 0.88%
```

10.4.2 Discrete Fourier Transform Method (Unknown f)

The major challenge of using DFT when the frequency of the sinusoid is not known is that it is difficult to ensure that the N samples of $s[n]$ contain an integer number of sinusoidal periods. In these cases, the DFT can still be used to estimate the frequency and amplitude of the signal. However, the resulting frequency and amplitude estimates may not be accurate due to spectral leakage and aliasing effects.

To mitigate these effects, one approach is to use windowing techniques such as the Hamming window before computing the DFT (Harris 1978). Windowing reduces spectral leakage by tapering off the edges of input signals.

The Hamming window $h[n]$ is an extension of the Hann window and is of the form

$$h[n] = 0.54 - 0.46\cos\left(\frac{2\pi n}{N-1}\right), \quad \text{for } n = 0, 1, \cdots, N-1. \tag{10.62}$$

We can then compute the DFT of the windowed signal $s_w[n] = s[n]h[n]$ following the same procedure presented before. The price we pay for using the window is a small reduction in the SNR of $s[n]$.

The frequency of the sinusoidal signal can be determined by finding the index of the maximum magnitude in the Windowed DFT (WDFT) output. This index corresponds to the frequency bin that contains the sinusoidal signal. With the frequency being known, we can then find the amplitude and phase as described in Sect. 10.3.2.

Computational Implementation

Sample Code 10.5: Sample code of Discrete Fourier Transform method when frequency is unknown.

Zero-padding is implemented to find peak frequency and calculate signal amplitude at that frequency.

```
import numpy as np

# First we define the parameters of the sinusoidal signal s(t)
# this are the values we want to estimate!
f = 10.555                              # Signal frequency in Hz
A = 2                                   # Signal amplitude
phi = np.pi/4                           # Signal phase

# Now we simulate the measurement process with additive noise
N = 1000                                # Number of measurements
m = 10                                  # number of cycles (integer value)
fs = N/m*f                              # Sampling frequency, in Hz
Ts = 1/fs                               # Sampling period
t_n = np.array(range(N)) * Ts           # Time array in seconds, from 0 to (N-1)*Ts

w = np.random.normal(scale=0.1, size=len(t_n)) # Random noise
s_n = A * np.sin(2 * np.pi * f * t_n + phi) + w # sampled signal s[n]

# Apply Hamming window to the signal
window = [(0.54 - 0.46*np.cos(2 * np.pi * n / (N-1)))  for n in range(N)]
sw_n = s_n * window


# Find DFT peak iteralively
A_est = -np.inf

# first we create a vector of frequencies to serves as peak candidates
centralFreq = 10                        # center of the peak
stepSize=0.2                            # frequency step interval
Nsteps=10                               # Number of steps to each side

for dec in range(4):                    # we will repeat the process 4 times
  # vector with frequency candidates
  freq_interval=np.arange(centralFreq-Nsteps*stepSize, centralFreq+Nsteps*stepSize, stepSize)
  for f_temp in freq_interval:
    Omega_0 = 2 * np.pi / N             # fundamental angular frequency
    m= N * f_temp / fs
    k= m
    S_m = 0
    for n in range(N):
      S_m += sw_n[n] * np.exp(-1j * k *Omega_0 * n)
```

```

    # Extract A and phi from S[m]. Stores them in temporary
    # variables and check them agains the best stored values
    A_temp = 2 * np.abs(S_m) / (N*np.mean(window))
    phi_temp = np.angle(S_m) + np.pi / 2

    # next we check if A_temp is bigger than the latest best estimate
    if A_temp > A_est:
      # store the current estimates if A_temp > A_est
      A_est = A_temp;
      freq_est = f_temp
      phi_est = phi_temp

      # updates the central frequency of the interval
      centralFreq = freq_est

  # reduces the size of the interval of frequencies
  stepSize = stepSize / 10

print(f'Real frequency: {f:.4f}; Estimated peak frequency = {freq_est:.4f}; Error: {100*(freq_est-f)/f:.2f}%')
print(f'Real amplitude: {A:.4f}; Estimated amplitude: {A_est:.4f}; Error: {100*(A_est-A)/A:.2f}%')
print(f'Real phase: {phi:.4f}; Estimated phase: {phi_est:.4f}; Error: {100*(phi_est-phi)/phi:.2f}%')
```

Output of Sample Code 10.5

```
Real frequency: 10.5550Hz; Estimated peak frequency = 10.5540Hz; Error: -0.01%
Real amplitude: 2.0000; Estimated amplitude: 1.9948; Error: -0.26%
Real phase: 0.7854; Estimated phase: 0.7873; Error: 0.24%
```

10.5 Simulations and Discussion

Most of the EIT systems have a voltage-controlled current source with a factory preset fixed frequency. However, the stability of electronic equipment is affected by temperature, aging, and other factors, and thus current source frequency may drift over time from its original value. Then, an important question to pose is *how accurate are the methods presented above in the presence of frequency drift and measurement noise*? In this section, we analyze the robustness of the presented methods to estimate the parameters of the sinusoid if the actual frequency deviates from its original value.

The boxplots in the following figures show the distribution of the estimation error of amplitude, phase, offset, and frequency (when applicable) for different simulated scenarios. Boxplots are a standardized way of displaying the distribution of data. It is based on a five-number summary, which includes the minimum, first quartile (Q1), median, third quartile (Q3), and maximum. The box in the boxplot represents the interquartile range (IQR), which is the range between Q1 and Q3. The whiskers extend from the box to the minimum and maximum values, excluding outliers.

Outliers are data points that are more than 1.5 times the IQR away from the box. In all cases, a sinusoidal with amplitude $A = 2\,\text{V}$, phase $\phi = \pi/4$ rad, and offset $c = 0.5\,\text{V}$, added with zero mean and 0.1 V standard deviation Gaussian noise, was used as simulated sampled signal. The sampled data used in Figs. 10.1 and 10.2 was simulated with 3, 10, and 30 points per cycle, while in Figs. 10.3 and 10.4 it was obtained using 3 and 30 points per cycle. Figure 10.2 was obtained using a data length of 3 cycles only, while the other figures were obtained using data lengths of 3 and 10 cycles. For each case, 100 measurement simulations were made.

10.5.1 Data Length Influence on Error When f Is Known

Figure 10.1 presents the data length influence on the estimation error of amplitude, phase, and offset for the three methods (LSQ, DFT, and WDFT) when the difference between actual and reported frequency varies from -3% to $+3\%$. Estimations with LSQ and DFT methods were obtained using sample codes presented in Sect. 10.3, i.e., when the frequency is known a priori, and WDFT estimation was implemented with the windowing technique presented in Sect. 10.4.2. Comparing the estimation error for all 3 parameters when 3 and 10 cycles were used, it is possible to observe that estimation errors using a smaller number of points (3 cycles) are significantly smaller than when using a large number of points (10 cycles), if the informed frequency is not exactly the same of the actual frequency (columns 1, 2, 4, and 5). When the true frequency of the signal perfectly matches the informed frequency (column 3), a larger number of points give smaller estimation errors. Also, the larger number of points results in a smaller variability in estimations, regardless of the estimation error. A smaller number of points require a smaller sampling interval and a less computational time. This is an important point to consider if the demodulation is being computed by a microcontroller, specially to produce EIT images in real time.

Looking at the second line of Fig. 10.1, it is possible to conclude that the error in phase estimation is directly proportional to both the error in frequency and the data length; that is, the greater the error in frequency and the greater data length used, the greater the phase estimation error.

It is important to notice that, assuming the data is composed of complete cycles, LSQ and DFT methods resulted in parameters with identical results. However, the error in amplitude estimation with WDFT method is smaller than the DFT method when using a larger number of points (10 cycles). This behavior can be interpreted by looking at the implications of changing the frequency, in which the existence of a frequency error implies that complete cycles are not guaranteed, which is widely known as an error factor for the DFT method. No difference for offset estimation was observed among the three implemented methods.

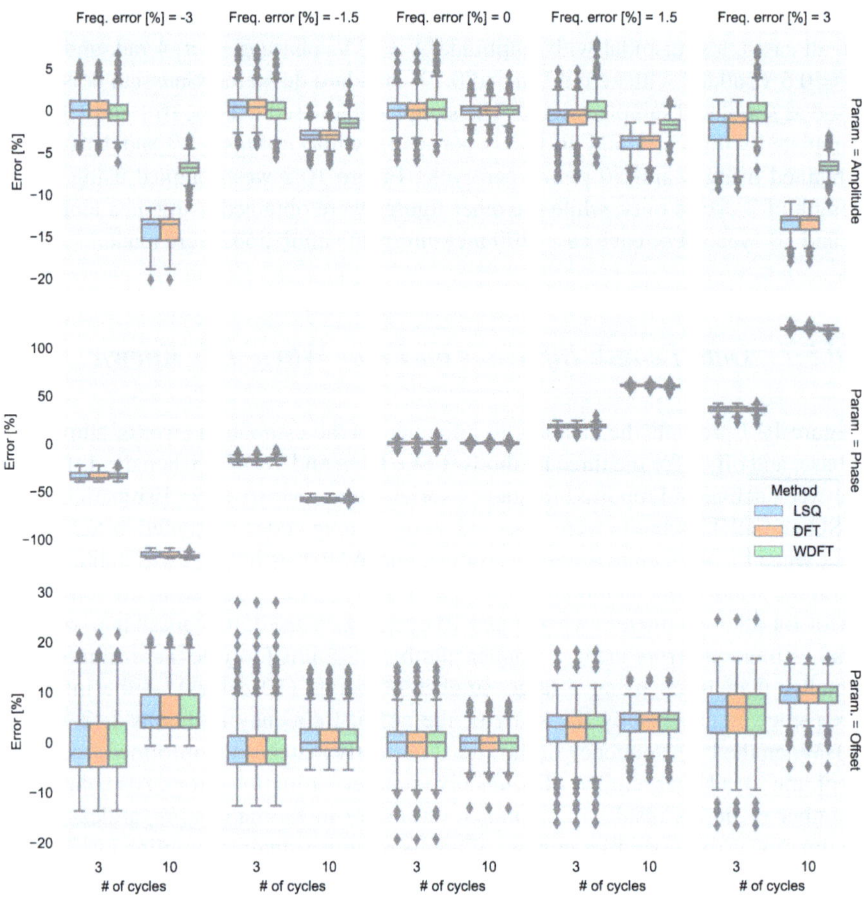

Fig. 10.1 Data length influence on error for LSQ, DFT, and Windowed DFT methods when f known, but not accurate. Simulation cases: the number of points per cycle = 3, 10, and 30; the number of cycles = 3 and 10; 100 measurement simulations for each number of points per cycle and number of cycles. Expected parameters: amplitude $A = 2.0$ V, phase $\phi = \pi/4$ rad, and offset $c = 0.5$ V

10.5.2 Sampling Frequency Influence on Error When f Is Known

Figure 10.2 presents an analysis similar to that presented in Fig. 10.1 but with respect to sampling frequency. The sampling frequency is represented by its normalized form: the number of points (samples) per cycle (a complete period) of a sinusoid.

Similarly to Fig. 10.1, the second line of Fig. 10.2 also shows that the phase error is directly proportional to the frequency error. It is possible to notice the same

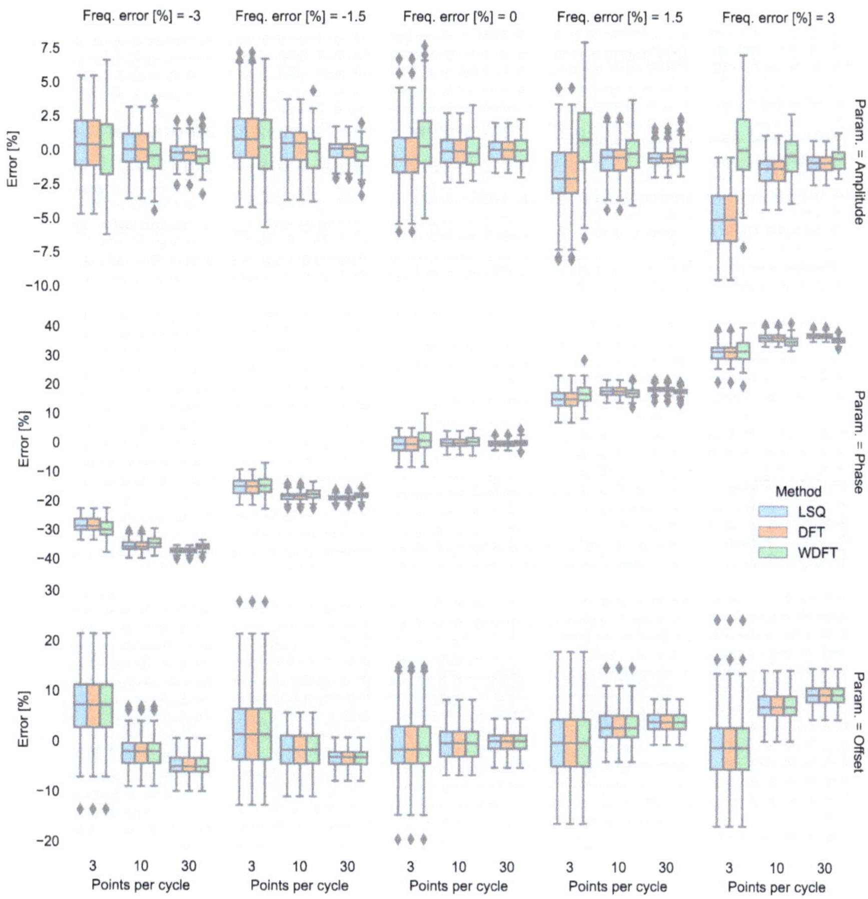

Fig. 10.2 Sampling frequency influence on error for LSQ, DFT, and Windowed DFT methods when f is known, but not accurate. Simulation cases: the number of points per cycle = 3, 10, and 30 with a fixed number of cycles = 3; 100 measurement simulations for each number of points per cycle and number of cycles. Expected parameters: amplitude $A = 2.0\,\text{V}$, phase $\phi = \pi/4$ rad, and offset $c = 0.5\,\text{V}$

tendency in which increasing the sampling frequency may result in a small increase in the phase estimation error when a frequency error is present.

The greater the sampling frequency, the smaller the error standard deviation. This does not necessarily mean that the parameter errors are smaller, just that there is less variability in the results obtained. That way, a bigger sampling frequency results in a bigger precision, but not necessarily a better accuracy. Analyzing separately the three studied parameters, a bigger sampling frequency resulted in bigger precision and accuracy for the amplitude; for the phase, it resulted in a bigger precision, but smaller accuracy; and for the offset, it resulted in a bigger precision and the accuracy depended on the frequency error percentage.

10.5.3 Sensitivity to Errors in the Linearization Point When f Is Unknown

Figure 10.3 presents the sensitivity to errors in the linearization point for the least-square method when f is unknown. Figure 10.4 presents the sampling frequency influence on errors for LSQ, DFT, and WDFT methods when f is also unknown.

Observing Fig. 10.3, the two most direct conclusions are (i) despite the percentage error considered in the frequency of the initial conditions, if a higher sampling frequency is used, the parameter estimation error is smaller and (ii) the

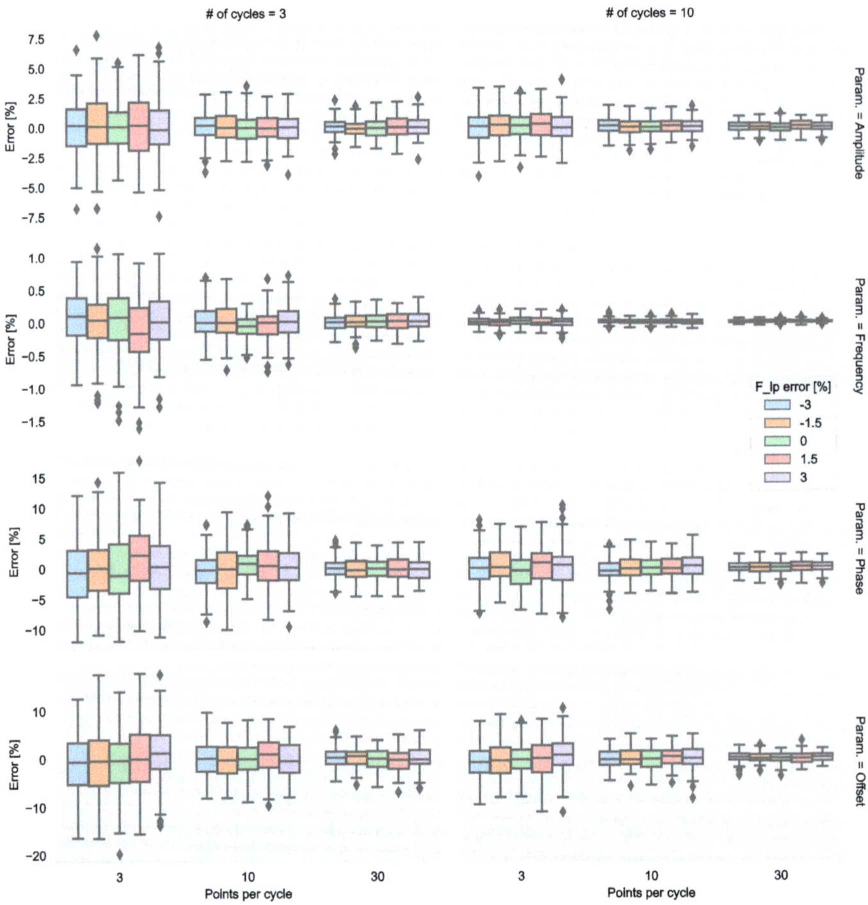

Fig. 10.3 Sensitivity to errors in the linearization point for the least-square method when f is unknown. Simulation cases: the number of points per cycle = 3, 10, and 30; the number of cycles = 3 and 10; 100 measurement simulations for each number of points per cycle and number of cycles. Expected parameters: amplitude $A = 2.0\,\text{V}$, phase $\phi = \pi/4$ rad, and offset $c = 0.5\,\text{V}$

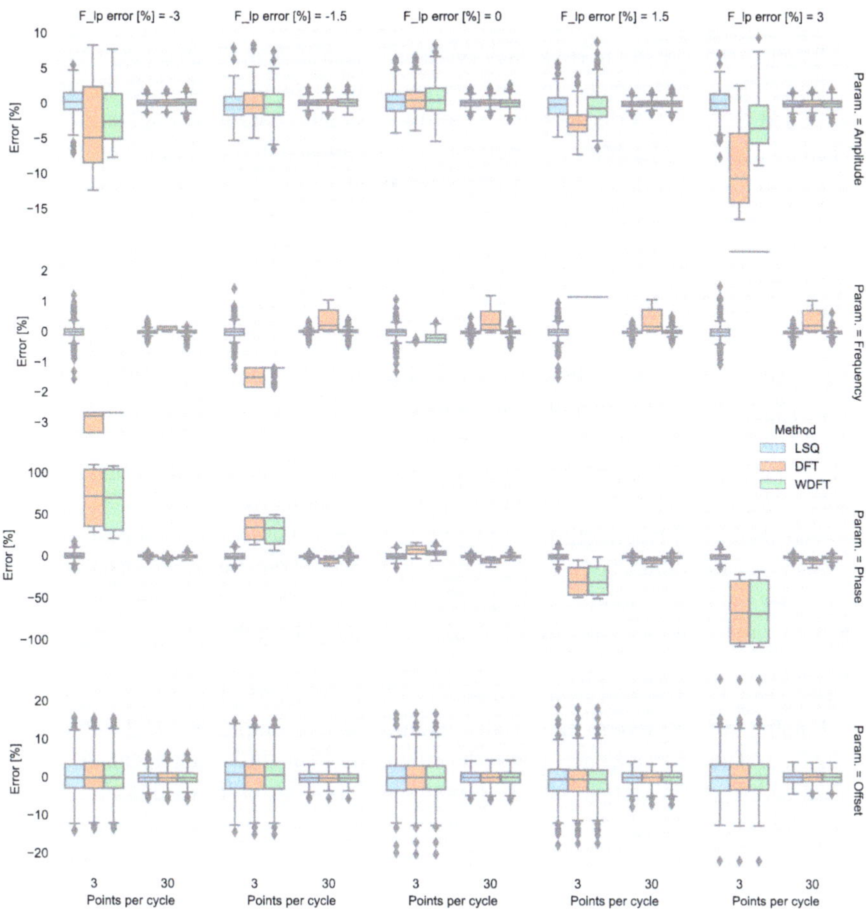

Fig. 10.4 Sampling frequency influence on errors for LSQ, DFT, and Windowed DFT methods when f is unknown. Simulation cases: the number of points per cycle = 3 and 30; the number of cycles = 3 and 10; 100 measurement simulations for each number of points per cycle and number of cycles. `F_lp` is the linearization point for the least square method and the initial condition for DFT and Windowed DFT methods. Expected parameters: amplitude $A = 2.0\,\text{V}$, phase $\phi = \pi/4$ rad, and offset $c = 0.5\,\text{V}$

same happens if a larger sample size is considered. Both of these interpretations are intuitive and straightforward.

It is important to say that the LSQ method starts to diverge if measurements with too many points (the number of cycles >10) are used in cases where there is error in the frequency. Since the expected frequency and the true frequency are different, a large number of points result in data chunks in phase and data chunks in counter-phase with the expected signal, which is similar to a beating phenomenon, making the method diverge. Furthermore, whenever the method presents a convergent behavior, regardless of the error in the initial linearization point, it will converge to approximately the same parameters. However, when the initial conditions present sufficiently large errors in frequency (greater than $>3\%$), the method also diverges.

Figure 10.4 shows that, for a smaller sampling frequency, the LSQ method has presented itself as a better method, resulting in smaller parameter estimation errors when compared to the other two methods. For a higher sampling frequency, the LSQ and WDFT methods have presented similar results.

10.6 Conclusion

We have shown two techniques for estimating the parameters of a sampled sinusoidal signal. Although these techniques can be used in other areas, estimating the amplitude, phase, offset, and frequency is a crucial task for the development of EIT hardware. Since EIT image estimation is an ill-conditioned problem, small-measurement errors may lead to large image estimation errors.

The theory of two methods was presented: one based on the Least-Squares method and the other based on the Discrete Fourier Transform. We discussed the particular case of assuming that the frequency is known or not known a priori in both cases. The use of a windowing technique with DFT was also discussed. Additionally, we performed several simulations with an artificial sampled data considering the effects of measurement noise and frequency drift.

Finally, we offered to the reader example codes implemented in Python 3 language. The source code of the algorithms used in this chapter can be freely downloaded from https://github.com/eitlab-ufabc/sinusoidDemodulation under GNU General Public License.

Acknowledgments This work was financed in part by the Coordenação de Aperfeiçoamento de Pessoal de Nível Superior—Brazil (CAPES)—Finance Code 001.

References

Bachmann, M Consuelo et al. (2018). "Electrical impedance tomography in acute respiratory distress syndrome". In: *Critical Care* 22.263. https://doi.org/10.1186/s13054-018-2195-6.

Castro Martins, Thiago de et al. (2019). "A review of electrical impedance tomography in lung applications: Theory and algorithms for absolute images". In: *Annual Reviews in Control* 48, pp. 442–471. ISSN: 1367–5788. https://doi.org/10.1016/j.arcontrol.2019.05.002. https://www.sciencedirect.com/science/article/pii/S1367578819300173.

Cooley, James W and John W Tukey (1965). "An algorithm for the machine calculation of complex Fourier series". In: *Mathematics of computation* 19.90, pp. 297–301.

Harris, Fredric J (1978). "On the use of windows for harmonic analysis with the discrete Fourier transform". In: *Proceedings of the IEEE* 66.1, pp. 51–83.

Nascimento, Milena S. et al. (2021). "Electrical impedance tomography in pediatric patients with COVID-19, the first reports". In: *BMC Pulmonary Medicine* 21.1, pp. 1–7. https://doi.org/10.1186/s12890-021-01716-y.

Perier, François et al. (2020). "Effect of positive end-expiratory pressure and proning on ventilation and perfusion in COVID-19 acute respiratory distress syndrome". In: *American journal of respiratory and critical care medicine* 202.12, pp. 1713–1717. https://doi.org/10.1164/rccm.202008-3058le.

Silva, Olavo Luppi et al. (2017). "Influence of current injection pattern and electric potential measurement strategies in electrical impedance tomography". In: *Control Engineering Practice* 58, pp. 276–286. ISSN: 0967-0661. https://doi.org/10.1016/j.conengprac.2016.03.003. https://www.sciencedirect.com/science/article/pii/S0967066116300351.

Smith, Julius Orion (2008). *Mathematics of the discrete Fourier transform (DFT): with audio applications*. W3K Publishing. ISBN: 978-0-9745607-4-8.

Part IV
Applied Technologies

Chapter 11
Basics of 3D Bioprinting Extrusion Process

Juliana Kelmy Macário Barboza Daguano, Andrea Cecilia Dorion Rodas, Karina Feliciano Santos, Camila Campos Santos, and Jorge Vicente Lopes da Silva

Abstract The extrusion-based bioprinting system is one of the most used techniques to potentially obtain functional tissues, being widely used both in the commercial sphere and in the research area. This technology allows the printing of biomaterials combined with living cells, bioactive molecules, or other agents, from deposition in filaments or continuous fibers, in order to build a three-dimensional structure, deposited layer by layer. 3D extrusion bioprinting has the possibility of operating with different types of materials for the formulation of hydrogels, with a wide range of viscosities. For these compositions to be considered a bioink, cells or biological agents are mandatory components. Furthermore, it is possible to include cells from different environments and shapes, such as unit cells, cell aggregates, spheroids, microtissues, and organoids. Additionally, it is possible to contain bioactive molecules, and other molecules containing DNA, RNA, small vesicles, or even biomaterials. On the other hand, compositions that do not contain these cells and biological agents cannot be qualified as bioinks but as biomaterial inks.

Keywords Additive manufacturing · 3D bioprinting · Extrusion · Biofabrication · Tissue engineering

J. K. M. B. Daguano · A. C. D. Rodas · K. F. Santos
Graduate Program in Biomedical Engineering, Federal University of ABC – UFABC Al. da Universidade, São Bernardo do Campo, SP, Brazil

Center for Information Technology Renato Archer, Campinas, SP, Brazil
e-mail: juliana.daguano@ufabc.edu.br

C. C. Santos
Graduate Program in Biomedical Engineering, Federal University of ABC – UFABC Al. da Universidade, São Bernardo do Campo, SP, Brazil

J. V. L. da Silva
Center for Information Technology Renato Archer, Campinas, SP, Brazil

C. B. Lombello, P. A. da Ana (eds.), *Current Trends in Biomedical Engineering*,
https://doi.org/10.1007/978-3-031-38743-2_11

Abbreviations

ALGMA	Methacrylated Alginate
AMP	Amorphous magnesium phosphates
BMSCs	Bone marrow stromal cells
ECM	extracellular matrix
Epi-SCs	Epidermal Stem Cells
GELMA	Gelatin Methacrylate
HAp	hydroxyapatite
Matrigel	is a reconstituted basement membrane derived from extracts of Engelbreth-Holm-Swarm mouse tumors
Lap	Laponite
PCL	Poly(ε-caprolactone)
PRP-Platelet	rich-plasma
SKps	Skin derived precursors
ZnO	Zinc oxide

11.1 Timeline of 3D Bioprinting

Biofabrication, as a set of technologies that encompasses 3D bioprinting, emerges as a powerful tool for the construction of biologically functional structures applied to the area of tissue engineering (TE) and regenerative medicine (Moroni et al. 2018). Groll et al. defined it as the automated generation of biological products with structural organization, aiming at making biological constructs from living cells, bioactive molecules, biomaterials, cellular aggregates, and microtissues to obtain organs and tissues in vitro. These constructions can be obtained through bioassembly or 3D bioprinting (Groll et al. 2016).

3D bioprinting is a variant of the 3D printing technology, also known as additive manufacturing (AM), that begins with the first printer created by Charles Hull, in 1986 (Karzyński et al. 2018; Su and Al'Aref 2018). He developed and patented the first device that is now known as stereolithography printing technique and which was able to build the first three-dimensional object. This is a manufacturing process in which thin layers (from polymeric resin) are built on top of each other to form a solid structure (Robles Martinez et al. 2018; Su and Al'Aref 2018). Soon after, Carl Deckard developed the selective laser sintering (SLS) printing technique as it is known today (Su and Al'Aref 2018). In 1989, Scott Crump created the polymer hot extrusion process known as fused deposition modeling (FDM) using filament as a feedstock. He founded the company called Stratasys and its process is the basis for most of the extrusion processes available today, including the one focused in this chapter. More recently, in 2003, the term 3D bioprinting was created when Tom Boland and collaborators carried out bioprinting using inkjet-based additive technology (Karzyński et al. 2018; Gu et al. 2020).

In 2006, there was the first discussion between researchers that led to the definition of 3D bioprinting and its technologies as we know it nowadays: *it can be defined as a set of additive manufacturing techniques, aiming to mimic organs and tissues and their functionalities through three-dimensional structures.* Thus, bioprinting can be seen as a tool that allows the use of cells, biomaterials and bioactive factors that, when combined, can result in advanced materials, often called bioinks (Mironov et al. 2006; Hospodiuk et al. 2017; Karzyński et al. 2018; Gargus et al. 2018).

From the development of new technologies for the obtainment of 3D structures, additive manufacturing (AM) has gained popularity not only in the biomedical field, but also in segments such as, pharmaceuticals, food, biotechnology, dentistry, biosensors, among others (Agarwal et al. 2020). To meet all this demand, there was an exponential growth of software (free and paid), hardware, and new models and types of printers. These printers range from a high sophistication profile to a simple profile (Ke et al. 2022).

According to the Tissue Engineering Global Annual Report, this field reached a turnover of around USD 9.9 billion in 2019. It is expected that it will exceed USD 31.23 billion in 2030, followed of 10.46% of Compound Annual Growth (CAGR). The United States (USA), China, South Korea, Germany, the United Kingdom (UK), and Canada scored as the most relevant countries where research on 3D bioprinting is currently ongoing. The USA has a significant participation of 30%, succeeded China, in the development of this niche technology (Tong et al. 2021). In respect a wide variety of bioprinters on the market, another question arises as to the cost. In this sense, companies (startups) have recently started to develop low-cost printers that have gained space in the commercial area. The available bioprinting techniques can be based on their operating principle as inkjet, extrusion, or laser (Ke et al. 2022). Among these processes, the most widespread is the extrusion based due to its greater commercial accessibility and the possibility of adapting its components. In addition, it has a lower cost when compared to other bioprinters systems as a laser-based bioprinter with highly sophisticated components, just to cite one, leading to a costly system (Tong et al. 2021).

Related to this area, a growing number of studies can be noted in the past 20 years (from 2002 to 2022), considering research and review papers published in scientific journals included in Scopus (Elsevier). A total number of 10,750 entries were reported on new technologies comprises keywords "*3B Bioprinting,*" "*Bioinks,*" and "*Tissue Engineering.*" Most of these studies have as their main approach to additive technologies and biomaterials applied in inks and bioinks development. Figure 11.1a shows the number of articles published since 2002–2022 and its expressive growth in recent years. Figure 11.1b highlight a clearly constant increase in number of publications on bioinks in the last 5 years. An extensive cleaning and deduplication process was subsequently performed, leading to 1839 selected articles, consisting of 1427 research articles about development of bioinks, printability, and printing parameters, while part of these are review works (412 publications), that intend to address types of biomaterials, classification, and their current application. Due to the progress of the field, it is worth noting that approximately 30% of these papers were published in 2022.

Despite 3D bioprinting, extrusion-based technique is more representative in commercial and academic niches, while in hospital and clinical environment, 3D

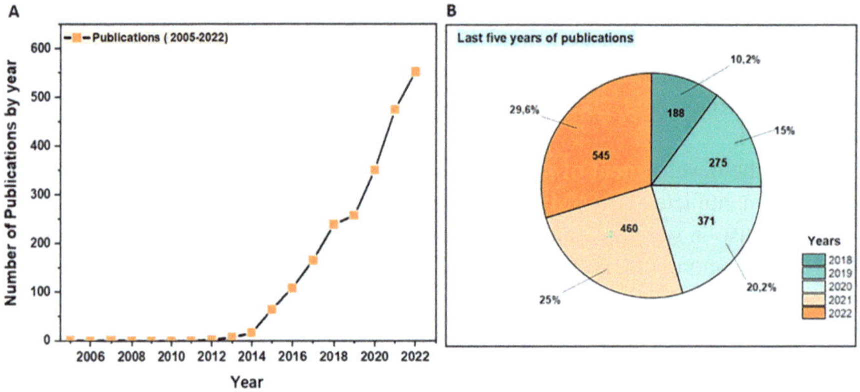

Fig. 11.1 Publications based on search in Scopus Database using keywords: *3D Bioprinting, Bioinks* and *Tissue Engineering*. (**a**) Number of publications by 2002–2022. (**b**) Number of publications in the last 5 years (total = 1839; 412 reviews × 1427 research articles)

bioprinting still remains unpredictable. A better comprehension of cell behavior and its possible sources, suitable bioinks, methodologies, adequate infrastructure, and bioethical and regulatory issues are still widely required (Kirillova et al. 2020).

11.2 Fundamentals of Extrusion-Based Bioprinting

The extrusion-based bioprinting technique has a robotic system as base of deposition (Naghieh and Chen 2021). Also known as Direct Ink Writing, in this technology, the materials (bioinks) are allocated in a disposable syringe or piston, in form of a gel, with the whole printing process automatically controlled by a computer system (Katja et al. 2016). According to the virtual project, printing parameters such as pressure, time, speed, flow rate, temperature are selected, which are previously determined by physical–chemical characteristics of each material (Naghieh and Chen 2021). After actuation (pneumatic, screw, or piston), a controlled head moves constantly in three-dimensional directions (x–y–z) and the bioink is deposited in form of cylindrical fibers (filaments) layer-by-layer on the substrate (Champeau et al. 2020). At the end of the process, its viscosity is partially recovered, creating a functional construct.

Independently of driving system (pneumatic, screw, or piston), these mechanisms present similar results, and each printing technique has its advantages and disadvantages (Fu et al. 2021). The pneumatic extrusion system uses pressurized air to produce a pressure to extrude the gel through a sterile syringe orifice. After coupling, a compressed air pump is used with a valve system that allows air control (Schwab et al. 2020). It is very recommended to print highly viscous gels (or ceramic composites) due to its different pressure gradients. However, as a limitation of this system, it requires work in an area with sterilization through air filters, to reduce potential microbiological contamination, but that make it more expensive. For mechanical systems (screw or piston) to dispense bioinks, motors are generally used to help material ejection. Thus, piston performs linear movements, and a volume is ejected proportionally to the piston displacement. The advantage of using mechanical systems is the ease and simplicity of finding their components on market (Schwab et al. 2020).

Nevertheless, a major advantage of extrusion-based bioprinting is a possibility of working with a wide range of biomaterials, whose viscosity varying between 30 and 6×10^7 $mPas^{-1}$ (Katja Hölzl et al. 2016; O'Connell et al. 2017). Appropriated viscosity is one of the main requirements for the success of the process. Materials that present high viscosity have greater structural support than gels with low viscosity (Cooke and Rosenzweig 2021). However, high viscosity can lead to disruption of the cell membrane when incorporated into bioinks due to the excessive shear stresses involved in the material extrusion.

11.3 Key Factors for Extrusion-Based Bioprinting of Bioinks

Adequate extrudability of bioinks is directly linked to the physicochemical characteristics; therefore, formulations must meet reproducible conditions and parameters. Understanding rheological behavior of gels not only influences the printability, but also impacts cell adhesion, proliferation, and differentiation within or on gel matrix (Blaeser et al. 2016). Main properties that must be standardized are viscosity, shear-thinning behavior, thixotropy, and cross-linking method that vary from material to material (Cooke and Rosenzweig 2021).

11.3.1 Viscosity

Viscosity describes physical property of a gel to resist flow induced by shear generated at syringe nozzle. This parameter governs gel printing capacity (printability), post-printing stabilization, and cell viability (Amorim et al. 2021). The printability can be translated as material extrudability. Variables such as concentration of polymeric solution, molar mass of components, presence of particles, and temperature can influence gel's quality. The higher the concentration of the polymeric solution and the molar mass, the greater is the gel viscosity (Temirel et al. 2022).

High viscosities prevent droplets formation due to surface tension, allowing the filaments to appear as a continuous fiber (Gargus et al. 2018). On the other hand, excessive viscosity requires greater force to expel material through nozzle, and generally, gels obtained present robust aspects, losing their shape fidelity (Gillispie et al. 2020; Schwab et al. 2020). In addition, higher pressure applied to extrude the material affect cell viability, since at the time of extrusion cells may experience damage caused by excessive pressure. In contrast, lower viscosity can lead to the instability of deposited structures and prevent the formation of filaments due to the greater predominance of a liquid phase (Champeau et al. 2020; Temirel et al. 2022).

Less viscous gels can be extruded more easily at a faster rate of time (Cooke and Rosenzweig 2021). Although it does not have an ideal viscosity value, studies suggest a range around 300 cP $< \mu < 0.4 \times 10^5$ and pressure $P > 5$ kPa, however, lower than $P < 35$ kPa for the extrusion technique (Montero et al. 2019).

11.3.2 Shear-Thinning Behavior

Shear-thinning refers to non-Newtonian behavior characterized by a decrease in viscosity as shear rate is increased. This property is critical for extrusion process, since the ability to gel flows is related to the viscosity of the fluid, which reduces as the shear increases, i.e., the more easily through a nozzle tip (Cooke and Rosenzweig 2021; Garcia-Cruz et al. 2021).

The reduction in viscosity occurs due to the elongation of polymeric chains that are organized in flow direction. After tension is removed, the structure of gels is reorganized, returning closely to its initial viscosity. As a requirement, viscosity must be high enough so that the material does not flow or collapse after printing, without need to apply excessive force (Gargus et al. 2018; Schwab et al. 2020). Another relevant point is yield stress (Champeau et al. 2020), which is the minimum stress applied for material to yield. Paxton et al. demonstrated that gels containing different concentrations of Polaxamer 407 had different flow limits as the polymer content increased. Due to the increase in polymeric concentration, an increase in viscosity was observed, thus influencing the flow limit (Paxton et al. 2017).

Some studies have used constitutive equations as guidance in design of bioinks. These proposed equations (rheological models) help to understand important parameters that relate shear stress, apparent viscosity with shear rate, involving the use of two to five parameters (Dávila and D'Ávila 2017). The Power-Law model equations (Eq. 11.1) relate key concepts such as (i) consistency index, (ii) shear-thinning index, (iii) shear stress, and (iv) viscosity (Jaishankar and McKinley 2013; Dávila and d'Ávila 2019). The constant (m) is related to the gel viscosity, referred to as apparent viscosity, and indicates extrudability of material (O'Connell et al. 2017; Fu et al. 2021).

$$\eta(\dot{\gamma}) = m\dot{\gamma}^{n-1} \tag{11.1}$$

where m is the consistency index and n is the power-law index; m is associated with the magnitude of the viscosity and n defines the viscosity behavior: shear-thickening if $n > 1$, Newtonian if $n = 1$, and shear-thinning if $n < 1$.

For beginners in bioprinting, Paxton et al. recommend following the printing parameters of Nivea® Cream that have reproducible coefficients ($m = 26.1$; and

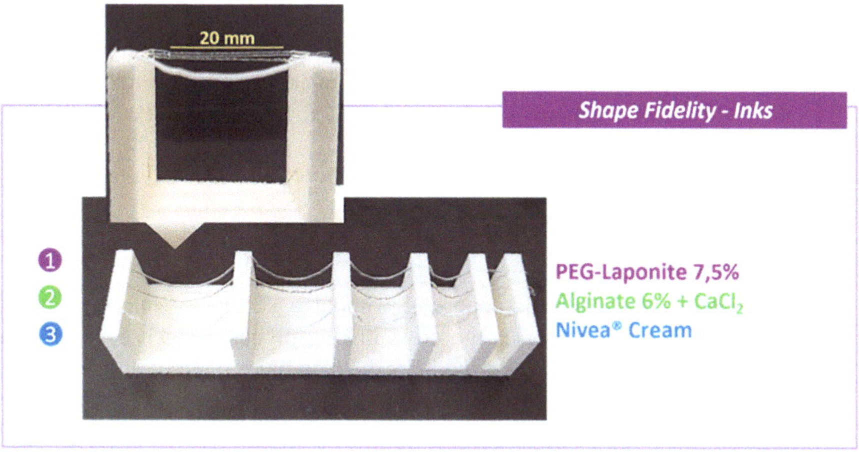

Fig. 11.2 Printed filaments deposited on pillars at different distances. Three different gels were tested for printability: 1 – PEG-Laponite 7.5%; 2 – Alginate 6% $CaCl_2$; 3 – Nivea® Cream

n = 0.552) (Paxton et al. 2017). Figure 11.2 compares good stability of Nivea® Cream to other gels, PEG-Laponite and Alginate 6% $CaCl_2$. This test may indicate the ability to withstand small distances and the occurrence of the filament deformation. In previous studies, these hydrogels demonstrated efficient rheological properties to be extruded, maintaining shape fidelity after 3D printing (Daguano et al. 2022; Dávila and d'Ávila 2019).

11.3.3 Thixotropy

Gels after being extruded need to recover their viscosity at a very fast rate of time so that they do not collapse. A tixotropic material is that its viscosity progressively decreases with time, under a constant shear rate, and subsequently has its viscosity gradually and completely recovered (Gao et al. 2019). Thixotropy is a characteristic that some shear-thinning materials have when subjected to constant stress in a given time. Thixotropy is considered one of the most important properties of the extrusion technique (Malda et al. 2013; Gao et al. 2019).

Although structural integrity is associated to viscosity, it cannot be considered as a single stability factor of the precursor gel after printing, thus it is necessary to understand the viscoelasticity relationships between loss modulus (G'') (viscous response) and storage modulus (G') (elastic response). These modules indicate a solid–liquid behavior. As viscosity directly influences the fidelity form of printed specimen, this parameter helps to keep formed filaments into their geometry without interpenetrating one another (Champeau et al. 2020). Therefore, gel-like behavior for bioinks exhibits better shape fidelity. Generally, this parameter is achieved completely through cross-linking as post-processing to ensure structural integrity. Thus, oscillatory tests such as stress or strain scanning allow understanding structures with preserved shape fidelity (Schwab et al. 2020; Temirel et al. 2022).

11.3.4 Cross-Linking Method

Most biomaterials are not self-supporting without a secondary intervention. In most cases, this secondary step is the cross-linking process, which consists of a physical or chemical procedure that modifies the material properties and this could interfere in the viability of cells. The cross-linking process avoids the dissolution of polymeric chains in aqueous media either by physical or chemical methods. Basically, these two processes differ in terms of type of bond formed between the polymeric chains of the gel (Ghavaminejad et al. 2020). In general, physically gelled bioinks can present electrostatic, ionic, hydrogen bonds, due to the nature of their interactions, then are reversible (Malda et al. 2013). In chemical process cross-linking, the interactions are covalent and irreversible, because of polymerization process and/or due to the addition of photosensitive agents (Ghavaminejad et al. 2020).

Regarding bioprinting, cross-linking is a crucial factor in shape fidelity, as it is responsible for tying the polymer chains, maintaining the structural integrity after the printing process, and the system viscosity. Although this step is often applied after extrusion, there are processes that can be done simultaneously with printing. Thus, the addition of a precrosslinker to the gel has been an option to improve the structural integrity, encapsulation of cells, and bioactive molecules (Ghavaminejad et al. 2020). On the other hand, an excessive level of crosslinking can detrimentally affect the printability of bioinks, as it becomes more challenging to precisely control, owing to the constrained mobility of polymeric chains during the extrusion process. Additionally, the potential for nozzle clogging may arise due to the presence of crosslinking agents, such as ionic species, or as a consequence of heightened viscosity, thereby exerting a notable impact on print quality (Remaggi et al. 2022).

Santos et al. (2022) demonstrated that PEG-Laponite-Alginate scaffolds were printed containing up to 10 layers and placed in a $CaCl_2$ solution to promote cross-linking of the alginate chains. Besides great printability, in vitro tests were performed in phosphate-buffered saline solution, showing that the dissolution of the composite hydrogel network was critical. Figure 11.3 illustrates the interactions formed by cross-linking process using 2% $CaCl_2$ for 15 min in alginate hydrogels (5% PEG-Laponite).

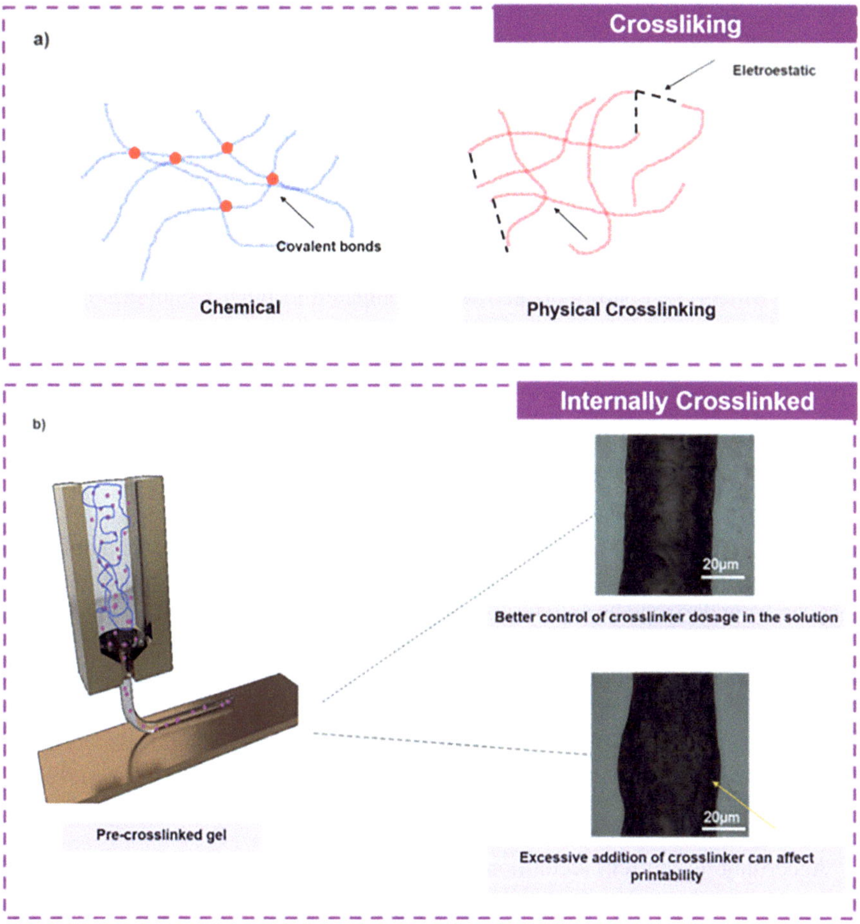

Fig. 11.3 (**a**) Representation of cross-linked polymeric chains by conventional chemical and physical methods. (**b**) Internally cross-linked alginate filaments printed. Excessive cross-linker when added to the solution can form aggregate or prevent flow

11.3.5 Printer Nozzles and Needles

Despite a great number of configurations for extrusion-based bioprinters, nozzles, or needles are used as a universal component. However, more attention must be paid to the fact that it can impact on cell viability, bioink spatial distribution, gel shear-thinning, generated shear stress, and shape fidelity (Blaeser et al. 2016; O'Connell et al. 2017). At the time of printing, the force applied is directly on the nozzle, generating an increase in pressure inside it, which can be felt by cells. The needles demonstrate more elongated shapes, so the material tends to be more sheared, producing more continuous filaments, but less cell viability. Therefore, nozzles are interested in terms of cell viability, as the pressures generated are <300 kPa (Emmermacher et al. 2020).

Additionally, nozzle/needle diameter can dictate the printed structures resolution that vary from 100 to 1000 μm (Hölzl et al. 2016; Miri et al. 2019). Lower resolution details are usually achieved using smaller diameter needles. For example, for tissues like vessels around 200 μm, cartilage due to its different zones 50–500 μm, bone 200–500 μm (Miri et al. 2019). However, although there is an approach to obtaining vessels by extrusion bioprinting, for some situations, such as blood capillaries (around 10 μm), it is not indicated when compared to other techniques (Hölzl et al. 2016).

11.4 Limitation and Advantage of Bioinks for Extrusion-Based Bioprinting

Advances in the cellular field have enabled the use of three-dimensional matrices as a cellular carrier to promote regeneration signals, creating tissue engineering (TE), that can be understood as an interdisciplinary field. TE applies engineering and life science concepts to the development of biological substitutes that aim to restore, maintain, or improve the function of injured tissues by new therapies and/or new biomaterials. For this, TE has as an approach the use of living cells, biocompatible biomaterials (scaffolds), and signaling mechanisms (growth factors). Recently, 3D bioprinting has been adopted in the field of TE, as one of its approaches, developing highly customized cell-loaded scaffolds to obtain healthy tissue from patients' own cells.

According to current definition, bioink is "a formulation of cells suitable for processing by an automated biofabrication technology that may also contain biologically active components and biomaterials" (J Groll et al. 2019 Biofabrication 11013001). Therefore, bioinks are formulations that have the presence of cells in their composition in a mandatory way. These formulations mimic similar structures from the mechanical and structural concepts of the extracellular matrix (ECM). Formulations that include thermoplastics supplemented with drug molecules, inorganic powders, or slurries that release bioactive ions cannot be considered as bioinks. In this way, groups of bioinks were proposed: (i) support bioinks are materials designed to support cell proliferation during delivery and act as an artificial ECM; (ii) fugitive bioinks are sacrificial materials that intend scape through a printed construct by internal voids or channels; (iii) structural bioinks provide mechanical performance to 3D structures over a relatively long timescale biodegradation; and (iv) functional bioinks provide biochemical, mechanical, or electrical cues to influence cellular behavior after a structure is printed (Groll et al. 2019).

In terms of composition, it is worth noting that other biological components may be contemplated in a bioink in addition to cells, such as ECM proteins, molecules, vesicles, or growth factors. They are mixed in different proportions depending on the final product application (Dzobo et al. 2019). This is partly due to the need for the bioink to provide physical and biochemical support to cells in the construct.

Although it is very common to designate all formulations that will receive cells as bioinks, there is terminological distinctness between inks made from biomaterials. Particularly, biomaterial inks do not contain cells in their preparation (Moroni et al. 2018; Groll et al. 2019). From a perspective of processing, these materials present greater manufacturing flexibility. After printing, cells are seeding on their surface after being consolidated, thus establishing their super-efficient adhesion and cell proliferation (Gungor-Ozkerim et al. 2016). Figure 11.4 illustrates the main differences between these two classes of materials.

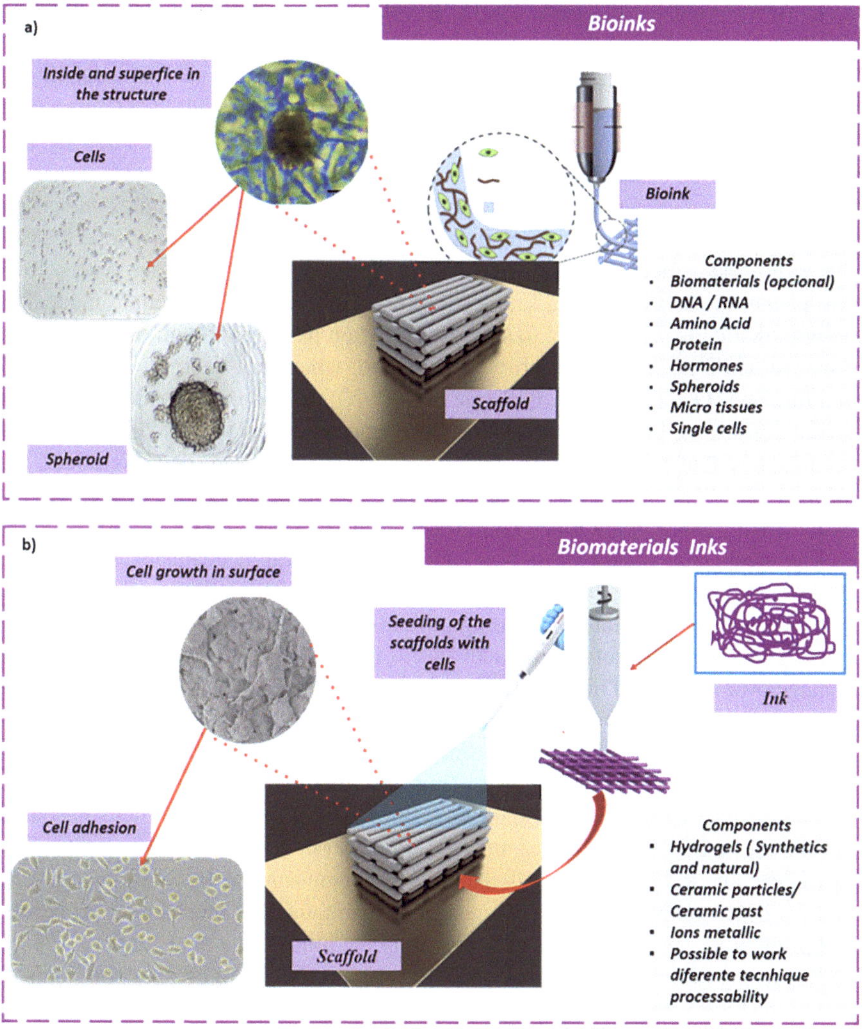

Fig. 11.4 Characteristics and components of (**a**) bioinks and (**b**) biomaterials inks

Bioinks can be subdivided into scaffold-free and cell-based scaffolds (Benwood et al. 2021). Those scaffold-free consist of structures based on cell suspension that try to mimic the formation of a neotissue from cell differentiation in a specific way. In this type of approach, it is employed structures such as spheroids or cell pellets that can be used as therapeutic agents, as they have tissue-specific mechanisms that many biomaterials cannot mimic (Murata et al. 2020).

On the other hand, the use of biomaterials in their composition is known as cell-based bioinks (Fantini et al. 2019; Benwood et al. 2021). As biomaterials perform their function, it opens space for cells to proliferate and occupy the free space to form biomimetic structures. Therefore, as a requirement, the biomaterials used must have adequate mechanical properties similar to the constructed tissue (Benwood et al. 2021). Another important condition, bioinks must present physical–chemical and biological compatibility characteristics inherent to the tissue, among others: sterility, facilitated cells growth, mechanical stability, and high shape fidelity of post-print during tissue reshaping (Smandri et al. 2020). Also, cell laden parameters interfere in the bioink integrity, such as cell type used in the bioink, cell density, and incubation period (Smandri et al. 2020).

Some companies disseminate the use of bioprinters based on complex and costly systems, claiming the need to guarantee the sterility of the process. Nevertheless, this is not the critical point of cell viability. Bioprinting taking place with a low-cost printer inside a biological safety cabinet already satisfies this requirement. The point is that cells are exposed to shear stress when passing through the nozzle and pressure while in the syringe prior to extrusion, which can decrease cell viability and function (Gao et al. 2019; Gillispie et al. 2020). Extrusion bioprinting produces robust constructs, but it is reported to typically promote up to 60% of cell damage immediately due to shear stress (Datta et al. 2019). To overcome this effect, the printing speed can be projected to decrease the pressure applied in extruded bioink, by increasing the printing time (Marques et al. 2019).

The mechanical stresses cause cell damage by several mechanisms, such as cell membrane rupture and generation of reactive oxygen species (ROS), which can lead to damage to cell structures, including proteins, lipids, and nucleic acids. The ROS can have temporal effects in the activation of signaling pathways, leading to cell damage over a period of time. Datta et al. have demonstrated that bioprinting process optimization as an option is not enough to avoid cell death in constructs, once cell rate viability decreases during time, limiting the impact on long-term functionality of the construct (Datta et al. 2019). Thus, to optimize the bioink should be one option to concern the long-time damage caused in post-printed. Another suggestion is utilizing antioxidants, for example, N-acetyl cysteine (NAC), as a component to reduce the ROS-induced cell damage, as demonstrated by NAC-incorporate bioinks based on MC3T3 cells-alginate that maintained its mechanical properties and printability, increasing cell viability along the time (Datta et al. 2019).

Regarding bioink optimization, one alternative is the use of cellular spheroids as part of the bioink because of its resistance capacities. Three-dimensional cell culture has enormous advantages to mimic the in vivo architecture and microenvironment of tissue and organs, and spheroids are currently the most attractive 3D model to produce uniform reproducible cell structures. The advantage of using spheroids in bioinks is related to the cell stability and surviving in biomaterials once the forces applied for extrusion in the AM damage the cell membrane, consequentially decreasing the cell viability detected. Moreover, spheroids may express molecules specialized in adhesion, particularly cadherin – main mediator of cell–cell interactions, which play a key role in intercellular interaction. Consequently, it promoted cell aggregation and compaction of 3D structures at advanced stages of development (Decarli et al. 2021).

As already mentioned, it is very challenging to design a bioink with encapsulating cells, from beginning to end. Before bioprinting, it is crucial to guarantee their survival while the process occurs, allowing homogeneous distribution of cells inside the engineered construct, since this is the final object resulting from bioprinting process (Valot et al. 2019). Figure 11.5 shows the challenges in the balance between printability and cell viability. Figure 11.5a demonstrated good 3D structure processability, however, inconsistent with cell viability inside this construct. In Fig. 11.5b, printability and cell viability are the best optimization, while in Fig. 11.5c, the printability and cell viability are the worst elaboration. Lastly, Fig. 11.5d has inconsistent processability of 3D construct, but good cell viability.

To balance the shear stress bioprinting and the cross-linking effects in cells, Good Laboratory Practices (GLP) of cell culture shall be applied. Physical and chemical conditions (temperature, pH, and UV irradiation) and material characteristics (biodegradability, viscosity, surface hydrophilicity, and stimulus responsiveness) should be previously determined (Li et al. 2021). In addition, satisfactory bioinks come from an adequate source of feedstock, and the supplier guarantee supported

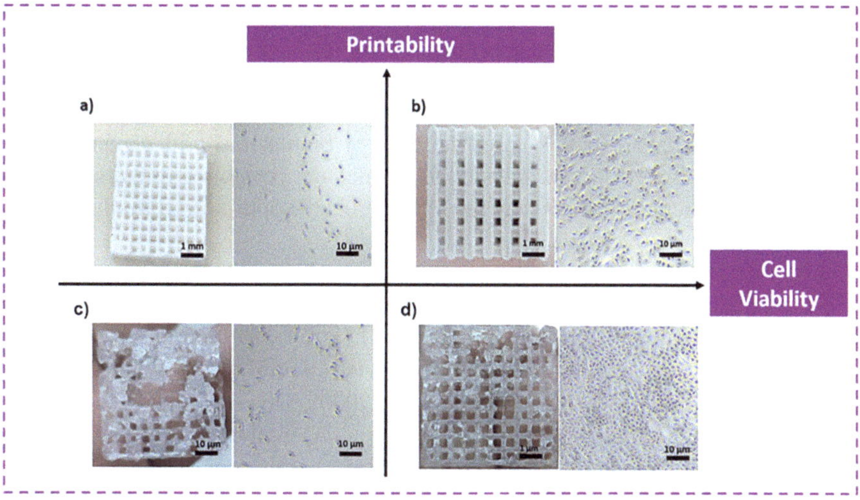

Fig. 11.5 Biofabrication window for designing bioinks. 3D constructs obtained by extrusion-based bioprinting (left) and their cell viability (right). Cell viability and printability are properties that need to be well balanced

by standards and routine tests. In this way, International Standard for biological evaluation of medical devices (e.g., ISO 10993-1 2018) gives parameters that corroborates to the GLP, once material information indicating how characterization process links to overall biological evaluation decision points to be taken when this biomaterial is applied to medicine, guiding what to look for bioink properties. These conditions, when considered to bioink design, help to guarantee the reproducibility (Gopinathan and Noh 2018).

The association of a well-prepared cell culture with appropriated biomaterial as substrate respecting the printability concept is the key of a successful bioink. Therefore, materials used to compose the bioink must attend to sterility rules to avoid bacterial or fungal contaminations (Valot et al. 2019). For sterility, the process of biomaterials used in bioink can be applied:

- Filtration if it is liquid enough, using a 0.22 μm filter.
- UVC irradiation for a transparent solution or hydrogel, but only if it cannot undergo unwanted UVA-cross-linking reactions.
- Autoclaving, only for compounds and solutions that are stable at high temperature, high pressure, and humidity (i.e., up to 130 °C with a humidity-saturated atmosphere).
- Pasteurization where material is placed in bath for long period, i.e., 1 h at 72 °C (Fantini et al. 2019).

Many studies have used natural hydrogels as bioink because of their improved bioactivity, water solubility under physiological conditions, and resemblance to the extracellular matrix both chemically and structurally (Montero et al. 2019). Chiefly, natural polymer have weaker mechanical properties, while synthetic hydrogels have less cost, are easily modified, and lack bioactive molecules for essential cell adhesion and proliferation (Gillispie et al. 2020). Synthetic materials preparation does not significantly interfere in cells of the bioink even when they use toxic solvents or have a melting point above the human body temperatures, which could compromise cell viability and make in situ cell encapsulation a difficult process (Montero et al. 2019). In this case, the type of biomaterial will depend on the application, cell type of organ intended to mimic, and possibly a mixture between natural and synthetic polymers may result in a probable optimized formulation with both benefits.

According to Santoni, bone is the tissue most studied in the topic of bioprinting; it corresponds to 31.7% of the total academic papers published, followed by vascularization (15.6%) (Santoni et al. 2022). For bone bioprinting, the most common used hydrogels are gelatin methacrylate (GelMa), polyethylene glycol (PEG), and Pluronic F-127® (Genova et al. 2020). However, there are various combinations with other materials and polymers. A combination of two or more biomaterials is the frequent choice to reach the best bioink properties.

Collagen (COL)-based bioinks are another commonly used bioinks with the intention to mimic cartilage, bone, and osteochondral tissue. COL bioinks can provide bindings for cell adhesion and reduce the risk of immune reaction; however, the property of this material alone is not completely stable due to its low viscosity. For this case, it is interesting blend COL with other materials to improve final mechanical properties. One example of hybrid material (natural plus synthetic biomaterial) is mixing COL with Pluronic®, which generates promising results in stable constructs 3D bioprinted, increasing cell number significantly after 24 h post-bioprinting (Marques et al. 2019).

Gelatin (GEL) is derived from denatured collagen and has bioactive properties, due to the presence of the amino acid sequence arginylglycylaspartic, which facilitates cell adhesion and allows cell remodeling in sites of degradation. This material, in addition to being thermoreversible, gels at temperatures below 25–35 °C, which is an important property for biomedical applications. The acrylate form, known as GelMa, has good mechanical properties and is more appropriate for cartilage tissues. However, it shows low biodegradation and low solubility in water, and depending on the cross-linking method, it can be toxic (Gao et al. 2019; Garcia-Cruz et al. 2021).

Polyethylene glycol (PEG) is a linear synthetic homopolymer obtained from the constituent monomers such as ethylene glycol, ethylene oxide, or oxyethylene. The advantages of PEG hydrogels over natural hydrogels are the control of their mechanical properties, biodegradability, among others with the insertion of chemical groups, associated with the type of cross-linking used. Due to its bioinert compatibility, it presents low adhesion and cell proliferation as a limitation to use as bioink. Consequently, it is necessary to merge its properties with those of other materials that are bioactive or to incorporate specific agents in its chain. In addition to the chemical modifications of PEG hydrogels, the physical properties are conditioned at the time of their reaction, as they directly influence the cross-linking density, swelling, elastic modulus, resulting in constructs with stiffness ranging from soft tissue (liver and skin) to high values of stiffness (articular cartilage and bone) (Li et al. 2021).

Pluronic® F127 has been used in the development of hydrogels for bioprinting due to its good pseudoplasticity. The properties are directly related to its molar mass, that is, between the ratio between the length of PEO and PPO blocks. The molar mass ratio between these blocks can be adjusted according to their purpose, so that it can directly influence in vivo applications and their interactions with cells and cell membranes. Thus, the material allows the control of its physicochemical characteristics, modifying the ratio of PEO and PPO. One explanation is associated with micellar formation that occurs at the critical micellization temperature because of dehydration of the PPO block. In this way, micellization significantly occurs with increasing temperature, due to its greater contact and there is no movement of the chains. Therefore, the formation of gels at high temperatures may be associated with the entanglement of these micelles (Genova et al. 2020).

Furthermore, PF-127 is a thermosensitive polymer showing inverse thermo-gelling properties, i.e., at low temperatures, the material exists in the form of a liquid. These gel–liquid transitions are often characterized by a lower critical solution temperature (LCST). Soluble in cold water (4–8 °C), its sol–gel transition occurs in the presence of heating. Consequently, higher temperatures provided characteristics of a soft gel in material, because of an increase in viscosity (Li et al. 2021).

Although bone and osteochondral bioprinting are the most studied topics, there is no final product in use and the research is still running. As many materials are developed to be used as bioink, there will be more options in the future for customized regenerative medicine. Some studies have searched for methods bedsides bioprinting, as presented in Table 11.1. An approach that has recently gained strength is direct printing of the bioink inside the defect, usually called in situ *bioprinting*. This new technique tries to realize the clinical application of bioprinting by resolving the limitations of conventional strategy that is bioprinting, bioreactor culture, and subsequent implantation.

Nonetheless, in situ bioprinting is an emerging field that requires materials and technologies with different characteristics compared with conventional bioprinting. This new branch is typically set up based on an open-loop three-axis motion platform, which implies that the printer can only perform calibrate-then-print operations and is only applicable to static target surfaces. Hence, no bioprinting approach can yet fully replicate the morphological, biochemical, and physiological properties of native tissues, particularly skin and its appendages (Chen et al. 2023; Zhao et al. 2023).

11.5 Accessible Hardware and Software for 3D Bioprinting

As FDM has a simple structural construction with low costs materials and components, flexibility for printing different feedstock, and flexibility of use, when the patent expired in 2009, many parts of the world started popping up alternative solutions to the FDM original process leading to a high dropping in price and increasing offer of analogous technologies. It was mainly related to open-source hardware and software collaboration for 3D printing. This initial 3D printing democratization movement was led about 2006, in parallel by Evan Malone in the USA at Cornell University with the Fab@Home project (Malone and Lipson 2007) and Adrian Bowyer in the UK at University of Bath with the RepRap project. At the Center for Information Technology Renato Archer, in Brazil, a similar open project was in development in 2005 with the purpose to reduce the amount of costly biomaterials as feedstock with flexibility and affordability. Immediately after the first publication on this area, CTI constructed its extrusion 3D printer based on Fab@Home project, focusing in biomaterials use as a multinational collaboration development with Cornell University (Lixandrão Filho et al. 2009). The strategy was the improvement of functionality, including the development of an interchangeable printing head and mini screw extrusion processes for polymer powder extrusion (Silveira et al. 2014).

Nowadays, there are a myriad of publications in the context of open-source hardware and software collaboration to adapt 3D printers into affordable 3D

Table 11.1 Most relevant bioprinting methods in the literature review

N°	Materials	Mimetic tissue	Model animal	Type of cells	Bioprinting	References
1	GelMa	Skin (hair follicle)	C57BL/B6 mice and BAL-B/c/nu/mice	Skin-derived precursors (SKps) and epidermal stem cells (Epi-SCs)	Robo-assisted	Chen et al. (2023)
2	PCL and Gelatin/Alginate and BMSCs bioink	Cranial defects	Canine (Beagle dog)	BMSC (bone marrow–derived mesenchymal stem cell)	Extrusion	Huan et al. (2023)
3	PRP/GelMa/AlgMa/Laponite	Bone (femoral condyle defect)	Rats	HUVECs and BMSCs	Bioploter	Cao et al. (2023)
4	Matrigel/Epi-SCs/SKps	Skin (hair follicle)	BALB/c nu/nu mice	C57BL/B6 neonatal mouse dorsal and Epi-SCs and SKPs	Robotic-printed	Zhao et al. (2023)
5	ECM/AMP	Bone	Rats	DPSCs (dental pulp stem cells)	Extrusion	Dubey et al. (2020)
6	PCL/ZnO/HAp	Bone	C57BL/6 mice	hMSCs	Extrusion	Mostafavi et al. (2021)
7	PCL	Tend-to-bone interface	Rats	Mesenchymal stem cells	Bioploter	Tarafder et al. (2020)
8	Fibrin/collagen	Excisional full-thickness wounds (Skin)	Porcine and Murine mice	Fibroblast and keratinocytes	Inkjet	Albanna et al. (2019)

bioprinters, beyond improvements of components, control, and flexibility (Tong et al. 2021; Ioannidis et al. 2020; Dávila et al. 2022a, b).

11.6 Future Perspectives

3D bioprinting is a technology that is still in consolidation and the terms related to products from this technology have been undergoing revisions that are converging to an international consensus. Extrusion-based bioprinting stands out as a promising future technique due to its versatility in obtaining biomimetic structures, in addition to facility of optimizing its construction parameters, as nozzle diameter and heating temperature.

The interaction of in vitro cultured cells with the extruded material is still a barrier in the commercial application of this technology. The final application of the product opens up a range of study possibilities, for example, imitation of skin, bone, and organs (liver, pancreas, and thyroid) requires different types of bioinks and the cells involved have unique characteristics that will interfere with the material. Yet, one may want the bioink to be degradable or not, the only convergent point is the viability of the cells in these materials.

Regenerative medicine is the biggest consumer of new technologies that benefit human health, it drives the development of new forms of treatment and allows research for customization of treatments, this means a flexibility of technology for individualized treatment with medicine focused on the real need of the patient, which has been bringing about new discussions in regulatory terms and expanding the reach of these studies.

The advancement of these researches has achieved the bioprinting of mini-organs aimed at analyzing the safety and efficacy of pharmaceutical products. Society's interest in not using animals for experimentation has driven research into the development of in vitro methodologies with more reproductive results. Much has already been characterized with cells cultivated in 2D environment, but it is known of the difference in results when cells are in 2D and 3D environments.

The cultivation of cells in 3D environment brings a better comparation of its results with those available from animal studies, these promising studies have boosted bioprinting to design compartmentalized organ models, which, together with microfluidics studies, are revolutionizing the diagnostics known as lab-on-a-chip.

Therefore, the applications of bioprinting are unlimited and the research group with multiple skills can make the knowledge about bioink, cell interactions, and product application go further in human life.

Funding Declaration/Acknowledgments The authors want to thank FAPESP for the grant (# 2019/11950-6) and the financial support of the CNPq/PCI program through the K. F. Santos fellowship.

References

Agarwal, S. et al. (2020) Current Developments in 3D Bioprinting for Tissue and Organ Regeneration – A Review. Frontiers in Mechanical Engineering vol 6: 58917 https://doi.org/10.3389/fmech.2020.589171.

Albanna, M. et al. (2019). In Situ Bioprinting of Autologous Skin Cells Accelerates Wound Healing of Extensive Excisional Full-Thickness Wounds. *Scientific Reports*, 9(1). https://doi.org/10.1038/s41598-018-38366-w.

Amorim, P.A. et al. (2021) Insights on shear rheology of inks for extrusion – based 3D bioprinting. Bioprinting vol 22, p. e00129. https://doi.org/10.1016/J.BPRINT.2021.E00129.

Benwood, C. et al. (2021) Natural biomaterials and their use as bioinks for printing tissues. *Bioengineering*, 8(2): 1–19. https://doi.org/10.3390/bioengineering8020027

Blaeser, A. et al. (2016) Controlling Shear Stress in 3D Bioprinting is a Key Factor to Balance Printing Resolution and Stem Cell Integrity, Advanced Healthcare Materials, 5(3): 326–333. https://doi.org/10.1002/adhm.201500677.

Cao, B. et al. (2023). 3D-printed vascularized biofunctional scaffold for bone regeneration. *International Journal of Bioprinting*, 9(3). Available at: https://doi.org/10.18063/ijb.702.

Champeau, M. et al. (2020) 4D Printing of Hydrogels: A Review. Advanced Functional Materials 30(31):1–22. https://doi.org/10.1002/adfm.201910606.

Chen, H. et al. (2023). Robot-assisted in situ bioprinting of gelatin methacrylate hydrogels with stem cells induces hair follicle-inclusive skin regeneration. *Biomedicine and Pharmacotherapy*. https://doi.org/10.1016/j.biopha.2022.114140.

Cooke, M.E. and Rosenzweig, D.H. (2021) The rheology of direct and suspended extrusion bioprinting, APL Bioengineering, 5(1): 1–20. https://doi.org/10.1063/5.0031475.

Daguano, J.K.M.B. et al. (2022) 'Shear-thinning sacrificial ink for fabrication of Biosilicate® osteoconductive scaffolds by material extrusion 3D printing', *Materials Chemistry and Physics*, 287. Available at: https://doi.org/10.1016/j.matchemphys.2022.126286.

Datta, S. et al. (2019) Bioink formulations to ameliorate bioprinting-induced loss of cellular viability. Biointerphases, 14(5): 051006. https://doi.org/10.1116/1.5111392.

Dávila, J.L. and D'Ávila, M.A. (2017) Laponite as a rheology modifier of alginate solutions: Physical gelation and aging evolution. Carbohydrate Polymers 151:1–8. https://doi.org/10.1016/j.carbpol.2016.09.057.

Dávila, J.L. and dÁvila, M.A. (2019) Rheological evaluation of Laponite/alginate inks for 3D extrusion-based printing. International Journal of Advanced Manufacturing Technology vol 101:675–686. https://doi.org/10.1007/s00170-018-2876-y.

Dávila, J.L., Manzini, B.M., d'Ávila, M.A., et al. (2022a) 'Open-source syringe extrusion head for shear-thinning materials 3D printing', *Rapid Prototyping Journal*, 28(8), pp. 1452–1461. Available at: https://doi.org/10.1108/RPJ-09-2021-0245.

Dávila, J.L., Manzini, B.M., Lopes da Fonsêca, J.H., et al. (2022b) 'A parameterized g-code compiler for scaffolds 3D bioprinting', *Bioprinting*, 27. Available at: https://doi.org/10.1016/j.bprint.2022.e00222.

Decarli, M.C. et al. (2021) 'Cell spheroids as a versatile research platform: Formation mechanisms, high throughput production, characterization and applications', *Biofabrication*. IOP Publishing Ltd. Available at: https://doi.org/10.1088/1758-5090/abe6f2

Dubey, N. et al. (2020). Extracellular Matrix/Amorphous Magnesium Phosphate Bioink for 3D Bioprinting of Craniomaxillofacial Bone Tissue. *ACS Applied Materials and Interfaces*, 12(21): 23752–23763. https://doi.org/10.1021/acsami.0c05311.

Dzobo, K., Motaung, K.S.C.M.; and Adesida, A. (2019) Recent trends in decellularized extracellular matrix bioinks for 3D printing. International Journal of Molecular Sciences 20(18):1–28. https://doi.org/10.3390/ijms20184628.

Emmermacher, J. et al. (2020) Engineering considerations on extrusion-based bioprinting: interactions of material behavior, mechanical forces and cells in the printing needle. *Biofabrication* 12(2): 025022. https://doi.org/10.1088/1758-5090/ab7553

Fantini, V. et al. (2019) Bioink composition and printing parameters for 3D modeling neural tissue. Cells 8(8):830. https://doi.org/10.3390/cells8080830.
Fu, Z. et al. (2021) Printability in extrusion bioprinting. Biofabrication 13(3): 033001. https://doi.org/10.1088/1758-5090/abe7ab.
Gao, T. et al. (2019) Optimization of gelatin-alginate composite bioink printability using rheological parameters: a systematic approach. Biofabrication 10(3):1–17. https://doi.org/10.1088/1758-5090/aacdc7
Garcia-Cruz, M.R. et al. (2021) Printability and bio-functionality of a shear thinning methacrylated xanthan-gelatin composite bioink. Biofabrication 13(3): 035023 https://doi.org/10.1088/1758-5090/abec2d.
Gargus, E.; Lewis, P.; Shah, R. (2018) Bioinks for 3D printing. In: 3D Bioprinting Regenerative Engineering, p 25.
Genova T, Roato I, Carossa M, Motta C, Cavagnetto D, Mussano F. (2020) Advances on Bone Substitutes through 3D Bioprinting. *International Journal of Molecular Sciences*; 21(19):7012. https://doi.org/10.3390/ijms21197012
Ghavaminejad, Amin et al. (2020) Crosslinking strategies for 3D bioprinting of polymeric hydrogels. Small 16 (35): 200293. https://doi.org/10.1002/smll.202002931
Gillispie, G. et al. (2020) Assessment methodologies for extrusion-based bioink printability. Biofabrication 12(2): 022003. https://doi.org/10.1088/1758-5090/ab6f0d.
Gopinathan, J. and Noh, I. (2018) Recent trends in bioinks for 3D printing. Biomaterials Research 22:1–15. https://doi.org/10.1186/s40824-018-0122-1.
Groll, J. et al. (2016) Biofabrication: Reappraising the definition of an evolving field. Biofabrication 8(1):013001. https://doi.org/10.1088/1758-5090/8/1/013001.
Groll, J. et al. (2019) A definition of bioinks and their distinction from biomaterial inks, Biofabrication, 11(1):013001. https://doi.org/10.1088/1758-5090/aaec52.
Gu, Z. et al. (2020) Development of 3D bioprinting: From printing methods to biomedical applications. Asian Journal of Pharmaceutical Sciences 15(5): 529–557. https://doi.org/10.1016/j.ajps.2019.11.003.
Gungor-Ozkerim, P.S. et al. (2016) Advanced Bioinks for 3D Printing: A Materials Science Perspective. Biofabrication 12(1): 631–634. https://doi.org/10.1007/s10439-016-1638-y.
Hölzl, K. et al. (2016) Bioink properties before, during and after 3D bioprinting. Biofabrication 8(3): 032002. https://doi.org/10.1088/1758-5090/8/3/032002.
Hospodiuk, M. et al. (2017) The bioink: A comprehensive review on bioprintable materials. Biotechnology Advances 35(2): 217–239. https://doi.org/10.1016/j.biotechadv.2016.12.006.
Huan, Y. et al. (2023) '3D bioprinted autologous bone particle scaffolds for cranioplasty promote bone regeneration with both implanted and native BMSCs', *Biofabrication*, 15(2): 025016. https://doi.org/10.1088/1758-5090/acbe21.
Ioannidis, K. et al. (2020) 'A Custom Ultra-Low-Cost 3D Bioprinter Supports Cell Growth and Differentiation', *Frontiers in Bioengineering and Biotechnology*, 8. Available at: https://doi.org/10.3389/fbioe.2020.580889.
ISO 10993-1. (2018, 08). Biological evaluation of medical devices – Part 1: Evaluation and testing with a risk management process. *International Standard, 5*. ISO.; Accessed in:5/9/2020.
Jaishankar, A. and McKinley, G.H. (2013) Power-law rheology in the bulk and at the interface: Quasi-properties and fractional constitutive equations. Proceedings of the Royal Society A: Mathematical, Physical and Engineering Sciences.
Karzyński, K. et al. (2018) Use of 3D bioprinting in biomedical engineering for clinical application. Medical Studies 34(1): 93–97. https://doi.org/10.5114/ms.2018.74827.
Katja, H. et al. (2016) Bioink properties before, during and after 3D bioprinting, Biofabrication, 8(3): 32002. http://stacks.iop.org/1758-5090/8/i=3/a=032002.
Ke, D., Niu, C. and Yang, X. (2022) Evolution of 3D bioprinting-from the perspectives of bioprinting companies, Bioprinting. https://doi.org/10.1016/j.bprint.2022.e00193.
Kirillova, A., Bushev, S., Abubakirov, A., & Sukikh, G. (2020). Bioethical and legal issues in 3D bioprinting. International Journal of Bioprinting, 6(3): 1–10. https://doi.org/10.18063/IJB.V6I3.272

Li, N.; Guo, R.; Zhang, Z. J. (2021) Bioink Formulations for bone tissue regeneration, Front Bioeng. Biotechnol. 9: 630488. https://doi.org/10.3389/fbioe.2021.630488
Lixandrão Filho, A. et al. (2009) Construction and adaptation of an open source rapid prototyping machine for biomedical research purposes—a multinational collaborative development, in *Innovative Developments in Design and Manufacturing*. CRC Press. Available at: https://doi.org/10.1201/9780203859476.ch73.
Malda, J. et al. (2013) 25th Anniversary Article: Engineering Hydrogels for Biofabrication. Advanced Materials 25(36): 5011–5028. https://doi.org/10.1002/adma.201302042.
Malone, E. and Lipson, H. (2007) Fab@Home: the personal desktop fabricator kit. *Rapid Prototyping Journal*, 13(4), pp. 245–255. Available at: https://doi.org/10.1108/13552540710776197.
Marques, C.F. et al. (2019) Collagen-based bioinks for hard tissue engineering applications: a comprehensive review. Journal of Materials Science: Materials in Medicine, 30(3)1–12. https://doi.org/10.1007/s10856-019-6234-x.
Miri, A.K. et al. (2019) Effective bioprinting resolution in tissue model fabrication. Lab on a Chip, 19(11): 2019–2037. https://doi.org/10.1039/C8LC01037D.
Mironov, V., Reis, N. and Derby, B. (2006) Review: Bioprinting: A Beginning. Tissue Engineering 12(4): 631–634. https://doi.org/10.1089/ten.2006.12.631.
Montero, F.E. et al. (2019) Development of a Smart Bioink for Bioprinting Applications. Frontiers in Mechanical Engineering. https://doi.org/10.3389/fmech.2019.00056.
Moroni, L. et al. (2018) Biofabrication: A Guide to Technology and Terminology. Trends in Biotechnology, 36(4): 384–402. https://doi.org/10.1016/j.tibtech.2017.10.015.
Mostafavi, A. et al. (2021) In situ printing of scaffolds for reconstruction of bone defects. *Acta Biomaterialia*, 127, pp. 313–326. https://doi.org/10.1016/j.actbio.2021.03.009.
Murata, D; Arai, K.; Nakayama, K. Scaffold-free bio-3D (2020) Printing using spheroids as "Bio-Inks" for tissue (Re-) construction and drug response tests. *Advanced Healthcare Materials*, (9):15 1901831, 2020. https://doi.org/10.1002/adhm.201901831.
Naghieh, S. and Chen, D. (2021) Printability – a Key Issue in Extrusion-based Bioprinting, Journal of Pharmaceutical Analysis 11(5): 564–579. https://doi.org/10.1016/j.jpha.2021.02.001.
O'Connell, C. et al. (2017) Characterizing Bioinks for Extrusion Bioprinting: Printability and Rheology. *3D Bioprinting: Principles and Protocols p111-133*. https://doi.org/10.1007/978-1-0716-0520-2.
Paxton, N. et al. (2017) Proposal to assess printability of bioinks for extrusion-based bioprinting and evaluation of rheological properties governing bioprintability, Biofabrication, 9(4):044107. https://doi.org/10.1088/1758-5090/aa8dd8.
Remaggi, Giulia et al. (2022) Alginate Self-Crosslinking Ink for 3D Extrusion-Based Cryoprinting and Application for Epirubicin-HCl Delivery on MCF-7 Cells. Molecules 27 (3): 882. https://doi.org/10.3390/molecules27030882.
Robles Martinez, P., Basit, A.W. and Gaisford, S. (2018) The history, developments and opportunities of stereolithography. 3D Printing of Pharmaceutical. https://doi.org/10.1007/978-3-319-90755-04.
Santoni, S., Gugliandolo, S. G., Sponchioni, M., Moscatelli, D., & Colosimo, B. M. (2022). 3D bioprinting: current status and trends—a guide to the literature and industrial practice. In Bio-Design and Manufacturing 5(1):14–42. https://doi.org/10.1007/s42242-021-00165-0.
Santos, K. et al. (2022). Estudo da reologia de hidrogéis compósitos de PEG-Laponita-alginato visando impressão 3D baseada em extrusão. *Matéria(Rio de Janeiro)*, v. 27. https://doi.org/10.1590/S1517-707620220002.1374.
Schwab, A. et al. (2020) Printability and Shape Fidelity of Bioinks in 3D Bioprinting. Chemical Reviews 120(19): 11028–11055. https://doi.org/10.1021/acs.chemrev.0c00084.
Silveira, Z. et al. (2014). Design development and functional validation of an interchangeable head based on mini screw extrusion applied in an experimental desktop 3-D printer. *International Journal of Rapid Manufacturing*, (4):1 p.49–65, 2014. https://doi.org/10.1504/IJRAPIDM.2014.062037.
Smandri, A. et al. (2020) Natural 3D-Printed Bioinks for Skin Regeneration and Wound Healing: A Systematic Review. *Polymers* 12(8):1782. https://doi.org/10.3390/polym12081782.

Su, A. and Al Aref, S.J. (2018) History of 3D printing, in 3D Printing Applications in Cardiovascular Medicine, p 1–10. https://doi.org/10.1016/B978-0-12-803917-5.00001-8.
Tarafder, S. et al. (2020). In situ tissue engineering of the tendon-to-bone interface by endogenous stem/progenitor cells. *Biofabrication*, 12(1). https://doi.org/10.1088/1758-5090/ab48ca.
Temirel, M., Dabbagh, S.R. and Tasoglu, S. (2022) Shape Fidelity Evaluation of Alginate-Based Hydrogels through Extrusion-Based Bioprinting. *Journal of Functional Biomaterials* 13(4). https://doi.org/10.3390/jfb13040225.
Tong, Anh et al. (2021) Review of low-cost 3D bioprinters: State of the market and observed future trends. Translating Life Sciences Innovation 26(4): 333–366. https://doi.org/10.1177/24726303211020297.
Valot, L. et al. (2019). Chemical insights into bioinks for 3D printing. Chemical Society Reviews, 48(15): 4049–4086. https://doi.org/10.1039/c7cs00718c.
Zhao, W., Hu, C. and Xu, T. (2023). In vivo bioprinting: Broadening the therapeutic horizon for tissue injuries. *Bioactive Materials*. KeAi Communications Co., pp. 201–222. https://doi.org/10.1016/j.bioactmat.2023.01.018.

Chapter 12
Optical Techniques for the Diagnosis and Monitoring of Oral Hard Tissue Lesions

Patricia Aparecida da Ana, Ilka Tiemy Kato Prates, Carolina Benetti, and Matheus Del-Valle

Abstract In association with biomedical optics, Biophotonics contributes to the field of health sciences through the development of optical methods for early disease diagnosis and monitoring, as well as assessing the effects of different therapies. Such methods are based on the knowledge of interactions between light and biological tissues, enabling the detection, quantification, and correlation of tissue abnormalities with several processes. Optical techniques have advantages, including the use of nonionizing radiation, fast and real-time detection of disease stages, imaging capabilities, and high sensitivity and specificity. Still, these techniques are noninvasive, providing comfort to the patients. In this chapter, the main optical techniques used for the early diagnosis of dental hard tissue pathologies will be presented, with emphasis on the demineralization of enamel and dentin, which is a common characteristic of caries and erosion lesions. Fluorescence imaging or spectroscopy techniques, Fourier transform infrared spectroscopy, Raman spectroscopy, optical coherence tomography, and transillumination are advantageous for these lesions. These methods present high resolution and allow the quantification of optical parameters that may be related to the disease progression. Some techniques are already in clinical use, while others are in broad scientific development, and very close to clinical reality.

Keywords Fluorescence · Raman spectroscopy · FTIR spectroscopy · Optical coherence tomography · Transillumination · Demineralization

P. A. da Ana (✉) · I. T. K. Prates · C. Benetti · M. Del-Valle
Graduate Program in Biomedical Engineering, Federal University of ABC – UFABC Al. da Universidade, São Bernardo do Campo, SP, Brazil
e-mail: patricia.ana@ufabc.edu.br

C. B. Lombello, P. A. da Ana (eds.), *Current Trends in Biomedical Engineering*,
https://doi.org/10.1007/978-3-031-38743-2_12

12.1 Introduction

Although optics is a branch of physics that basically studies light-related phenomena, the interaction of different light sources with biological tissues stands out in the health sciences, and this is the focus of Biophotonics. This area provides great biotechnological development for both diagnosis and treatment of different clinical conditions. The great advantage of using optical techniques for diagnostics and monitoring is the possibility of the early detection of small tissue alterations, which may be morphological, optical, or even biochemical. Also, with the development of collimated lasers and LEDs, the construction of low-cost equipment with great potential for long-term monitoring has been observed, as it does not involve ionizing radiation.

The optical techniques are widely used in several specialties of medicine, such as ophthalmology, oncology, dermatology, and cardiology, as well as in regenerative medicine since it is possible to construct and longitudinally monitor the interaction of scaffolds and implants with the biological tissues in real time (Chen et al. 2023). Among the main optical techniques that are used in medicine, we can highlight the vibrational spectroscopy techniques, which let optical biopsy procedures and, thus, are widely used in the early diagnosis of malignant neoplasms (Meza-Ramirez et al. 2023). Imaging techniques, such as fluorescence imaging and infrared thermography, can detect inflammatory and/or infectious processes even at an initial stage, as well as allow a quick and real-time visualization of the magnitude of one or more lesions simultaneously (Schiavon et al. 2021). Technologies, such as optical coherence tomography, have revolutionized the field of ophthalmology, among others, by the visualization of tissues in depth with high spatial resolution (Kim et al. 2023). Other techniques can be cited such as laser Doppler flowmetry (Kralj and Lenasi 2023), laser speckle contrast imaging (Konovalov et al. 2023), and a number of other methods. In this chapter, focus will be given to the main techniques used for diagnosing hard tissue pathologies, that is, enamel, dentin, and bone.

Dentistry is one of the health areas that benefit from optical techniques for early diagnosis, and the most prevalent pathologies that can be supervised are those related to dental hard tissues, such as caries and erosion lesions. Although the physiopathology of these lesions is widely known by clinicians and the scientific community, their diagnosis at a very early stage is not always possible using conventional techniques, but it is strictly necessary as it enables clinical interventions that can reverse them. Also, monitoring the treatment performed is an important step for the patient's health.

The search for more precise, objective, and quantitative diagnostic methods has become constant in recent decades, and transillumination techniques (FOTI –Fiber Optics Transillumination and DIFOTI – Digital Imaging Fiber Optics Transillumination), fluorescence (DIAGNOdent, DELF – dye-enhanced laser fluorescence and QLF – quantitative light fluorescence), and tomographic (OCT – optical coherence tomography, PS-OCT – polarization-sensitive optical Coherence Tomography) monitoring, among others, were developed and continue to be tested. In this chapter, these techniques will be briefly described.

12.2 General Aspects of Dental Lesions

In summary, dental caries is an infectious, transmissible, chronic, multifactorial, and slow-progressing disease that can affect deciduous and permanent teeth (Machiulskiene et al. 2020). The clinical aspect is the presence of a demineralization process, which is the result of the dissolution of the inorganic matrix of dental hard tissues (composed mainly of carbonated hydroxyapatite) by acids produced by oral bacteria during the metabolism of fermentable carbohydrates. As it is a multifactorial disease, the development and progression of carious lesions depend on several factors, such as the type of substrate, type of diet, and the composition of saliva, among others. The onset and the progression of carious lesions are related to the imbalance between demineralization and remineralization.

The initial appearance of caries lesions is that of a white spot, which corresponds to the change in the refractive index of the hard tissue due to the loss of subsurface minerals with the maintenance of the intact surface (Fejerskov and Nyvad 2015). At this stage, the lesions can have a depth of up to 200 μm in enamel, or up to 1.5 mm in enamel and dentin (Ko et al. 2005). As these lesions have an intact surface, they can be remineralized, with no need to remove tissue to restore the tooth. At this moment, the correct diagnosis is essential, which often does not occur because such lesions can appear in interproximal regions (between the teeth) which are not detected with the naked eye. Furthermore, traditional diagnostic methods, such as interproximal radiography, do not have the in-depth resolution to detect such small lesions but are effective in detecting lesions deeper than 50 μm (Wijesinghe et al. 2016).

The progression of carious processes in enamel and dentin leads to the occurrence of lesions with a series of histological alterations, many of which cannot be identified macroscopically (Silverstone et al. 1988). For this reason, not only is early diagnosis important, but discrimination of the tissue condition in a chronic cavitated lesion is also required. When it is necessary to treat the caries lesion, the current minimal intervention protocol recommends preserving the structure of the dental element as much as possible, and it is desirable that the minimum amount of tissue is removed (Horowitz 1996). Thus, the optical diagnostic methods can differentiate the courses of lesions, as well as identify the possibility of their progression and arrest in real time, with low cost and high specificity and sensitivity.

Tooth erosion is another multifactorial prevalent lesion that can be present in enamel and dentin. It is characterized by the dissolution of the dental hard tissue by acids of non-bacterial origin, when the saliva is undersaturated in relation to the tooth mineral (Schlueter et al. 2020). The acids involved in the etiology of dental erosion are those from soft drinks, gastroesophageal reflux, acidic foods and fruits, or even gaseous acids or another chemical breathes (Kanzow et al. 2016). Clinically, the dissolution of enamel or dentin promotes the loss of texture, a glossy appearance, the flattening of the occlusal surface, or concavities on smooth surfaces. However, sometimes the early diagnosis of erosion lesions can be difficult since they are accompanied by few signs or symptoms and, nowadays, there are no clinical devices that allow the specific tooth erosion diagnosis and monitoring (Joshi et al. 2016). Some studies have shown the potential of some optical techniques in the discrimination of hard tissues in the initial stages of demineralization, considering that demineralization increases the scattering and changes the light polarization (Cara et al. 2014; Maia et al. 2016; Pereira et al. 2018). In this way, it is possible to expand the potential of these techniques in the early diagnosis of carious and erosion lesions.

12.3 Basic Principles of Biophotonics

All optical methods for diagnostic and monitoring are based on the detection of optical changes in tissues during the course of the disease and their subsequent quantification. It should be considered, on the other side, that the optical techniques applied to dental tissues are inherently complex due to the non-homogeneity and anisotropic nature of the biological materials present. To understand how to discriminate healthy from diseased tissue, the four main interactions of photons with biological tissues are basically considered (Prasad 2004): reflection, absorption, scattering, and transmission (Fig. 12.1a).

Reflection occurs when the incident light returns after interacting with the surface of a medium. This phenomenon occurs when light, which is propagating in one medium, falls on another medium with a distinct refraction index. Absorption occurs when the medium attenuates the intensity of the incident light, as the molecules of the medium and the energy of the incident photons are in resonance. The molecules that carry out light absorption are called chromophores. The main chromophores present in biological tissues are water and hemoglobin in its oxygenated form. The region between 650 and 1300 nm is called "biological window," because the absorption is minimized, prevailing the scattering.

Scattering occurs when the vibration frequencies of light and molecules in the medium are different, and the direction of the light propagation is changed, together with the probable loss of light intensity. The transmission is given by the portion of light that does not undergo these interactions, crossing the medium. The biological tissues are highly heterogeneous, with different molecules and structures; in this way, all these light-tissue interactions can occur simultaneously, but in general, absorption and scattering are the interactions that occur with greater intensity. The proportions of occurrence vary according to the wavelength of the incident light and the tissue in question. The optical parameters of the tissues that characterize these interactions are defined as absorption coefficient (μ_a) and scattering coefficient (μ_s) (Jacques 2013).

Some characteristics concerning the light source are important to induce a detectable interaction with the biological tissue. One characteristic that some light

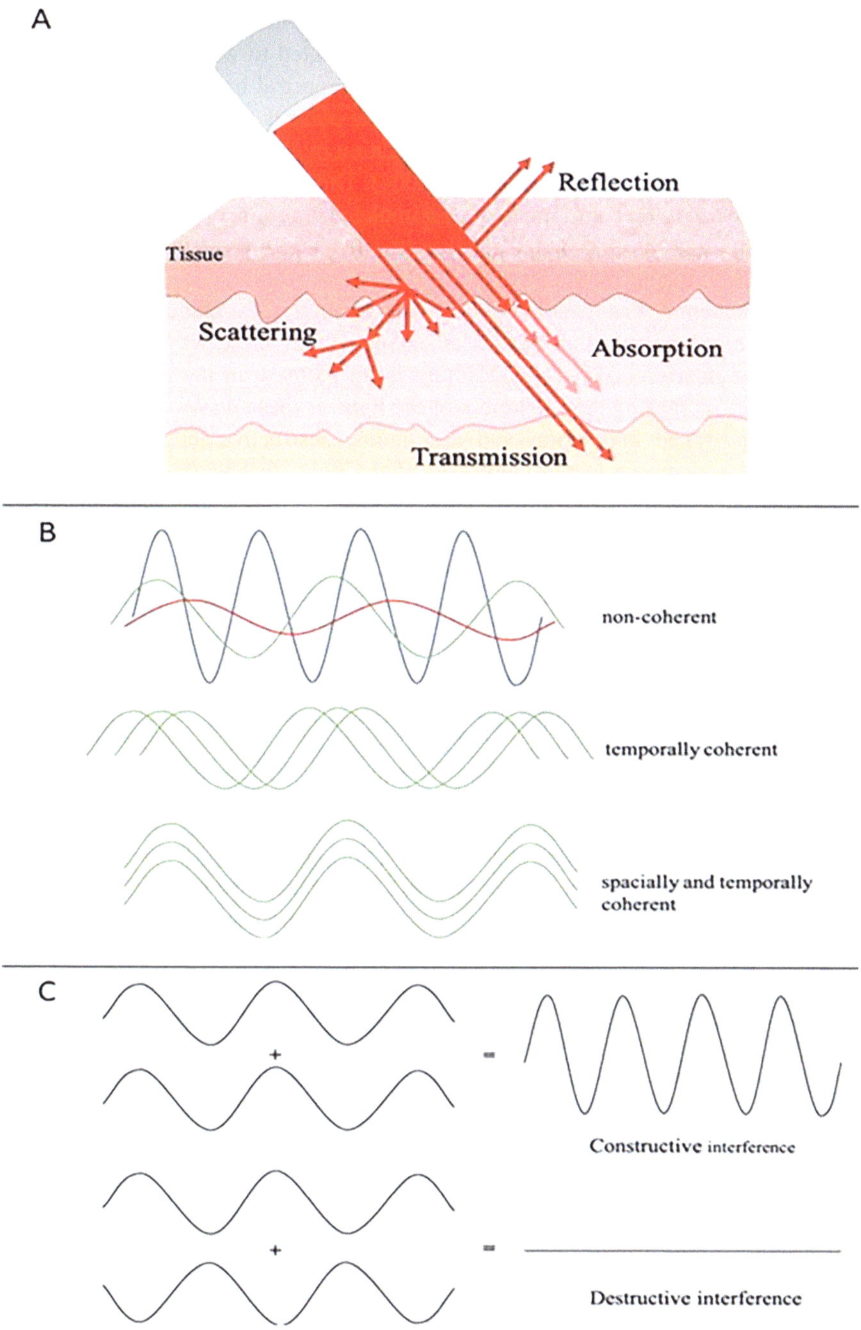

Fig. 12.1 (**a**) Scheme of the four main interactions of photons with the skin; (**b**) Scheme of temporal and spatial coherence; (**c**) Scheme of constructive and destructive interference

sources have is coherence. The concept of coherence involves two types: temporal and spatial (Fig. 12.1b). Temporal coherence is related to waves having the same wavelength or close wavelengths, that is, narrow spectral bandwidth. This type of light is called monochromatic, in which a single color is observed in its emission. The spatial coherence is related to the propagation in phase, where the maxima and minima of the waves are aligned (Prasad 2004). These characteristics are present in lasers (acronym of light amplification by stimulated emission of radiation) and some LEDs (light emitting diode), and for this reason, these are the main light sources used in optical diagnostic devices.

A phenomenon that can be understood through the coherence property is interference, where two light waves are combined. If these waves are in phase, constructive interference occurs, resulting in an increase in the amplitude of the light, that is, increasing the brightness of the light. If these waves are not in phase, that is, misaligned maximums and minimums, destructive interference occurs, resulting in a decrease in the amplitude of the light, that is, decreasing or even annulling the brightness of the light (Fig. 12.1c). The interference phenomenon is used in some optical devices such as FTIR spectroscopy and OCT (Brezinski 2013). Next, we will summarize some fundamental aspects related to the main optical technologies currently used for the diagnosis of dental hard tissues.

12.4 Fluorescence-Based Methods

Fluorescence is the radiation emitted by a molecule due to the absorption of electromagnetic radiation (Prasad 2004). For the occurrence of this phenomenon, the molecule must absorb a photon with higher energy (thus, with shorter wavelength) and emit another one with lower energy, thus having a longer wavelength. Therefore, a molecular excitation only occurs when photons of some specific wavelengths are absorbed by this molecule. For example, the excitation in the ultraviolet region causes the fluorescence emission ranging from violet to blue; blue excitation results in blue, green, or yellow emission; and red excitation results in near-infrared fluorescence.

For the diagnosis of dental hard tissue diseases, methods based on fluorescence have been available in clinical practice since the 1980s (Lussi et al. 2004). They include light-induced fluorescence (LIF), fluorescence imaging, and fluorescence spectroscopy alone or combined with imaging.

LIF quantification methods are based on the autofluorescence of dental hard tissues. In this way, the demineralization promotes the loss of this natural fluorescence, which can be quantified through the differentiation between signals from healthy and demineralized regions. The exact cause of dental fluorescence is still not fully understood and may result from the combination of the inorganic matrix with organic absorber molecules (fluorophores). Autofluorescence of carious dentin, for instance, occurs at various excitation and emission wavelengths and can be attributed to multiple potential fluorophores, including collagen and elastin, amino acids (tryptophan and tyrosine), porphyrin molecules (such as protoporphyrin IX, coenzymes such as NAD(P)H and flavin molecules), and plasma proteins (Kwak et al. 2001; Stordelder et al. 2004). Some examples of LIF devices are DIAGNOent and DIAGNOdent Pen.

A fluorescence imaging system is more complex because it comprises an objective lens, a scientific CCD camera, a computer, an illumination system to induce fluorescence, and a spectral selection element (Lins 2009). For fluorescence image processing, it is necessary to detect differences in intensity between a healthy and a demineralized region; based on this, processing is used to improve the captured images, for example, by means of optical filters. The QLF technique (quantitative light-induced fluorescence) is also a commercial device developed to detect and quantify caries lesions using a CCD camera with a filter that only allows the transmission of longer wavelengths than 540 nm (Amaechi 2009). In this way, it is possible to monitor demineralization lesions over time with higher precision. The QLF system uses an argon laser to excite the tissues and, due to its higher cost, other devices were further developed using LED sources for the same objective such as Soprolife.

The same principle can be applied to scientific instruments, but using LEDs as the excitation source. This approach is very interesting since it enables the construction of efficient and low-cost devices. In Fig. 12.2, a fluorescence imaging of a sound and a carious dentin sample is shown, revealing a decrease in the green intensity in carious dentin, which is characteristic of mineral loss and, thus, loss of fluorescence signal. In this example, a blue LED light ($\lambda = 405 \pm 30$ nm) was used for excitation, along with a broadband filter with a wavelength cutoff at 450 nm. Due to this fact, the fluorescence of dentin molecules is emitted in the green color (Ana et al. 2015).

Fluorescence spectroscopy consists of a quantitative and non-invasive technique in which a tissue is excited with monochromatic light with subsequent capture of the signal resulting from tissue interaction, by optical fibers (Amaechi 2009). A detector converts the optical into a digital signal, which is visualized in the form of spectra (wavelength versus signal strength). The spectra are differentiated between

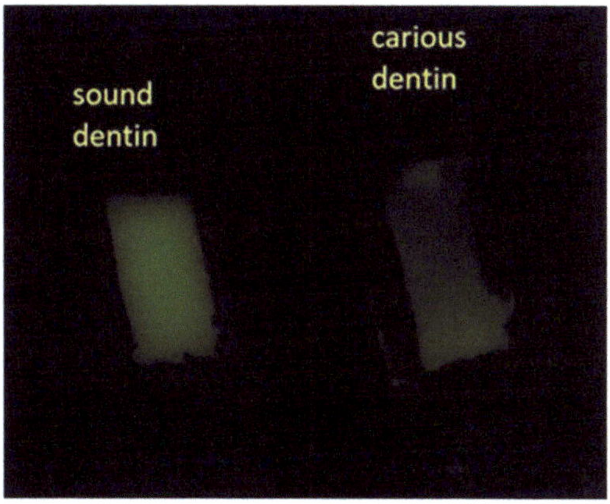

Fig. 12.2 Fluorescence imaging of a sound and a demineralized dentin sample

healthy and diseased tissue. For dental purposes, the technique allows the detection of mineral changes with high sensitivity and specificity. There are several excitation sources that can be used to distinguish tissues (Ribeiro Figueiredo et al. 2005; Zezell et al. 2007). It should be noted that the fluorescence emission peaks may vary depending on the transmission curve of the filters, the detector used, as well as the group of teeth studied (Buchalla 2005). The analysis of the obtained spectra is also important in the interpretation of the results and, although several methodologies are described in the literature, there is still no consensus (Ribeiro Figueiredo et al. 2005; Zezell et al. 2007). What is known is that the variation in spectral properties can provide important information about the origin of the lesion, its stage, and structure, thus allowing its monitoring.

The development of techniques that combine spectroscopy with fluorescence imaging provides an important advantage in the analysis of optical signals. These techniques provide simultaneous spatial and spectral information, enabling analysis of the two-dimensional distribution of intensities at the same time (Abdel Gawad et al. 2019). They represent a great advantage when compared to traditional spectroscopy techniques, which only allow the study of a very small area of tissue. These techniques are more sensible to different demineralization stages and allow the quantification of small biochemical changes.

Of all-optical techniques, those based on fluorescence are the most clinically used for diagnosing demineralization of dental hard tissues. Currently, the aid of machine learning techniques has improved its sensitivity, especially when used in association with hyperspectral imaging (Wang et al. 2023). Although it is not yet in clinical use, this is a very promising strategy for an increasingly early and assertive diagnosis.

12.5 Optical Coherence Tomography (OCT)

Optical coherence tomography is a non-invasive technique that can provide images in sections (as the X-ray tomography), presented for the first time by Huang et al. in 1991. The OCT device has a light source emitted in the near-infrared region (NIR) of the electromagnetic spectrum, allowing penetrations up to 3 mm and resolution up to 15 μm in biological tissues. These characteristics depend on the instrumentation of the equipment and the tissue to be imaged (Drexler and Fujimoto 2015).

The operation of OCT can be seen as an analog of ultrasonography. In the case of ultrasonography, the sound (mechanical waves) strikes the tissue, the intensities of the reflected waves are detected, and the time delay between the incident and reflected wave is calculated by an electronic system to obtain the depth related to each intensity. In OCT, light strikes the tissue, and the intensities of backscattered waves are detected, but as the speed of light is much greater than that of sound (in vacuum, 300,000 km/s of light against 0.3 km/s of the sound), measuring the time delay by a current electronic system is impossible (Brezinski 2013).

The operational principle of OCT is based on a Michelson interferometer, a device used to analyze the interference of two beams. In a simplified time-domain scanning system, the OCT operation begins with the light coming from the source being divided into two parts by a filter (beam splitter). One part is directed to a reference mirror (movable mirror) and the other part to the sample. Since the light source used in OCT is in the NIR region and, thus, within the biological window, it will be prevalently scattered. Part of this beam will return to the filter by backscattering, that is, 180° scattering. Simultaneously, the beam incident on the reference mirror will be reflected to the filter. The reflected and backscattered beams will recombine at the filter, resulting in an interference pattern. By positioning the mirror where the most intense interference pattern occurs (coherent beams), it is possible to identify the depth of the signal returned from the sample (Drexler and Fujimoto 2015).

OCT allows the following acquisitions: one-dimensional (1D), two-dimensional (2D), and three-dimensional (3D) imaging. In 1D acquisition, the profile of signal intensities is plotted along a single depth line, called A-scan. In 2D acquisition, multiple A-scans are combined, forming a B-scan image (B-scan), where an A-scan represents a column of the B-scan. In 3D acquisition, all B-scans of a region are grouped together forming a volumetric image (Kraus et al. 2012). These acquisition methods are not only important for visualizing biological tissue but are also useful in quantifying features that may be related to the pathophysiology of diseases.

Figure 12.3 shows images from an in vitro study carried out by our research group, and it is noted that the OCT images are different when performed in distinct regions of a chronic caries lesion. The characteristics observed in these images allow the differentiation between active and inactive lesions, as well as the identification of caries-affected dentin and infected dentin. Such diagnosis assists the clinician in determining the necessity of tissue removal during restorative procedures.

The use of OCT for early diagnosis of caries lesions has been suggested since the 1990s (Colston et al. 1998) and, for this purpose, distinct devices were used, and wavelengths from 780 to 1550 nm were successfully tested (Nakagawa et al. 2013; Maia et al. 2016; Pereira et al. 2018; Moraes et al. 2021; Zhu et al. 2023). Different analytical methodologies were also proposed to quantify demineralization in OCT images, which is essential to monitor the emergence, progression, or arrestment of caries lesions in both enamel and dentin. For erosion lesions, methods have been developed to quantify the depth of the lost surface (Pereira et al. 2018; Romero et al. 2021) and to assess the roughness and thickness of the remaining enamel in an erosive process. Recently, an analytical methodology has been proposed to quantify the activity of caries lesions, including secondary caries, by OCT (Chang et al. 2022).

12.6 Vibrational Spectroscopy

Vibrational spectroscopy devices have proven to be very efficient in the qualitative and semi-quantitative characterization of biological tissues. Because they are non-destructive techniques and require relatively simple sample preparation, they have

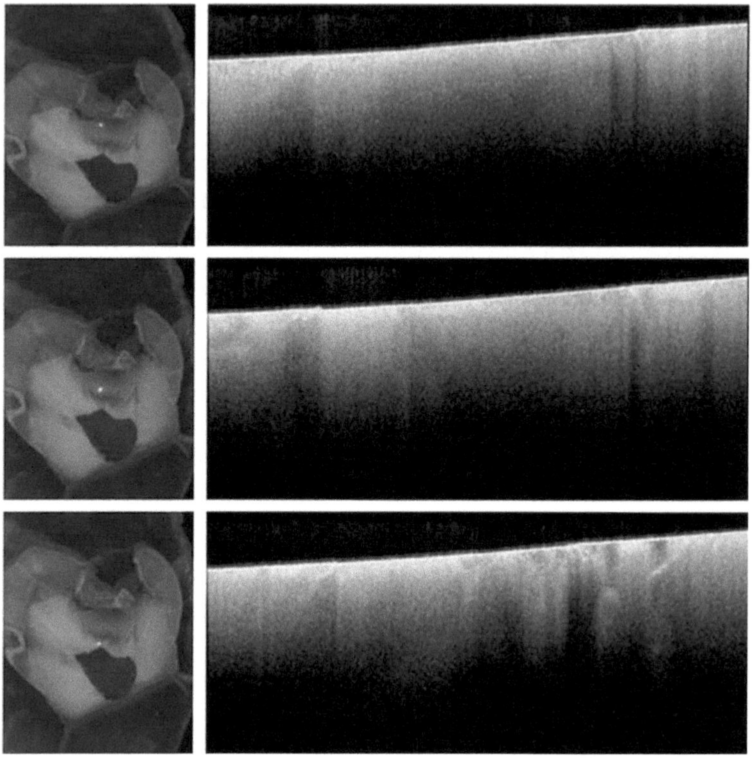

Fig. 12.3 Two-dimensional OCT images (B-scans) obtained in different regions (white dots) of a carious lesion in a molar tooth ex vivo. Different aspects of the in-depth images are noticed, revealing characteristics of the lesion that are not seen in an evaluation from the surface

great potential for clinical diagnosis and their use in mineralized tissues has been growing over the years. In these, the Fourier Transform Infrared (FTIR) and Raman spectroscopies have already proved to be efficient in obtaining information on the mineral content, crystallinity, and maturity of collagen, indicating the potential of the technique for differentiating dental lesions (Ana et al. 2014; Zezell et al. 2015; Pereira et al. 2018; Dias-Moraes et al. 2021).

In infrared absorption spectroscopy, a tissue is exposed to radiation from a continuous light, typically in the spectral region of 2.5–25 μm or 4000–400 cm^{-1}. The transmitted or reflected radiation is compared to the reference (transmitted or reflected radiation in the absence of a sample) and the spectrometer records the result as an absorption band. The absorption of these low-energy photons only increases the amplitude of the molecules' vibrations of the tissue, whose signal is detected by the equipment (Skoog et al. 2007). The most sophisticated measurement technique currently involves the use of a Michelson interferometer and the subsequent Fourier transform of the obtained interferogram (FTIR).

Raman spectroscopy, on the other hand, uses a laser of a specific wavelength to interact with the tissue being evaluated; the scattering signal resulting from this interaction is detected and converted into a spectral signal. The generated scattering is also characteristic of each molecule present in the tissue; in this way, it is possible to detect it (Skoog et al. 2007).

Considering the interference of water in the signal obtained by FTIR, Raman spectroscopy has emerged as a promising technique for the molecular diagnosis of soft tissue pathologies (Beeram et al. 2023). Also, Raman equipment can be easily miniaturized for clinical use, as the scattering signal can be obtained with collimated light sources of different wavelengths. However, the Raman scattering signal is inherently weak; thus, FTIR spectroscopy can be more advantageous, mainly for the analysis of hard tissues. Therefore, it can be stated that vibrational spectroscopy techniques are complementary (Gieroba et al. 2020), and both can provide precise and highly sensitive information about the analyzed tissue.

Concerning the diagnosis of demineralization, both FTIR and Raman spectroscopy provide information about the proportion of mineral content and organic matrix, as well as parameters such as mineral density, collagen denaturation, water content increase, and hydroxyapatite crystallinity. These techniques allow assessing the progression of lesions, monitoring remineralization process, and evaluating other hard tissue pathologies, such as fluorosis (Seredin et al. 2020). Furthermore, both FTIR and Raman spectroscopy are also effective in assessing the impact of different treatments on enamel and dentin, as well as highly effective in detecting the occurrence of secondary caries.

Figure 12.4 shows an example of Raman bands detected in an ex vivo molar decayed tooth. It is noticed that the spectrum of carious dentin presents important differences when compared to the spectrum of healthy dentin, and the quantification of such changes is important to detect the presence and the activity of a caries lesion.

12.7 Transillumination

In general, transillumination is understood as the phenomenon of the transmission of radiation with predefined parameters by an obstacle; in this way, the transmitted light becomes a carrier of information about the object. For diagnostic of oral lesions, it is frequently used near-infrared radiation (NIR) to interact with the tissues considering the weak scattering and absorption in this region, which allows imaging structures up to 2.5 mm depth (Staninec et al. 2011).

It was reported that optical transillumination can be similar to other projection imaging modalities like conventional X-rays, however, the contrast image arises from changes in the dispersion of tissues as opposed to variations in tissue density (Jones et al. 2003). Furthermore, it was stated that, from this point of view, this method may be more sensitive than X-rays for the detection of early caries lesions. NIR transillumination is cited as having considerable potential as a routine screening

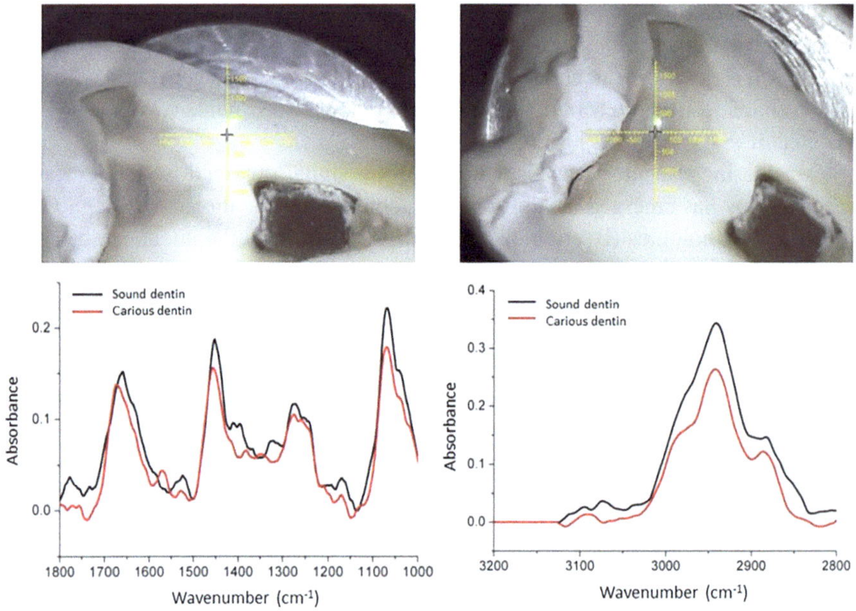

Fig. 12.4 Raman spectra of carious and sound dentin in ex vivo molar teeth

tool for diagnosing carious lesions throughout the dentition, since it does not involve ionizing radiation (Bühler et al. 2005). The main limitation of this method is the high cost of the imaging technology originating from InGaAs sensors; however, it is probable that, in the near future, the NIR technology will be available at a lower cost or that alternative systems will be developed operating at 830 or 1550 nm, balancing cost, sensitivity, and performance.

Some NIR measurements can be used to show the structure, depth, and severity of natural caries lesions on the occlusal surface and have the potential for nondestructive assessment of the severity of caries lesions (Staninec et al. 2011). The reliability of transillumination in the evaluation of interproximal carious lesions was also demonstrated. However, depending on the instrumentation used, transillumination may have less accuracy when used for visual assessment, followed by laser fluorescence and digital radiography techniques (Kühnisch et al. 2016; Berg et al. 2018). With the development of new higher resolution infrared cameras, it was demonstrated a significant improvement in the diagnosis of interproximal enamel caries lesions, with high sensitivity and overall accuracy in comparison with digital radiography (Ortiz et al. 2020; Edrees et al. 2023). These findings demonstrate the viability of a diagnostic system that allows the monitoring of lesions without worrying about the harm that can be caused by the use of ionizing radiation.

Figure 12.5 shows a scheme of a transillumination system when tested in a molar tooth. It can be noticed that the transmission of NIR radiation through the

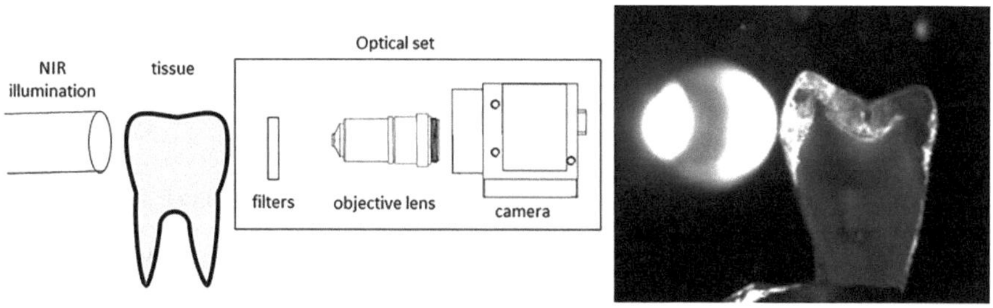

Fig. 12.5 Scheme of the basic instrumentation of a transillumination system applied for scientific purposes

enamel shows characteristics in the enamel image compatible with the presence of demineralization.

12.8 Conclusion

Optical methods have emerged as promising techniques for the diagnosis of oral hard tissue diseases. The use of advanced optical instruments, such as lasers and LEDs, coupled with appropriate filters and detecting systems, has enabled the development of efficient dental diagnosis devices. Moreover, optical methods use nonionizing radiation, offering non-invasive and painless diagnostic procedures that are well-received by both patients and practitioners. Currently, fluorescence-based techniques and transillumination have been integrated into dental practice for the detection of demineralized tissue. Ongoing technological advancements hold a great potential for further improving the sensitivity and diagnostic capabilities of these techniques. Additionally, Fourier transform infrared spectroscopy, Raman spectroscopy, and optical coherence tomography can provide valuable information regarding mineral density, crystallinity, organic composition, depth, and extent of demineralization. These techniques demonstrate great potential for clinical implementation in the diagnosis of demineralized lesions in oral hard tissue. As research in this field continues to advance, optical methods are expected to assume an increasingly important role in providing accurate and specific characteristics of various oral lesions. This information will enable dental professionals to make more informed decisions regarding appropriate treatment strategies and preventive measures for optimal oral health.

Funding Declaration/Acknowledgments The authors would like to thank to FAPESP (2017/21887-4), CAPES (PROCAD-CAPES 88881.068505/2014-01) and CNPq (INCT # 465763/2014-6 – Photonics National Institute for Science and Technology – INFO – and 440228/2021-2 National System of Photonics Laboratories – Sisfoton-MCTI).

References

Amaechi BT (2009) Emerging technologies for diagnosis of dental caries: the road so far. J Appl Phys 105:10247-110247-9.
Ana PA, Kauffman CMF, Bachmann L, Soares LES, Martin AA, Gomes ASL, Zezell DM (2014) FT-Raman spectroscopic analysis of Nd:YAG and Er,Cr:YSGG laser irradiated enamel for preventive purposes. Laser Physics 24:035603.
Abdel Gawad AL, El-Sharkawy Y, Ayoub HS, El-Sherif AF, Hassan MF (2019, March) Classification of dental diseases using hyperspectral imaging and laser induced fluorescence. Photodiagnosis Photodyn Ther 25:128–135. https://doi.org/10.1016/j.pdpdt.2018.11.017
Ana PA, Brito AMM, Zezell DM, Lins ECCC (2015) Characterization of caries progression on dentin after irradiation with nd:YAG laser by FTIR spectroscopy and fluorescence imaging, Biophotonics South America [Preprint]. https://doi.org/10.1117/12.2181080.
Beeram R, Vepa KR, Soma VR (2023) Recent Trends in SERS-Based Plasmonic Sensors for Disease Diagnostics, Biomolecules Detection, and Machine Learning Techniques. Biosensors (Basel) 13:328.
Berg SC, Stahl JM, Lien W, Slack CM, Vandewalle KS (2018) A clinical study comparing digital radiography and near-infrared transillumination in caries detection. J Esthet Restor Dent 30:39–44.
Brezinski ME (2013) Optical coherence tomography: Principles and applications. Amsterdam: Elsevier.
Buchalla W (2005) Comparative fluorescence spectroscopy shows differences in noncavitated enamel lesions. Caries Res 39:150–156.
Bühler CM, Ngaotheppitak P, Fried D (2005) Imaging of occlusal dental caries (decay) with near-IR light at 1310 nm. Opt Express 13: 573–82.
Cara ACB, Zezell DM, Ana PA, Maldonado EP, Freitas AZ (2014) Evaluation of two quantitative analysis methods of optical coherence tomography for detection of enamel demineralization and comparison with microhardness. Lasers Surg Med 46:666–671.
Chang NN, Dillas T, Fried D (2022) Assessing lesion activity of secondary lesions on extracted teeth by thermal dehydration measurement and optical coherence tomography. Proc SPIE Int Soc Opt Eng 11942:1194204.
Chen Z, Cheng Q, Wang L, Mo Y, Li K, Mo J (2023) Optical coherence tomography for in vivo longitudinal monitoring of artificial dermal scaffold. Lasers Surg Med 55:316–326.
Colston BW, Sathyam US, DaSilva LB, Everett MJ, Stroeve P, Otis LL (1998) Dental OCT. Opt Express 3:230.
Drexler W and Fujimoto JG (2015) Optical Coherence Tomography Technology and applications. Cham: Springer International Publishing.
Edrees A, Hassanein O, Shaalan O, Yassen A (2023) Accuracy of high definition near infrared transillumination camera in detection of hidden proximal caries. J Clin Exp Dent 15:e1–e8.
Fejerskov O and Nyvad B (2015) Dental caries: The disease and its clinical management. Chichester: Wiley/Blackwell.
Gieroba B, Krysa M, Wojtowicz K, Wiater A, Pleszczyńska M, Tomczyk M, Sroka-Bartnicka A (2020) The FT-IR and Raman Spectroscopies as Tools for Biofilm Characterization Created by Cariogenic Streptococci. Int J Mol Sci 21:3811.
Horowitz AM (1996) Introduction to the symposium on minimal intervention techniques for caries. J Public Health Dent 56:133–4.
Huang D, Swanson EA, Lin CP, Schuman JS, Stinson WG, Chang W, Hee MR, Flotte T, Gregory K, Puliafito CA, et al. (1991) Optical coherence tomography. Science 254:1178–1181.
Jacques SL (2013) Optical properties of biological tissues: a review. Phys Med Biol 58:R37–R61.
Jones RS, Huynh G, Jones G, Fried D (2003) Near-infrared transillumination at 1310-nm for the imaging of early dental decay. Opt Express 11:2259–2265.
Joshi M, Joshi N, Kathariya R, Angadi P, Raikar S (2016) Techniques to Evaluate Dental Erosion: A Systematic Review of Literature. J Clin Diagnostic Res 10: ZE01–ZE07.

Kanzow P, Wegehaupt FJ, Attin T, Wiegand A (2016) Etiology and pathogenesis of dental erosion. Quintessence Int 47:275–8.

Kim DH, Kim SW, Hwang SH (2023) Efficacy of optical coherence tomography in the diagnosing of oral cancerous lesions: systematic review and meta-analysis. Head Neck 45:473–481.

Ko AC, Choo-Smith LP, Hewko M, Leonardi L, Sowa MG, Dong CC, Williams P, Cleghorn B (2005) Ex vivo detection and characterization of early dental caries by optical coherence tomography and Raman spectroscopy. J Biomed Opt 10:031118.

Konovalov A, Gadzhiagaev V, Grebenev F, Stavtsev D, Piavchenko G, Gerasimenko A, Telyshev D, Meglinski I, Eliava S (2023) Laser Speckle Contrast Imaging in Neurosurgery: A Systematic Review. World Neurosurg 171:35–40.

Kralj L, Lenasi H (2023) Wavelet analysis of laser Doppler microcirculatory signals: Current applications and limitations. Front Physiol 13:1076445.

Kraus MF, Potsaid B, Mayer MA, Bock R, Baumann B, Liu JJ, Hornegger J, Fujimoto JG (2012) Motion correction in optical coherence tomography volumes on a per a-scan basis using orthogonal scan patterns. Biomed Opt Express 3:1182–1198.

Kühnisch J, Söchtig F, Pitchika V, Laubender R, Neuhaus KW, Lussi A, Hickel R (2016) In vivo validation of near-infrared light transillumination for interproximal dentin caries detection. Clin Oral Invest 20:821–829.

Kwak ES, Kang TJ, Vanden Bout DA (2001) Fluorescence lifetime imaging with near-field scanning optical microscopy. Anal Chem 15: 3257–62.

Lins ECCC (2009) Espectroscopia da Fluorescência Na Citricultura. Available at: https://teses.usp.br/teses/disponiveis/76/76132/tde-09102009-095928/publico/EmeryCleitonCabralCorreiraLinsDOcorrigida.pdf (Accessed: 12 May 2023).

Lussi A, Hibst R, Paulus R (2004) Diagnodent: an optical method for caries detection. J Dent Res 83:80–3.

Machiulskiene V, Campus G, Carvalho JC, Dige I, Ekstrand KR, Jablonski-Momeni A, Maltz M, Manton DJ, Martignon S, Martinez-Mier EA, Pitts NB, Schulte AG, Splieth CH, Tenuta LMA, Ferreira Zandona A, Nyvad B (2020) Terminology of Dental Caries and Dental Caries Management: Consensus Report of a Workshop Organized by ORCA and Cariology Research Group of IADR. Caries Res 54:7–14.

Maia AMA, de Freitas AZ, de L. Campello S, Gomes ASL, Karlsson L (2016) Evaluation of dental enamel caries assessment using Quantitative Light Induced Fluorescence and Optical Coherence Tomography. J Biophotonics 9: 596–602.

Meza Ramirez CA, Greenop M, Almoshawah YA, Martin Hirsch PL, Rehman IU (2023) Advancing cervical cancer diagnosis and screening with spectroscopy and machine learning. Expert Rev Mol Diagn 23:375–390.

Moraes MCD, Freitas AZ, Castro PAA, Pereira DL, Ana PA, Zezell DM (2021) Assessment of the preventive effects of Nd:YAG laser associated with fluoride on enamel caries using optical coherence tomography and FTIR spectroscopy. PLoS One 16:e0254217.

Nakagawa H, Sadr A, Shimada Y, Tagami J, Sumi Y (2013) Validation of swept source optical coherence tomography (SS-OCT) for the diagnosis of smooth surface caries in vitro. J Dent 41:80–89.

Ortiz MIG, Alencar CM, De Paula BLF, Magno MB, Maia LC, Silva CM (2020) Accuracy of near-infrared light transillumination (NILT) compared to bitewing radiograph for detection of interproximal caries in the permanent dentition: A systematic review and meta-analysis. J Dent 98:103351.

Pereira DL, Freitas AZ, Bachmann L, Benetti C, Zezell DM, Ana PA (2018) Variation on molecular structure, crystallinity, and optical properties of dentin due to Nd:YAG laser and fluoride aimed at tooth erosion prevention. Int. J. Mol. Sci. 19: 1–14.

Prasad PN (2004) Introduction to Biophotonics. Hoboken: Wiley.

Ribeiro Figueiredo AC, Kurachi C, Bagnato VS (2005) Comparison of fluorescence detection of carious dentin for different excitation wavelengths. Caries Res 39:393–396.

Romero MJRH, Bezerra SJC, Fried D, Yang V, Lippert F, Eckert GJ, Zero DT, Hara AT (2021) Cross-polarization optical coherence tomographic assessment of in situ simulated erosive tooth wear. J Biophotonics 14:e202100090.

Schiavon G, Capone G, Frize M, Zaffagnini S, Candrian C, Filardo G (2021) Infrared Thermography for the Evaluation of Inflammatory and Degenerative Joint Diseases: A Systematic Review. Cartilage 13:1790S–1801S.

Seredin P, Goloshchapov D, Ippolitov Y, Vongsvivut J (2020) Development of a new approach to diagnosis of the early fluorosis forms by means of FTIR and Raman microspectroscopy. Sci Rep 2020 10:20891.

Schlueter N, Amaechi BT, Bartlett D, Buzalaf MAR, Carvalho TS, Ganss C, Hara AT, Huysmans MDNJM, Lussi A, Moazzez R, Vieira AR, West NX, Wiegand A, Young A, Lippert F (2020) Terminology of Erosive Tooth Wear: Consensus Report of a Workshop Organized by the ORCA and the Cariology Research Group of the IADR. Caries Res. 54(1):2–6. https://doi.org/10.1159/000503308

Silverstone LM, Hicks MJ, Featherstone JD (1988) Dynamic factors affecting lesion initiation and progression in human dental enamel. The dynamic nature of dental caries. Quint Int 19: 683–710.

Skoog DA, Crouch SR and Holler FJ (2007) Principles of instrumental analysis. London: BrooksCole.

Staninec M, Douglas SM, Darling CL, Chan K, Kang H, Lee RC, Fried D (2011) Non-destructive Clinical Assessment of Occlusal Caries Lesions Using Near-IR Imaging Methods. Lasers Surg Med 43:951–959.

Stordelder A, Buijs JB, Bulthuis J, Gooijer C, Zwan G (2004) Fast-gated intensified charge-coupled device camera to record time-resolved fluorescence spectra of tryptophan. Appl Spectrosc 58:705–710.

Wang C, Zhang R, Wei X, Wang L, Xu W, Yao Q (2023) Machine learning-based automatic identification and diagnosis of dental caries and calculus using hyperspectral fluorescence imaging. Photodiagnosis Photodyn Ther 41:103217.

Wijesinghe RE, Cho NH, Park K, Jeon A, Kim J (2016) Bio-photonic detection and quantitative evaluation method for the progression of dental caries using optical frequency-domain imaging method. Sensors 16:2076.

Zezell DM, Ribeiro AC, Bachmann L, Gomes AS, Rousseau C, Girkin J (2007) Characterization of natural carious lesions by fluorescence spectroscopy at 405-nm excitation wavelength. J Biomed Opt 12:064013.

Zezell DM, Benetti C, Veloso MN, Castro PAA, Ana PA (2015) FTIR Spectroscopy Revealing the Effects of Laser and Ionizing Radiation on Biological Hard Tissues. J Braz Chem Soc 26: 2571–2582.

Zhu Y, Kim M, Curtis D, Wang J, Le O, Fried D (2023) Active Surveillance of Root Caries in Vivo with CP-OCT. Diagnostics (Basel) 13:465.

Chapter 13
3D Electrical Mapping of the Heart

João Loures Salinet, Ilija Uzelac, Jimena Gabriela Siles Paredes, Vinicius Silva, Ítalo Sandoval Ramos de Oliveira, Tainan Cerqueira, Idágene Aparecida Cestari, and Marcelo Mazzeto

Abstract Cardiovascular diseases are the main causes of death worldwide and among them, heart diseases are particularly important with an incidence rate of 4% of the Brazil population with a cost of 2% of GDP. Many of these patients can present abnormalities related to the heart electrical conduction system, resulting in arrhythmic episodes. Over the last decade, new technologies have been developed

J. L. Salinet (✉)
Graduate Program in Biomedical Engineering, Federal University of ABC – UFABC, São Bernardo do Campo, SP, Brazil

HeartLab, Center for Engineering, Modeling and Applied Social Sciences (CECS), Federal University of ABC – UFABC, São Bernardo do Campo, SP, Brazil

Graduate Program in Biotechnoscience, Federal University of ABC – UFABC, São Bernardo do Campo, SP, Brazil
e-mail: joao.salinet@ufabc.edu.br

I. Uzelac
Georgia Institute of Technology, School of Physics, Atlanta, GA, USA

J. G. S. Paredes
HeartLab, Center for Engineering, Modeling and Applied Social Sciences (CECS), Federal University of ABC – UFABC, São Bernardo do Campo, SP, Brazil

Graduate Program in Biotechnoscience, Federal University of ABC – UFABC, São Bernardo do Campo, SP, Brazil

V. Silva · T. Cerqueira
HeartLab, Center for Engineering, Modeling and Applied Social Sciences (CECS), Federal University of ABC – UFABC, São Bernardo do Campo, SP, Brazil

Í. S. R. de Oliveira
Graduate Program in Biomedical Engineering, Federal University of ABC – UFABC, São Bernardo do Campo, SP, Brazil

HeartLab, Center for Engineering, Modeling and Applied Social Sciences (CECS), Federal University of ABC – UFABC, São Bernardo do Campo, SP, Brazil

I. A. Cestari · M. Mazzeto
Bioengineering Department, Heart Institute (InCor), Hospital das Clínicas da Faculdade de Medicina, Universidade de São Paulo, São Paulo, Brazil

C. B. Lombello, P. A. da Ana (eds.), *Current Trends in Biomedical Engineering*,
https://doi.org/10.1007/978-3-031-38743-2_13

to improve the diagnosis and classification of arrhythmic mechanisms, in order to provide more effective treatments, and consequently, increase patients' quality of life. The heart's electrical activity can be acquired by electrical and optical mapping (functional imaging). The electrical mapping typically comprises a set of electrodes in contact with the heart (endocardium or epicardium), termed invasive, or throughout electrodes spread in the body torso, for noninvasive measurement. These two distinct techniques are commonly referred to as body surface mapping or electrocardiographic imaging. Optical mapping uses optical cameras to acquire changes in the emitted fluorescence from the bound to the cardiac cell membrane, correlated to the cell transmembrane voltage change. These methods differ from each other in how electrical activity is acquired. This chapter presents the principles and advantages of each method and its contributions to the clinical outcome.

Keywords Electrophysiology · Cardiac arrhythmia · Diagnosis · Treatment · Mapping systems · Biomedical engineering

13.1 Introduction

The heart can be characterized as an electrical-mechanical pump responsible for transporting essential nutrients, metabolites, O_2, CO_2, hormones, and antibodies to various tissues and organs throughout the body (Tortora and Derrickson 2006). It stands out as one of the main organs of the body responsible for the maintenance of vital functions in all cells. Thus, any changes in the heart's ability to eject blood effectively can contribute to dysfunction and irreversible damage to organs, and potentially lead to death.

The heart is composed of four chambers, the upper atrial and the lower ventricles. The synchronous contraction starts in the right atrium through the electrical depolarization of the pacemaker cells in the sinoatrial node. Then it spreads to the other regions of the atria and reaching to the atrioventricular node. The depolarization of the ventricles starts from the right and left branches of the Purkinje system, toward their base, then spreads to the remaining areas of the ventricle's Purkinje fibers (Tortora and Derrickson 2006). The propagation of these action potentials (APs) to the rest of the heart involves complex intracellular mechanisms that rely on exchange of ions and intracellular molecules through various ion channels and transporters on the cellular level (Nerbonne 2004). Differences between APs may be a consequence of changes in the size, shape, and structure of heart cells, contributing to changes in cell conductivity and reshaping the AP (Nerbonne 2004).

Various factors such as genetics, aging, poor dietary habits, a sedentary lifestyle, and excessive alcohol consumption can lead to changes in the anatomy and physiology of the heart, contributing to the development of heart diseases (Winkel 2010), which might be associated with cardiovascular diseases, the leading causes of death in the world, being the major structural ones associated with cardiac arrhythmia (Bonny et al. 2020). Moreover, ischemic heart disease is the leading cause of morbidity and mortality, with the majority of deaths due to arrhythmias, ranging from brady to tachyarrhythmias in the atria or ventricles and atrioventricular block (Shah et al. 2021). The identification of arrhythmic regions in the heart is typically accomplished through electrophysiological examinations, which can involve invasive procedures and carry some risks to the patient. These procedures can also be time-consuming and costly, which may be a barrier for some patients or for the general population.

Atrial fibrillation (AF) and flutter prevalence is 50% in heart failure and rheumatic heart diseases (Shah et al. 2021). AF is the most prevalent cardiac arrhythmia encountered in clinical practice affecting 1–2% of the general population (Bollmann and Lombardi 2006). This condition leads to uncoordinated (i.e., irregular) and rapid electrical activation of the atria and reduces the mechanical ability of the atria to pump blood effectively. During AF, the ventricles beat irregularly and rapidly if the conduction is intact. The symptoms of AF include palpitations, tiredness, shortness of breath, dizziness, and chest pain. Since the mechanical pumping ability of the atria is compromised, the resulting pooling of blood in the atria increases the long-term risk of ischemic stroke by fivefold (Connolly 2011).

AF is a worldwide public health problem imposing huge costs on healthcare systems, accounting for 33% of all hospitalizations for arrhythmia (Fuster et al. 2006). This arrhythmia is associated with high morbidity and mortality and has become a chronic noninfectious cardiovascular epidemic, with a significant burden on healthcare resources. Antiarrhythmic (AA) drugs are commonly used to treat AF or to control the atria rhythm or ventricles rate. Rhythm control is the option of choice but has low efficacy and side effects on patient health, thereafter clinicians commonly choose the ventricles rate control, especially in the elderly population.

It is thought that the limited efficacy of AA is due to various factors, such as tissue fibrosis caused by cellular apoptosis and genetic disorders that affect ion channels of the cardiac cell's membrane, which reduce the atrial refractory period. Radiofrequency (RF) ablation is a more effective method than AA drugs for rhythm control of AF patients (Nault et al. 2010). During RF ablation procedure, a catheter containing one or more electrodes is inserted into the heart, via the femoral vein, and RF cauterization is performed on the endocardial areas identified as responsible for AF, in order to disrupt the mechanisms leading to AF. However, it's important to note that the results of the ablation procedure can vary, particularly in patients with long-standing atrial fibrillation and cardiac remodeling. In some cases, repeated ablation procedures may be necessary if AF recurs in the short term. The long-term results are not always encouraging. Moreover, one of the main current challenges in the clinic is the determination by the electrophysiologist of the endocardial target areas, avoiding cauterization of areas not related by triggering and AF maintenance, giving maximum efficacy and safety.

To improve the AF treatment, it is necessary first to fully understand the mechanisms of AF. Over the years, multiple mechanisms have been proposed to explain the initiation and maintenance of AF, based on experimental, clinical, and numerical simulation studies. These mechanisms include multiple wavelets, focal sources, mother rotor fibrillation, multiple stable or meandering rotors, double layer hypothesis, and micro-anatomic intramural reentry, among others. Although each mechanism can initiate and maintain AF independently, it is believed that their occurrence varies individually among patients, with multiple mechanisms coexisting simultaneously, or interspersed with each other, favoring the complex heart rhythm associated with AF (Nattel and Dobrev 2017).

For the past few decades, several invasive and noninvasive systems have been developed, to aid in improving our current knowledge and success of arrhythmias treatments. These systems allow the recreation of a three-dimensional (3D) geometry of the heart chamber(s) or torso and based on the electrical potentials recorded using high spatial density of electrodes, these potentials are then overlaid onto the 3D geometry shell and color-coded to generate 3D mappings. Clinicians can visualize and manipulate the maps to examine potential values, as well as other characteristics, such as activation frequency, wavefront time propagation, conduction velocity, and phase. The tools are based on the representation of the intrinsic variables as color maps on an anatomically accurate 3D representation of the atrium.

In certain cases, the interpretation of signals acquired by the above electroanatomic mapping systems and its respective 3D maps during AF is complex and uncertain (Allessie and Groot 2014; Narayan and Jalife 2014; Zaman et al. 2015), and the location of arrhythmogenic sources can pose a challenge for electrophysiologists. Mechanisms elucidation and improvements from current knowledge of AF have been achieved through the advent of high spatial resolution obtained by optical mapping (Berenfeld et al. 2011), considered the gold standard in the mapping of cardiac electrical activity. In these systems, voltage-sensitive dyes alter their emission spectrum according to changes in the transmembrane voltage of the myocardial cells. Thus, changes in the amount of fluorescent light emission of the cells are proportional to the changes in transmembrane voltage, allowing noninvasive measurements of the electrical activity in a high spatiotemporal resolution, contributing to a reliable characterization of the complex patterns of AF (Berenfeld et al. 2011). In vivo and in vitro studies allowed to identify and study the mechanisms of AF described, which were later also observed in patients.

Figure 13.1 presents the main aid systems utilized in clinic and research in the diagnosis of triggering and maintenance mechanisms for cardiac arrhythmias.

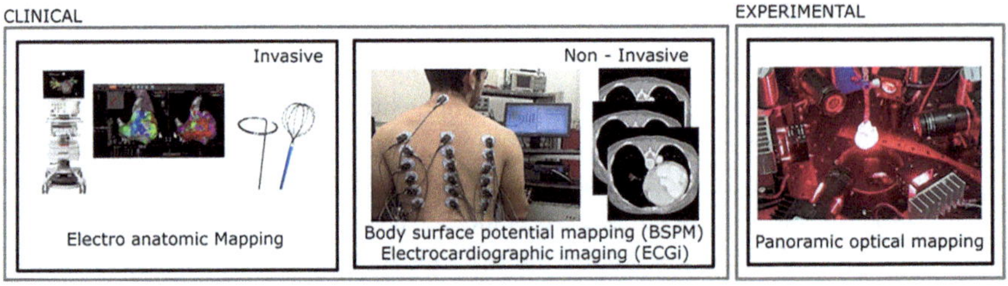

Fig. 13.1 Methods to map the cardiac electrical activity. Clinical: 3D invasive (electroanatomic mapping) and noninvasive (BSPM and ECGi) mapping systems. Experimental: panoramic optical mapping

Measuring and modeling the genesis and propagation of the electrical activity in the heart is important to improve our understanding and treatments of heart arrhythmias.

13.2 Electroanatomic Mapping

Since the advent of RF ablation in the 1980s, several techniques and new technologies have been incorporated to advance the diagnosis of AF mechanisms, improve the efficacy of ablation, and minimize its complications. Auxiliary imaging methods to aid in the identification of AF mechanisms and areas to be ablated, such as 3D electroanatomic mapping, are considered the standard treatment worldwide. The electroanatomic mapping systems allow acquisition of electrograms providing visualization of the spatiotemporal propagation of the cardiac cells AP. Signals can be captured using several types of catheters localized inside the patient's heart chamber and in connection with the electroanatomic mapping system. The catheters differ in terms of number of electrodes, arrangement, inter-electrodes spacing, and coverage. Typical examples are the Afocuss II, Lasso, and PentaRay. They are also more global electrode arrangements to cover simultaneously a wider area of the heart such the termed high-density baskets, such as the HD grid, the Intellamap Orion 64 electrodes, and Constellation basket catheter (Coveney et al. 2022).

Cardiac electrophysiology mapping is used to create a 3D reconstruction contributing significantly to the reduction of fluoroscopy time and radiation exposure to avoid stochastic or deterministic radiation risks to both patients and the entire electrophysiology laboratory staff (Borlich and Sommer 2019). At the mapped points, electrical information is recorded and can be displayed in various ways, such as color-coded activation mapping or displaying unipolar/bipolar electrograms. Advanced modules allow clinicians to infer detailed analysis of the atria substrate. The CARTO (Biosense-Webster), with the coherent mapping analysis, can globally map the propagation of the electrical activity and also localizes areas of slow conduction. Recently, EnSite (Abbott) visualizes the direction of the wavefront propagation, allowing the identification of collisions and conduction blocks. The Lumipoint (Rhytmia, Boston Scientific) also maps the atria globally, highlighting areas with fractionated signals and slow conduction. Acute Medical AcQMap allows to localize delayed activation areas with high resolution (Coveney et al. 2022).

Many other mapping systems are available in the market, with functions that move forward the characterization of the patient's cardiac activity patterns and tissue substrate, to allow identification of the arrhythmogenic areas that maintain the arrhythmia, and target of ablation, including systems from the following companies: Medtronic, MicroPort Scientific Corporation, EP Solutions SA, Acutus Medical, Philips, Biotronik; Catheter Precision, CoreMap, BioSig Technologies, APN Health, Kardium, and EPmap-System. The objective of these systems is to visually guide the operator in understanding the anatomical complexity of the atria and provide access to intracardiac electrograms. These systems are considered invasive, because intracavitary catheters with multiple electrodes are introduced into the heart

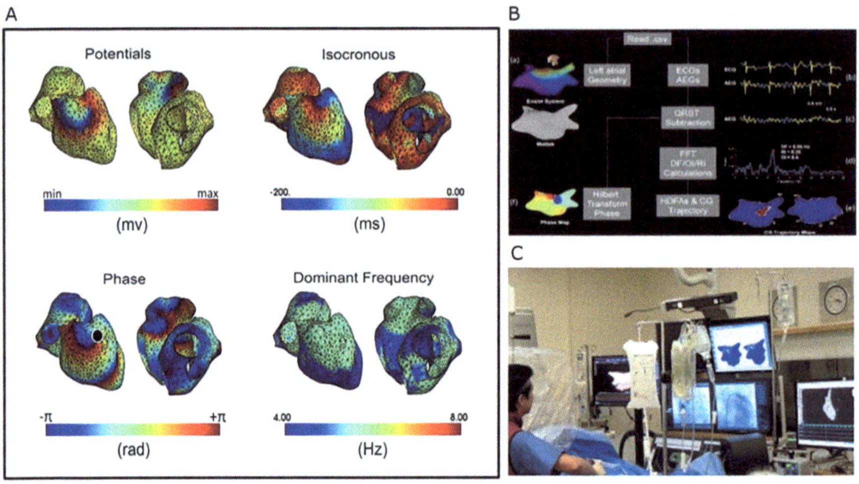

Fig. 13.2 (**a**) The main maps used by basic and clinical research for the study of AF: isopotential, isochronous, phase, and dominant frequency; (**b**) the signal processing methods applied in the USURP-GUI Platform; and (**c**) the USURP-GUI Platform being used prospectively in clinic to guide catheter ablation in a patient with persistent AF

chambers to map the electrical activity of the endocardium, with the acquisition of electrograms in real time for morphological and temporal analysis of its activation and repolarization phases, as well as the projection of its electrophysiological characteristics on 3D maps of the atria. The main maps used by basic and clinical research for the study of AF and its mechanisms through the methods of capturing cardiac electrical activity described above are isopotential, isochronous, phase, and frequency maps (Fig. 13.2a).

Isopotential maps are generated from the voltage captured by the electrodes at a given time, grouped into voltage ranges, and represented them in the 3D geometry of the torso or heart using a color-coded voltage range. The analyses of these maps involve the evaluation of the areas with maxima and minima voltage values, how they propagate, and distribution in consecutive time frames during the cardiac cycle and/or ongoing AF.

Isochronous maps are generated from the selection of a certain time interval during AF. In the segment chosen by the operator, the moment in time when the electrical activation of the myocardium occurs for each electrode is identified. The activation time can be obtained by identifying the maximal positive derivative of the signal. The electrodes with equal activation time values are connected to each other forming the isochronous line maps at equidistant intervals. All the generated lines are represented simultaneously, with a color code indicating the instant in which they appeared, thus forming a 3D map for the study and evaluation of fibrillation activation patterns. This map can be used for the calculation of the driving speed, and the sequence and the timing in which electrical activation occurs in the torso or heart.

The speed of cardiac electrical activation can be determined by analyzing the spacing between isochronous lines, allowing identification of the earliest activity and the related patterns of propagation of the ongoing AF, providing important quantitative electrophysiological information about the underlying tissue substrate, and identifying mechanisms of arrhythmogenesis (Cantwell et al. 2015). Both the speed and the sequence of electrical activation qualitatively describe the cardiac electrical phenomenon. However, the isochronous line maps also provide quantitative data on the time elapsed for the occurrence of the electrical phenomenon in each electrode. By analyzing the sequence of atrial activation times in the electrodes, the behavior of electrical activation can be characterized.

Phase maps provide a way to quantify the dynamics of excitation profiles, aiding in the identification of singularity points (SPs, i.e., points where the phase around them progresses in a complete cycle from $-\pi$ to $+\pi$) and reentrant circuits identified in noninvasive maps (Salinet et al. 2017a). SPs have been demonstrated as pivot areas of functional reentrant circuits and may or may not be anchored to nonexcitable obstacles. The identification of these circuits is often found in areas of the heart partially disconnected from the tissue or electrically isolated due to extensive areas of fibrosis. Ablation of these circuits has been shown to improve treatment rates for patients with AF. To obtain the phase maps, the Hilbert transform is applied to the signals, and the instantaneous phase is calculated through the inverse tangent between the imaginary and real components of the Hilbert transformed signal. Thus, for each sample in time, the phase range is between $-\pi$ and $+\pi$ and can be visualized using a color scale (Salinet et al. 2017a). Extending the analyzing to all signals simultaneously collected, 3D sequential maps can be created, providing visualization of the spatiotemporal dynamics of both the activation and repolarization episodes.

Frequency maps are generated by applying the Fourier transform to the signals to locate points whose dominant frequency (DF) of the spectrum, within the physiological range of the AF, is related to foci responsible for AF maintenance (Salinet et al. 2014). Additionally, high-frequency signals from certain regions of the heart suggest that AF could be originated in these places and propagate to the other areas of the heart. This map can help identify patterns that, after ablation, contribute to AF recurrence when sinus rhythm is restored (Li et al. 2017). The regions of higher frequencies in the DF maps can be isolated and mapped consecutively over time, allowing the investigation of arrhythmic behaviors and episodes (Salinet Júnior et al. 2013; Salinet et al. 2014). The organization index is a related metric that allows identify how important DF is in relation to the calculated spectrum, that is, the higher this index is, the more prominent the DF along the power spectrum is. Low values give indications of the presence of multiple competing peaks with DF, suggesting the hypothesis of the confrontation of multiple wavefronts (Li et al. 2017).

The patterns of AF mechanisms observed through these maps present intrinsic characteristics that provide better characterization. For example, the ectopic activity observed in isochronous maps, calculated from the electrograms of invasive methods and panoramic optical mapping, present a typical pattern, with this area being activated early compared to the others, whose wavefront emerges from the myocardial cells of this site propagating radially (or centrifugal) to the other areas with a uniform or anisotropic pattern. The frequency map shows that the region with the highest frequency of activation is the myocardial area with ectopic activity, with a frequency gradient for neighboring regions in the atrium. In the body surface potential mapping (BSPM) method, the phase maps present the ectopic activity with a transient and unstable rotational pattern seen in the torso, and large areas of higher frequency, with values close to the frequency of activation of the ectopic activity of the atrium.

The rotor pattern observed in both isochronous and phase maps, obtained from electrograms collected via invasive and noninvasive methods, reveals a rotating activity where myocardial cells surrounding a nonexcitable nucleus have a linear progression of the phases starting at activation until repolarization when the cycle restarts. The rotor exhibits spatiotemporal stability patterns located in the atrium, in addition to a cyclic behavior. The regions with the highest frequency were found to correlate with the rotor core. In the BSPM method, the phase maps display a spiral reentrant pattern in the torso with some spatiotemporal stability, and large areas of higher frequency, with values close to the rotation frequency of the atrium rotor.

The mechanism of multiple and continuous intra-atrial reentries is generally characterized by "chaotic" patterns, in contrast to the ectopic and rotor mechanisms. The collision of the reentry circuits results in unstable and short-lived propagation behaviors in phase and isochron maps, sometimes leading to the identification of spurious rotors and ectopic activity. Furthermore, these collisions between wavefronts can generate high-frequency values in certain areas of the atria, failing to represent the true areas responsible for the maintenance of AF.

Several platforms have been developed over the years to enable scientists and translational researchers to investigate AF mechanisms throughout the 3D mapping off-line, in an attempt to improve catheter ablation efficacy lately. A customized software platform, the USURP-GUI Platform, that allows the measurement of different spatiotemporal biomarkers, after ventricular far-field cancellation (Salinet et al. 2013), was first developed for off-line analysis (Fig. 13.2b) and later applied prospectively in the clinic (Fig. 13.2c) to guide catheter ablation of persistent AF patients (Li et al. 2017). Other platforms were later developed including Directed-Graph-Mapping (Van Nieuwenhuyse et al. 2022) and openEP software (Williams et al. 2021). ElectroMap (O'Shea et al. 2019) and a package of software (Map 3d, PFEIFER. The Forward/Inverse toolkit and UncertainSCI) developed by the NIH Center for Integrative Biomedical Computing from University of Utah Scientific Computing and Imaging Institute (SCI).

13.3 Body Surface Potential Mapping (BSPM)

The 12-lead electrocardiogram is the method used worldwide to evaluate the excitation and electrical propagation characteristics of the heart. By analyzing changes in the morphological characteristics and times of the P-QRS-T segments, clinicians can identify potential regions of the heart that might be "altered" and determine the appropriate clinical conduct to be used. This diagnostic test is low cost, noninvasive and accessible, and highly reproducible. However, the limited number of electrodes hinders the accurate diagnosis of numerous heart diseases, leading to the "silent" progression of the disease (Petruzzeli 2000).

The BSPM is an advanced system utilizing a dense array of electrodes, from 32 to 300, to collect the electrical activity of the heart across areas of the torso in real time, being able to detect minute changes in the heart rhythm (Marques et al. 2020). These electrodes are distributed throughout the torso and provide a 3D perspective of the chest capturing electrocardiograms (Fig. 13.3a). The voltage signals obtained are also projected through a 3D spatial perspective of the torso on consecutive maps in time. There are several BSPM systems, including ones that use flexible electrode garments, and others with traditional disposable passive electrodes of silver chloride with conventional electrode paste (skin-electrode coupling) attached in rubber and fixed in place with double-sided adhesives (Lux 2011).

In recent years, a 64-lead BSPM system was developed (Salinet et al. 2017b) (Fig. 13.3a), featuring a hardware organized in four modules each containing 16 channels, which perform analog amplification and filtering. Each channel has an overload protection circuit consisting of a neon lamp to absorb the discharge and Zener diodes to limit the input voltage. The signals are then passed to an input buffer for impedance matching and amplified with a gain of 30 times (29.54 dB), followed by band-pass filtering between 0.5 and 150 Hz. A second amplifier with programmable gain of 10 or 30 times is applied. Each module contains a sample hold circuit, a 16 channels multiplexer, and a high-voltage isolating

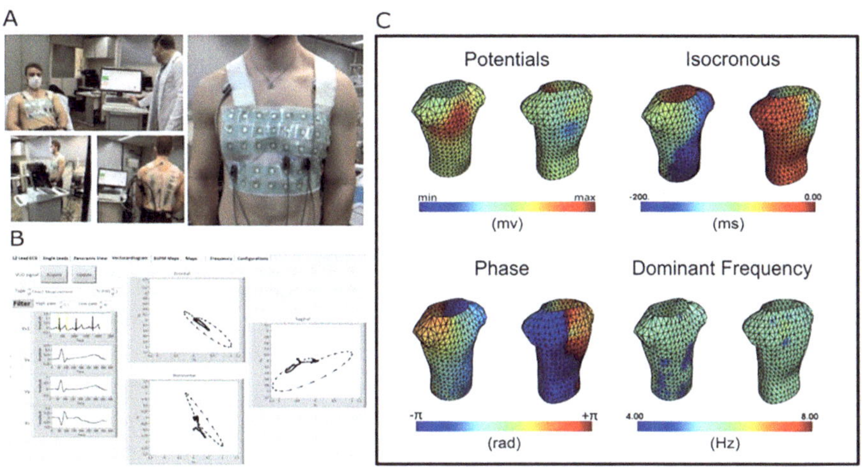

Fig. 13.3 (**a**) The 64-lead BSPM system showing both hardware and vest; (**b**) The VCG tab developed in the CardioMap-GUI platform; and (**c**) 3D BSPM isopotential, isochronous, phase, and dominant frequency maps during AF

amplifier. A control board sends 16 kHz pulses to a National Instruments (NI) USB-6363 data acquisition system (DAQ), which performs the analog to digital (AD) conversion providing digital signals with 16 bits of precision. Each module containing 16 multiplexed BSPM signals corresponds to a channel from the DAQ. The demultiplexing is made via software. The reference for all channels is connected to the patient's right leg. A customized platform runs on an executable file generated in LabVIEW™ (version 14.0.1f3,32-bit). The CardioMap-GUI is divided into tabs, in which several features of the system are distributed. It also controls the source of the data, between two possible modes: patient examination and case review.

The digitalized surface potentials are demultiplexed in 64 channels with 1 kHz sampling rate. To ensure accurate signal processing, standard international recommendations are followed for the digital filters used (Kligfield et al. 2007). A second-order Savitzky-Golay low-pass filter with a cut-off frequency at 150 Hz is used to remove high-frequency components (Christov et al. 2017). The notch filter is used to eliminate the 60 Hz powerline interference (Zahradnik 2013), with a notch bandwidth of 6 Hz and passband attenuation of −0.5 dB. A second-order Lynn's lowpass filter is used to remove baseline frequencies below 0.26 Hz, with its output subtracted from the delayed input (Kozumplík and Provaznik 2017). All filters have a linear phase, which is essential to avoid signal distortion waveform. Preprocessing is carried out in pseudo real-time configuration, with buffers utilized.

The system allows waveform visualization from each of the 64 BSPM leads, as well as the 12-Lead ECG, and the Vectocardiograms extracted based on Frank's lead system (Frank 1956). The orthogonal VCG leads (Vx, Vy, and Vz) can also be obtained indirectly through a weighted average on selected leads of the 12-lead ECG (i.e., Inverse Dower, Uijen and Willems methods) (Macfarlane et al. 2010). The platform presents the orthogonal VCG leads (Vx, Vy, and Vz) and the spatial VCG's projection on the three planes of the body (X, horizontal; Y, vertical Y; and Z, antero-posterior) representing the electrical heart's cycle phenomenon. A fixed time window around a detected R peak is used to define P-QRS-T boundaries for the VCG (Fig. 13.3b).

In addition to the Vectocardiogram maps, which do not require 3D representation, the platform allows visualization of the isopotential, isochronous, and among others in 3D. To achieve this, a standardized 3D geometry of the torso was obtained from the online education and research ECGSIM software (Van Oosterom and Oostendorp 2004). The torso includes 410 vertices, each with its own X, Y, and Z coordinates, allowing accurate mapping of the electrical activity onto the 3D geometry of the torso, providing a more complete understanding of the heart's electrical behavior. Figure 13.3c presents the typical isopotential, isochronous, phase, and DF 3D maps during AF.

13.4 Electrocardiographic Imaging

The noninvasive electrocardiographic imaging (ECGi) method is a medical tool for the estimation of the electrical activity in the heart noninvasively. This technique has been widely used in the study of various heart diseases, due to its ability to analyze the heart's electrical activity in a manner comparable to invasive techniques with catheters, but with a greater spatial resolution. ECGi offers greater versatility, and the possibility to study and plan therapeutic strategies before surgical treatment (Salinet et al. 2021).

To perform ECGi, two sets of data are required: the potentials around the torso surface and the 3D geometries of the heart and torso surfaces (Fig. 13.4a). These data can be acquired completely noninvasively, with the potentials in the torso acquired by a high number of electrodes distributed to simultaneously capture their potentials every 1 ms, using the BSPM system. Geometric parameters can be obtained through anatomical images of computed tomography or magnetic resonance imaging. This provides the epicardial geometry and the positions of the electrodes on the torso in the same image. The recorded torso potential and geometric information are used as input data for the ECGi algorithm, which reconstructs the epicardial/endocardium potentials of intrinsic 3D maps (Fig. 13.4b) (Salinet et al. 2021). In this method, ventricular far-field cancellation (Salinet et al. 2013) is needed prior to estimation of the epicardium/endocardium signals noninvasively.

The underlying physics of the ECGi is governed by Laplace's equation (Eq. 13.1), in which the electric field φ generated by heart's excitation is propagated to the body surface passing through a volume of passive electrical conduction (Salinet et al. 2021). Where $\phi = \phi_T$ is on the torso surface, $\phi = \phi_E$ is on the epicardium, and ∇^2 is the Laplacian operator.

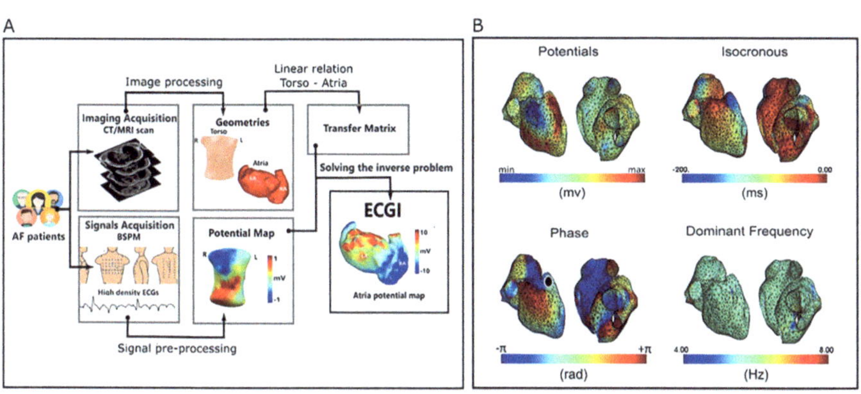

Fig. 13.4 (**a**) ECGi methodology scheme; (**b**) Epicardial potentials, isochronous, phase and dominant frequency, 3D maps reconstructions

$$\nabla^2 \phi = 0 \tag{13.1}$$

This method involves the calculation of potentials in the epicardium (ϕ_E) by solving the Laplace equation within the volume of the torso, using the electrical potentials acquired noninvasively in the torso (ϕ_T) and the geometric relationship between the 3D surfaces of the epicardium and torso (A). The discretization of the 3D surfaces of the epicardium and torso in triangular elements, which can be obtained through the use of the method of contour elements, is necessary for this purpose, allowing the relationship of ϕ_T and ϕ_E through a linear relationship matrix A, which contains the geometric information that relates both surfaces (Salinet et al. 2021), with ϕ_T and ϕ_E represent in vectors the electrical potentials of the surface of the torso and epicardium, respectively.

$$\phi_T = A\, \phi_E \tag{13.2}$$

The objective of ECGi is to solve the inverse problem of electrocardiography, that is, to obtain the epicardial potentials (ϕ_E) from the potentials measured in the torso (ϕ_T). The inverse problem is challenging due to potential errors (i) in the measurement of torso signals (ϕ_T) (i.e., noise and/or electrode position) and (ii) in the 3D geometry of the torso/heart, which can result in significant errors in the calculation of ϕ_E (Rudy 2013). This is due to the noninvertible character of the transference matrix A. To address this issue, regularization methods can be applied (Rudy 2013), and one of the most used is the Tikhonov method (Tikhonov 1963), which imposes limits on ϕ_E delimiting the amplitudes or derivatives of epicardial potentials in space, time, or both, and can be used for limiting these parameters to be within the electrophysiological and electric field limits of the heart (Rudy 2013). In this method, the smoothest approximate solution, compatible with the observation data, is determined for a certain noise level (Salinet et al. 2021).

$$\phi_E = A^+\, \phi_T \tag{13.3}$$

13.5 Panoramic Optical Mapping

A panoramic optical mapping system can map the entire epicardium, providing researchers the opportunity to study the wavefront's propagation and patterns during complex arrhythmias (Kay et al. 2004; Lou et al. 2011). The basis for this experimental system is to use dyes that bind to or interact with cell membranes, along with optics and photodetectors to record this interaction (Arora et al. 2003). This method is used in experiments with isolated hearts (in vivo) and retrograde heart perfusion using the Langendorff technique to maintain heart viability but can be applied to other models, such as transthoracic open chest. This technique is used to investigate the contractile and electrical conduction properties of the heart in physiological, pathological, and pharmacological research (Bell et al. 2011).

Optical mapping studies are based on the use of calcium- and voltage-sensitive dyes, allowing measurement of transmembrane action potentials and calcium transients with high spatial resolution without physical contact with the tissue (Lou et al. 2011; Laughner et al. 2012). Changes in the intensity of fluorescent light emission of the cells are proportional to the changes in transmembrane voltage, allowing noninvasive assess and high spatiotemporal resolution of the electrical activity. Optics include the excitation source that has an appropriate excitation wave corresponding to the dye absorption spectrum (Laughner et al. 2012) and commonly are high-power excitation light sources composed of light-emitting diodes (LEDs) (Marina-Breysse et al. 2021), filters, and condenser lens to orient the light as much as possible over the studied area. Photodetectors quantify the amount of emitted fluorescence, previously filtered and focused using specific filters and lenses, respectively (Siles-Paredes et al. 2022). A video camera with high sensitivity and low noise acts as an optical sensor for high-speed capturing of the photons emitted by the dye attached to the cell membrane. To avoid distortion of the image acquisition, the heart contractility (i.e., mechanical movement) is immobilized through contraction uncouplers.

To generate the 3D surface of the heart to 3D mapping, a circuitry is needed, allowing performing a full 360° rotation sequentially (Siles et al. 2022) for later extraction of the heart's silhouette (Qu et al. 2007). In addition to traditional maps, specified in the previous section (see section Electrophysiology Mapping), optical mapping can also generate other types of maps and analyses such as action potential duration (APD), repolarization maps, and optical action potential upstroke analysis (Laughner et al. 2012).

The fluorescent dyes used in optical mapping experiments are toxic and cannot be used in the in situ human heart (Clayton and Nash 2015). However, in the use of animal models in translational research, optical mapping has become of extreme importance to clinical research (Laughner et al. 2012) and currently is considered the gold standard method for mapping cardiac electrical activity (Lee et al. 2019). Multisite contact mapping used in clinics presents technical limitations associated with amplification, gains, sampling rates, signal-to-noise ratio, and the inability to see signals during high-voltage shocks. On the other hand, microelectrode techniques that record action potentials cannot record from several sites simultaneously (Arora et al. 2003). Using optical mapping, more sites can be imaged and information on repolarization can be obtained even during tachyarrhythmias when the T-wave is not visible in electrical recordings. Additionally, optical mapping is not affected by artifacts from electrical stimuli (Kay et al. 2004).

Although single optical mapping provides more information compared to traditional methods, the mapped region is limited to the field of view (Qu et al. 2007). One way to overcome this limitation is to use mirrors or multiple sensors to obtain multiple views of the heart, spaced together around the heart. The main objective is to obtain fluorescent images of the electrophysiological activity of the entire epicardial surface of the heart, allowing the study of the spatiotemporal propagation

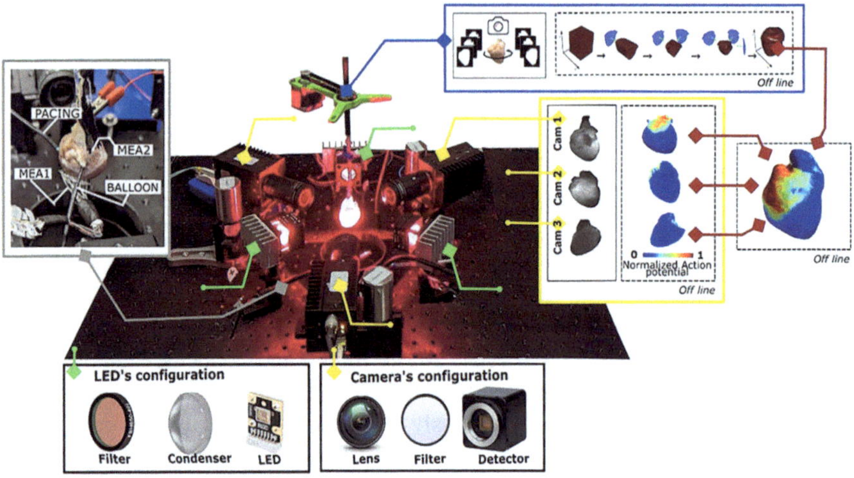

Fig. 13.5 Experimental animal setup for panoramic optical mapping and electrical mapping developed in Brazil (São Paulo). The light sources are highlighted in green, and their principal components are shown (filter, condenser, and LED). The camera configuration is highlighted in yellow, and its principal components are shown (lens, filter, and detector). On the right side, the offline postprocessing results and the final 3D heart model are presented. The electrical mapping setup is highlighted in grey and shows the MEAs, pacing electrode, and balloon

of the electrical activity responsible for the maintenance of cardiac arrhythmias. Furthermore, some studies have utilized the benefits of optical mapping to improve traditional clinical systems where both are simultaneously recorded, providing insights into the normal electrophysiology of the heart and pathophysiological mechanisms (Di Diego et al. 2013; Berenfeld et al. 2011).

Figure 13.5 shows an experimental animal setup for panoramic optical mapping and electrical mapping developed in Brazil (São Paulo) (Siles et al. 2022). The optical mapping system includes an excitation source composed of three high-power deep-red LEDs, an aspheric condenser lens, and a band-pass filter (LED configuration in Fig. 13.5, highlighted in green). For fluorescence acquisition, the emitted fluorescence passes through a C-mount lens and a filter before being detected by a camera (camera configuration in Fig. 13.5, highlighted in yellow). The resulting fluorescence map after pre- and postprocessing is presented in a dotted square on the left and it was generated off-line in MATLAB. A customized three-pointed 3D printed adapter (highlighted in blue) and a stepper motor are positioned at a fixed distance. One optical camera records a sequence of heart images while the adapter rotates 360 degrees. The offline process for acquiring the 3D surface generation is presented in another dotted square. The resulting 3D surface with the optical maps generated is used to obtain the final 3D model of the heart. Additionally, an electrical mapping system (highlighted in grey) records electrical signals through two multielectrode arrays (MEA1 and MEA2) under stimulation (pacing) with the help of a balloon to allow the heart cavities to expand.

Funding Declaration/Acknowledgments This chapter is supported by grant no. 2018/25606-2, São Paulo Research foundation (FAPESP) and CNPq INCT INTERAS. call 58/2022. J.G. Siles-Paredes is supported by grant #2020/03601-9, FAPESP. I. Sandoval is supported by Coordenacão de Aperfeiçoamento de Pessoal de Nível Superior – Brasil (CAPES).

References

Allessie M, Groot N (2014) CrossTalk opposing view: Rotors have not been demonstrated to be the drivers of atrial fibrillation. The Journal of Physiology, 592(15): 3167–70.

Arora R, Das MK, Zipes DP et al. (2003) Optical mapping of cardiac arrhythmias. Indian Pacing and Electrophysiology Journal, 3(4), 187.

Bell RM, Mocanu MM et al. (2011) Retrograde heart perfusion: the Langendorff technique of isolated heart perfusion. Journal of Molecular and Cellular Cardiology, 50(6), 940–950.

Berenfeld, O. et al. (2011) Time- and frequency-domain analyses of atrial fibrillation activation rate: the optical mapping reference. Heart Rhythm 8, 1758–1765.

Bollmann A, Lombardi F. (2006) Electrocardiology of atrial fibrillation: Current knowledge and future challenges. IEEE Engineering in Medicine and Biology Magazine, 25(6): 15–23.

Bonny A, Ng A, Sliwa K, Kengne AP, Chin A, Mocumbi AO, Ngantcha M, Ajijola OA, Bukhman G (2020) A Systematic Review of the Spectrum of Cardiac Arrhythmias in Sub-Saharan Africa. Global Heart. 15(1): 37.

Borlich M, Sommer P (2019) Cardiac mapping systems: rhythmia, topera, ensite precision, and CARTO. Cardiac Electrophysiology Clinics, 11(3), 449–458.

Cantwell CD, Roney CH, Ng FS, Siggers JH et al. (2015) Techniques for automated local activation time annotation and conduction velocity estimation in cardiac mapping. Computers in Biology and Medicine, 65, 229–242.

Christov I, Neycheva T, Schmid R, Stoyanov T et al. (2017) Pseudo-real-time low-pass filter in ecg, self-adjustable to the frequency spectra of the waves. Medical Biological Engineering Computing 1–10.

Clayton RH, Nash MP (2015). Analysis of cardiac fibrillation using phase mapping. Cardiac electrophysiology clinics, 7(1), 49–58.

Connolly SJ (2011) Atrial fibrillation in 2010: advances in treatment and management. Nature Reviews Cardiology 8:67–68.

Coveney S, Cantwell C, Roney C (2022) Atrial conduction velocity mapping: clinical tools, algorithms and approaches for understanding the arrhythmogenic substrate. Med Biol Eng Comput 60, 2463–2478.

Di Diego JM, Sicouri S, Myles RC, Burton FL et al. (2013). Optical and electrical recordings from isolated coronary-perfused ventricular wedge preparations. Journal of Molecular and Cellular Cardiology, 54, 53–64.

Frank E (1956). An accurate, clinically practical system for spatial vectorcardiography. Circulation 13(5), 737–749.

Fuster V, Ryden et al. (2006) ACC/AHA/ESC 2006 Guidelines for the management of patients with atrial fibrillation–executive summary: a report of the american college of cardiology/american heart association task force on practice guidelines and the european society of cardiology committee for practice guidelines (writing committee to revise the 2001 guidelines for the management of patients with atrial fibrillation): developed in collaboration with the European Heart Rhythm Association and the Heart Rhythm Society. Circulation 114, 700–752.

Kay MW, Amison PM, Rogers JM (2004) Three-dimensional surface reconstruction and panoramic optical mapping of large hearts. IEEE Transactions on Biomedical Engineering, 51(7), 1219–1229.

Kligfield P, Gettes LS, Bailey JJ, Childers R et al. (2007) Recommendations for the standardization and interpretation of the electrocardiogram. Circulation 2007; 115(10), 1306–1324.

Kozumplík J, Provazník I (2017). Fast time-varying linear filters for suppression of baseline drift in electrocardiographic signals. Biomedical Engineering Online 16(1), 24.
Laughner JI, Ng FS, Sulkin MS, Arthur RM, Efimov IR (2012). Processing and analysis of cardiac optical mapping data obtained with potentiometric dyes. American Journal of Physiology-Heart and Circulatory Physiology, 303(7), H753–H765.
Lee P, Quintanilla JG, Alfonso-Almazan JM, Galán-Arriola C et al. (2019). In vivo ratiometric optical mapping enables high-resolution cardiac electrophysiology in pig models. Cardiovascular Research, 115(11), 1659–1671.
Li X, Salinet JL, Almeida TP, Vanheusden FJ, Chu G, Ng GA, Schlindwein FS (2017) An interactive platform to guide catheter ablation in human persistent atrial fibrillation using dominant frequency, organization and phase mapping. Computer Methods and Programs in Biomedicine 141:83–92. https://doi.org/10.1016/j.cmpb.2017.01.011
Lou Q, Li W, Efimov IR (2011). Multiparametric optical mapping of the Langendorff-perfused rabbit heart. Journal of Visualized Experiments, 55, e3160.
Lux RL (2011) Body Surface Potential Mapping Techniques. Em Comprehensive Electrocardiology. Editors Peter Macfarlane, Adrian Van Oosterom,Olle Pahlm, Paul Kligfield, Michiel Janse, Jhon Camm. Second Edition, v. 3, 1362–1372, Springer-Veralog London Limited.
Macfarlane PW, Van Oosterom A, Pahlm O, Kligfield P, Janse M, Camm J (2010) Comprehensive Electrocardiology. Springer Science & Business Media.
Marina-Breysse M, García-Escolano A, Vila-García J, Reale-Nosei G et al. (2021). A complete and low-cost cardiac optical mapping system in translational animal models. Frontiers in Physiology, 12, 696270.
Marques VG, Rodrigo M, Guillem MS, Salinet J (2020) Characterization of atrial arrhythmias in body surface potential mapping: A computational study. Computers in biology and medicine, 127, 103904.
Narayan SM, Jalife J (2014) CrossTalk proposal: Rotors have been demonstrated to drive human atrial fibrillation. J Physiol 592(15), 3163–6.
Nattel S, Dobrev D (2017) Controversies about atrial fibrillation mechanisms: aiming for order in chaos and whether it matters. Circulation Research, 120(9), 1396–1398.
Nault I, Miyazaki S, Forclaz A, Wright M, Jadidi A, Jaïs P, Hocini M, Haïssaguerre M (2010) Drugs vs. ablation for the treatment of atrial fibrillation: the evidence supporting catheter ablation. European Heart Journal 31(9), 1046–1054.
Nerbonne JM (2004) Heterogenous Expression of Potassium Channels in the Mammalian Myocardim. Cardiac Electrophysiology: from cell to bedside.
O'Shea C, Holmes, AP, Yu TY et al. (2019) ElectroMap: High-throughput open-source software for analysis and mapping of cardiac electrophysiology. Sci Rep 9, 1389.
Petruzzeli SEA (2000) Limitations of ECG in Diagnosing Pulmonary Embolism. U.S. National Library of Medicine, 113(2), 559.
Qu F, Ripplinger CM, Nikolski VP, Grimm C, Efimov IR (2007) Three-dimensional panoramic imaging of cardiac arrhythmias in rabbit heart. Journal of biomedical optics, 12(4), 044019-044019.
Rudy Y (2013) Noninvasive Electrocardiographic Imaging of Arrhythmogenic Substrates in Humans. Circulation Research 2013, 112(5), 863–874.
Salinet JL, Madeiro JPV, Cortez PC, Stafford PJ et al. (2013) Analysis of QRS-T subtraction in unipolar atrial fibrillation electrograms. Medical & Biological Engineering & Computing, 51, 1381–1391.
Salinet Júnior JL, Oliveira GN, Vanheusden FJ, Comba JLD et al. (2013) Visualization of Intracardiac Atrial Electrograms of Patients with Atrial Fibrillation using Spectral Analysis. Computing in Science & Engineering, 15, 79–87.
Salinet JL, Tuan JH, Sandilands AJ, Stafford PJ et al. (2014) Distinctive patterns of dominant frequency trajectory behavior in drug-refractory persistent atrial fibrillation: preliminary characterization of spatiotemporal instability. Journal of Cardiovascular Electrophysiology, 25(4), 371–379.

Salinet J, Schlindwein FS, Stafford P, Almeida TP et al. (2017a) Propagation of meandering rotors surrounded by areas of high dominant frequency in persistent atrial fibrillation. Heart rhythm, 14(9), 1269–1278.

Salinet JL, Marques VG, Mazzetto M, Camargo EDLB et al. (2017b) A 64-lead Body Surface Potential Mapping System. In: 44th annual scientific meeting of Computing in Cardiology (CinC) Rennes. Computing in Cardiology 2017; Vol 44:1:4. ISSN: 2325-887X.

Salinet J, Molero R, Schlindwein FS, Karel J et al. (2021) Electrocardiographic imaging for atrial fibrillation: a perspective from computer models and animal experiments to clinical value. Frontiers in Physiology, 12, 653013.

Shah JA, Naz F, Kumar R, Hassan M, Shah G, Ahmed K, Hussain J, Abid K, Karim M (2021) Incidence of Cardiac Arrhythmias in Acute Myocardial Infarction Patients Undergoing Primary Percutaneous Coronary Intervention and Associated Outcomes During the First 24 Hours. Cureus. 13(1), e12599.

Siles-Paredes JG, Crowley CJ, Fenton FH, Bhatia N, et al. (2022) Circle Method for Robust Estimation of Local Conduction Velocity High-Density Maps from Optical Mapping Data: Characterization of Radiofrequency Ablation Sites. Frontiers in Physiology, 1358.

Siles JG, Uzelac I, Sandoval I, Silva V et al. (2022) Panoramic Optical Mapping setup in a Langendorff Perfused Rabbit Heart for the Study of Atrial Fibrillation. IX Latin American Congress on Biomedical Engineering (CLAIB) & XXVIII Brazilian Congress.

Tikhonov AN (1963) On the solution of ill-posed problems and the method of regularization. Doklady Akademii Nauk SSSR 1963; 151: 501–504.

Tortora GJ, Derrickson B (2006) Chapter 20 – The cardiovascular system: The heart. In: B Roesch, C Stone, editors. Principles of anatomy and physiology, 11th ed., John Wiley & Sons, Inc.

Van Nieuwenhuyse E, Hendrickx S, Abeele RVd et al. (2022) DG-Mapping: a novel software package for the analysis of any type of reentry and focal activation of simulated, experimental or clinical data of cardiac arrhythmia. Med Biol Eng Comput 60, 1929–1945.

Van Oosterom A, Oostendorp T(2004) Ecgsim: an interactive tool for studying the genesis of qrst waveforms. Heart, 90(2):165–168.

Winkel T (2010) Risk factors and outcome of new-onset cardiac arrhythmias in vascular surgery patients. American Heart Journal, v. 159, n. 6, p. 1108–1115.

Williams SE, Roney CH, Connolly A, Sim I et al. (2021) OpenEP: A Cross-Platform Electroanatomic Mapping Data Format and Analysis Platform for Electrophysiology Research. Front. Physiol. 12:646023.

Zahradnik P (2013) Notch filtering suitable for real time removal of power line interference. Radioengineering, (1):186–193.

Zaman JAB, Baykaner T, Schricker AA, Krummen DE, Narayan SM (2015) Mechanistic targets for the ablation of atrial fibrillation. Global Cardiology Science & Practice, 67(1): e201707.

Chapter 14
Fully Bioresorbable Vascular Stents

Sônia Maria Malmonge and Camila Cliquet

Abstract People from all around the world are annually affected by the arterial coronary disease. Nowadays, the clinical procedure to restore normal blood flow and avoid the other critical consequences of vessel narrowing are the coronary interventions, including angioplasty and stent placement, that represent the most performed invasive medical procedures in cardiology. The clinical use of coronary stents has a long history, as they have been used since the 1980s and have become a milestone in the treatment of coronary vascular diseases. Coronary stents save many lives every day, however, these same stents have caused multiple complications to patients, requiring innovation and new developments in the area. Thus, the fully bioresorbable vascular stents, also called vascular scaffolds, are intended to completely upgrade cardiovascular medicine due to their tissue engineering and regenerative medicine approach. However, with the upcoming scaffolds also came several doubts on whether their benefits would overcome the risks discovered so far. This chapter summarizes the history of the fully bioresorbable stents and discusses the future of these devices based on the results of clinical trials available and the main outcomes.

Keywords Biomaterials · Bioresorbable vascular stent · Tissue engineering

14.1 Introduction

Cardiovascular diseases (CVDs) are the leading cause of death worldwide. In 2019, more than 8.9 million people died from such diseases, accounting for 16% of global deaths (WHO 2021). According to Oliveira et al., authors of the Cardiovascular Statistics – Brazil 2021, more than 17 million global deaths are caused by CVD,

S. M. Malmonge (✉) · C. Cliquet
Graduate Program in Biomedical Engineering, Federal University of ABC – UFABC, São Bernardo do Campo, SP, Brazil
e-mail: sonia.malmonge@ufabc.edu.br

C. B. Lombello, P. A. da Ana (eds.), *Current Trends in Biomedical Engineering*,
https://doi.org/10.1007/978-3-031-38743-2_14

while in Brazil 30% of deaths from noncommunicable chronic diseases (NCDs) are caused by CVD (Oliveira et al. 2022).

Coronary artery disease (CAD) consists of a group of clinical conditions characterized by the narrowing of the lumen of the artery due to plaque deposited under the endothelium, which reduces blood flow to the myocardium and an inadequate supply of nutrients and oxygen to the cardiac muscle causing myocardial ischemia and ultimately myocardial infarction (Cassar et al. 2009).

Today, the clinical procedure to restore normal blood flow and prevent other critical consequences of vascular stenosis are coronary interventions, including angioplasty and stent placement, which are the most performed invasive medical procedures in cardiology (Laçin and Utkan 2014; Oliveira et al. 2022).

Coronary angioplasty, a nonsurgical solution to coronary arteries restoration, requires the use of balloon catheters (Centemero et al. 1993). A small balloon is introduced into the blocked coronary artery and then inflated, which relieves the blockage. To ensure that the artery does not recede, an endoprosthesis is installed – a stent or scaffold (Hospital Israelita Albert Einstein 2020).

Stents are expansible tubular structures whose purpose is to restore blood flow in an arterial segment that has suffered from stenosis, i.e., narrowing of the diameter of the vessel (Pant et al. 2011). From a structural point of view, stents are small, complex cylindrical hollow structures formed into a sequential ring structure consisting of multiple struts and connecting elements. Stent placement became the standard of care for percutaneous coronary intervention (PCI). Figure 14.1 illustrates typical stent geometries and Fig. 14.2 shows a deflated balloon and the expanded stent fixed in place, which increases the lumen for blood flow recovery in a vascular restoration procedure.

The clinical use of coronary stents has a long history, being used since the 1980s, and has become a milestone in the treatment of several vascular diseases. Since the 1980s, the treatment of CAD has been carried out with the implantation of metallic cardiovascular stents – bare metal stents (BMS), but the incidence of thrombosis and restenosis after implantation has been high. The metals used in the manufacture of BMS have suitable properties to provide the necessary support to the lesion site while maintaining the shape and integrity of the device, but their non-degradability in physiological medium leads to several post-intervention complications such as a high incidence of thrombosis and restenosis (Scafa Udrişte et al. 2021).

To reduce the risk of restenosis, drug-eluting stents (DES), a polymer-coated metal stent that releases antiproliferative drugs that reduce neointimal hyperplasia has emerged. However, these drugs also inhibit endothelialization, which increases the risk of thrombosis (Sigwart et al. 1987; Tomberli et al. 2018). The first-generation DES used stainless steel as the structural material covered by sirolimus and paclitaxel as antiproliferative drugs. In the second-generation DES, the support base was changed to a cobalt-chromium or platinum-chromium alloy, which made it possible to reduce the thickness of the support and increase flexibility. The safety and efficacy of second-generation DES have been improved with limus-

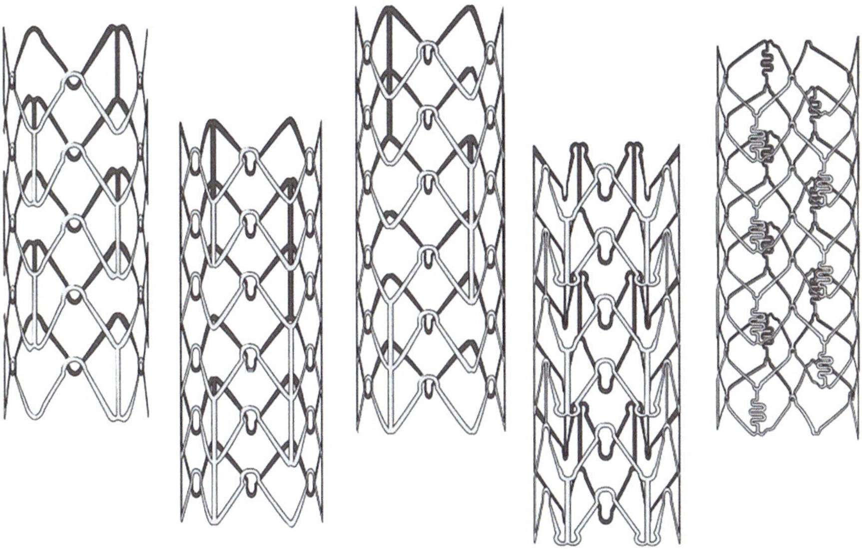

Fig. 14.1 Illustration of typical stent geometries. The figure was partly generated using Servier Medical Art, provided by Servier, licensed under a Creative Commons Attribution 3.0 unported license

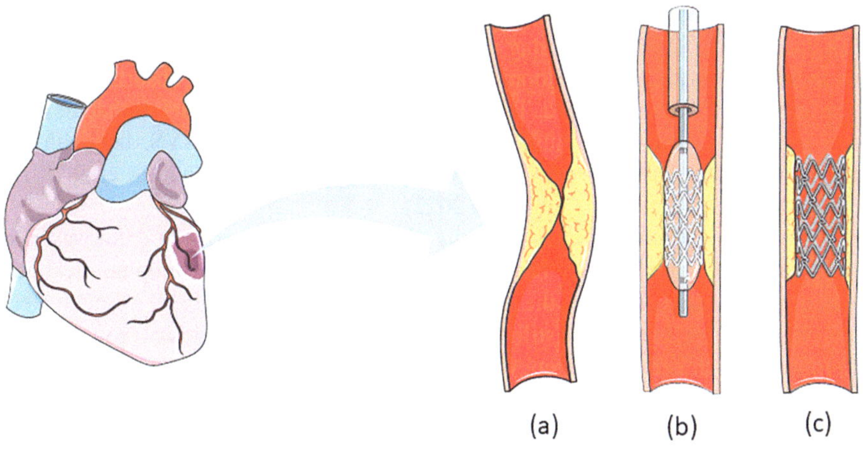

Fig. 14.2 Illustration of a "vascular restoration" in the case of coronary artery disease (CAD) with stent placement: (**a**) atherosclerosis, (**b**) stent placement with balloon catheter angioplasty, and (**c**) expanded stent placed in the narrowed vessel. The figure was partly generated using Servier Medical Art, provided by Servier, licensed under a Creative Commons Attribution 3.0 unported license

family antiproliferative drugs (zotarolimus, everolimus, and novolimus) with faster elution with earlier endothelial coverage (Scafa Udrişte et al. 2021). However, DESs have the same disadvantages as BMSs related to non-resorption in the human body, that is, DESs are a permanent device with limitations in terms of adaptive vascular remodeling because the metal structure remains attached to the vessel wall causing an abnormal coronary vasomotion and unspecified antiproliferative drug interactions (Indolffi et al. 2016; Scafa Udrişte et al. 2021).

In the light of this scenario and considering the concept of "vascular restoration" used in coronary interventions, the fully bioresorbable polymeric stent or bioresorbable vascular scaffold (BVS) technology appeared, representing an important revolution in interventional cardiology (Tomberli et al. 2018). The term scaffold is sometimes used because it implies that the device is a temporary implant that supports the arterial wall as opposed to the permanent vessel cage that is created with metal stents. BVSs are made of resorbable polymeric biomaterials that provide temporary mechanical support to the vessel wall, allowing early revascularization and preventing vessel collapse. Therefore, these devices must maintain mechanical properties and integrity for at least 6–12 months before resorption, when the treated vessel becomes a healthy endothelium and normal vasomotion (Indolffi et al. 2016; Jezewski 2019).

Since 2010, clinical trials of BVS have been conducted, and although the results are not conclusive, expectations for BVS have encouraged manufacturers to invest in research aimed at improving the performance of BVS (Indolffi et al. 2016; Jezewski 2019; Jinnouchi et al. 2019; Omar and Kumbhani 2019; Azzi and Shatila 2021).

In parallel with the development of BVS for 2000 years, absorbable metal scaffolds (AMS) also appeared. These stents are designed to use metals that dissolve and are reabsorbed in the human body in the same way as BVS, but with the mechanical properties that provide greater structural strength (Indolffi et al. 2016; Jezewski 2019; Drelich and Goldman 2022).

Contributing to the promising results of the fully bioresorbable cardiovascular stent, the healthcare industry became interested and eager to develop an ideal and functional device. Over time, the understanding of fully bioresorbable vascular stent has changed a lot, but there is still no clear answer about their viability or which path to follow, which is the next step, whether BVS or AMS.

Figure 14.3 shows a timeline with major milestones for coronary angioplasty and coronary stenting. The development of percutaneous coronary intervention (PCI), whose main steps of which were the first balloon angioplasty performed on patients by Andreas Grüntzig in 1964, the introduction of BMS into clinical practice, DES to prevent restenosis, and fully bioresorbable stents, which are considered an important innovation in interventional cardiology (Indolffi et al. 2016; Canfield and Totary-Jain 2018; Jinnouchi et al. 2019; Peng et al. 2020; Azzi and Shatila 2021).

In this chapter, we present an overview of the development of fully bioresorbable stents, BVS and AMS stents, which are currently in progress or have already concluded clinical trials, and the most important results.

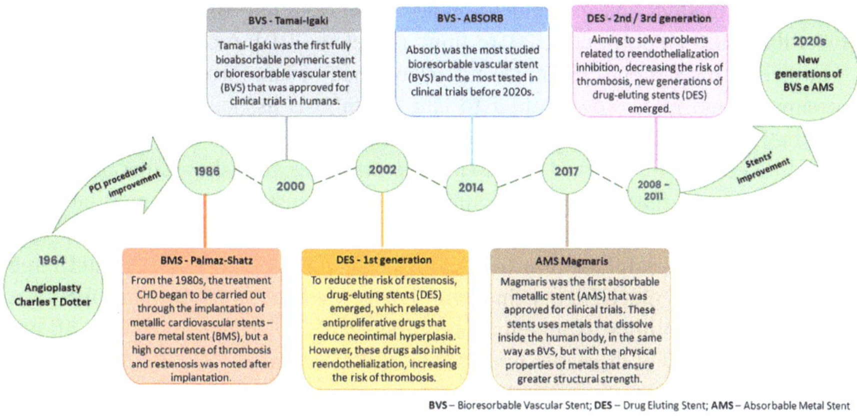

Fig. 14.3 Landmarks in coronary angioplasty and stent development

14.2 Bioresorbable Vascular Stents (BVS) in CAD Interventions

Since stent thrombosis usually happens due to the damaged endothelium of the metallic stent, BVS appeared to be a promising candidate for improving vascular healing after CAD interventions. Stent resorption can potentially lead to favorable remodeling of coronary arteries, sparing vasoreactivity, and preventing luminal narrowing.

The first fully bioresorbable stent that was tested in humans was the BVS Igaki-Tamai (Kyoto Medical Planning, Japan), using poly (L-lactic acid) (PLLA). Although the clinical trials of this device were not entirely satisfactory, the positive aspects encouraged the development of BVS to continue. Over the 2000 years, several BVSs have reached the clinical trial stage (Tamai et al. 2000; Indolffi et al. 2016; Jezewski 2019; Jinnouchi et al. 2019; Peng et al. 2020; Azzi and Shatila 2021).

Indolfi et al. provide a list of BVS on the market or in the clinical trials or development until 2016 with most important informations to facilitate model comparison. These authors also described and discussed the main limitations of BVS, the basis of restenosis (recurrence of vessel narrowing), thrombosis (formation of blood clots) in the stent site, and duration of dual antiplatelet therapy (DAPT) (Indolffi et al. 2016). Abbott Vascular, the manufacturer of Absorb, conducted several clinical trials with different versions of this BVS around the world, starting in 2014–2015, but the results were not very encouraging. Absorb consisted of PLLA coated with a layer of poly (DL-lactic acid) (PDLLA) to deliver the antiproliferative drug everolimus. This BVS was the most clinically evaluated. Jinnouchi et al. (2019) and Azzi and Shatila (2021) also presented details and discussed clinical trials of different generations of Absorb BVS stents. A preliminary clinical trial of this BVS showed a very high incidence of in-stent and in-segment binary restenosis, like BMS levels, but much higher than DES (Indolffi et al. 2016; Jinnouchi et al. 2019; Azzi and Shatila 2021).

In theory, BVS should eliminate the occurrence of late thrombosis, but this did not happen, and rates were similar to late thrombosis associated with the use of DES. In the case of BVS, late thrombosis is associated with stent under expansion or discontinuation of long-term DAPT. However, the literature shows that until 2016, the complications that appeared in clinical trials are just a setback and that future technological developments are the solution (Indolffi et al. 2016).

Two years after implantation of Absorb II, the rates of thrombosis and target lesion failure were much higher than with Elixir Medical's PLLA BVS DESsolve, which carries novolimus as an antiproliferative drug. In 2019, Jezewski et al. presented an overview of available first- and second-generation BVS and summarized the results of clinical trials using them. It discusses explanations for adverse results, suggested improvement of techniques, and a potential niche market for BVS use. According to these authors, the ratio of stents to implant site diameters has a direct effect on target lesion failure, because the higher the value, the less symmetrical is the session, and BVS – especially at the Absorb group – depends on a perfect implantation (Jezewski 2019).

In 2019, Jinnouch et al. also presented a review, focusing mainly on clinical trials for Absorb and detailing clinical trials and the results of different BVS tested until 2019. The authors describe in detail the steps involved in the study and the results related to the development of various BVS, and, according to these authors, until 2019, Absorb was the most advanced BVS studied in clinical trials, providing important information on the structure thickness and the influence of vessel wall coating on thrombosis. In addition, the authors discussed issues regarding the time of bioresorption and important factors related to implantation techniques and their influence on post-stenting (Jinnouchi et al. 2019).

Since 2019, expectations for BVS have been moderated by the publication of long-term clinical trials, especially for second-generation BVS, such as Absorb II and DESolve. Although the 5-year analysis was promising because BVSs were expected to be superior to DESs in terms of safety, they were not superior in randomized trials. Complications found in BVS after clinical trials included thrombosis – direct and late –the target lesion failure, restenosis, and local ischemia (Jinnouchi et al. 2019).

During the twenty-first century, several fully bioresorbable stents were developed by different manufacturers and clinically tested, which broadened the knowledge about the performance of the device and the interaction mechanisms between the stent and vasculature, as well as the main problems identified. Several authors published reviews reported and analyzed existing data to draw conclusions and established guidelines for the development of these technologies (Indolffi et al. 2016; Jinnouchi et al. 2019; Peng et al. 2020; Kumar et al. 2021; Azzi and Shatila 2021).

In 2020, Peng et al. presented a review that summarized the research results and clinical evidence of various fully bioresorbable stents and discussed the factors that may be associated with an increased risk of stent thrombosis (ScT), which has limited its clinical use. These authors also provide brief summaries of the called new generations of fully bioresorbable stents, which use resorbable polymeric biomaterials with stronger mechanical properties with the aim of providing thinner struts (Peng et al. 2020).

In 2021, a new study was published that evaluated the safety of BVS use in relation to acute myocardial infarction. Although the sample size was not very large in the study proposed by Kumar et al., the findings were again encouraging and based on the specific implantation technique, the success of BVS is possible and the frequency was considered acceptable (Kumar et al. 2021).

Azzi and Shatila (2021) presented a comprehensive review of real-world BVS data and concluded that "thinner structures, newer design characteristics, appropriate selection of patients and standardized techniques of implantation may lead to better outcomes and improve the care of patients." The authors also discussed aspects of the "vascular restoration" process of fully bioresorbable stents and noted that implantation of polymeric BVS requires careful implantation technique. BVSs are much more demanding in terms of implant technique and require a special protocol called "PSP technique" (Prepare the lesion, Size adequately, Post-dilate) (Azzi and Shatila 2021).

Based on the results confirmed in clinical trials, efforts have been made to improve the performance of fully bioresorbable stents through adaptations of material and design adjustments, especially to reduce thickness without compromising the radial strength, since surface characteristics and a low value of radial strength seems to be critical. Therefore, bioresorbable metals have the advantage of being more mechanically resistant.

14.3 Absorbable Metal Stents (AMS) in CAD Interventions

The main advantages of AMS over BVS are good radial strength, low values of recoil, and good adaptability to vascular anatomy. Although metal strength is often considered as a critical property, ductility, corrosion rate, toxicity, and biocompatibility are also important features in the property profile. AMS should provide spatial and temporal mechanical stability with a resorption rate like that of regenerative tissue. Corrosion products must be biocompatible and promote certain biological effects such as anti-inflammation, anti-platelet activation, and anti-smooth muscle cell hyperplasia (Indolffi et al. 2016; Jinnouchi et al. 2019; Wlodarczak et al. 2019; Toong et al. 2020; Peng et al. 2020; Scafa Udrişte et al. 2021; Drelich and Goldman 2022).

Magnesium (Mg), Iron (Fe), Manganese (Mn), Zinc (Zn), and its alloys are the most commonly used metals in AMS structure. The use of these pure metals is not recommended. Mg lacks structural strength and degrades quickly; Fe has the required structural strength but degrades slowly and its biocompatibility is unclear. As for Zn, its structural strength is lower than that of Mg, but its degradation time is intermediate when compared to Mg and Fe. The properties of Fe alloys make it possible to produce a stent with smaller structural thickness, in addition, they contribute to the binding and transport of oxygen through the blood and to the growth and differentiation of cells. The addition of metals such as Manganese (Mn), Silicon (Si), and Palladium (Pd) reduces the degradation time of the Iron stent (Toong et al. 2020).

As for Zn, its interaction with biological processes such as gene expression ensures its viability as a biomaterial. If the Zn is used individually, the tension required to produce the stents will not be sufficient. The formation of alloys through the mixture of Zn with other metals becomes necessary to dissipate its tension (Toong et al. 2020). The use of the Zn–Cu–Fe (zinc–copper–iron) alloy was suggested as viable for the development of AMS (Yue et al. 2017).

TOONG et al. presented a study that demonstrates the formulation of various Fe and Zn alloys and also highlights the main properties obtained from the alloy and compared the degradation time of each (Toong et al. 2020).

The first AMS to appear was the Magmaris from Biotronik (Rapetto and Leoncini 2017; Drelich and Goldman 2022). This AMS consists of magnesium ($Mg > 90\%$, rare earth metals $<10\%$), and its high concentration in coronary arteries can additionally lead to greater cell proliferation (Jezewski 2019) and reduce cases of arrhythmia, in addition to reducing ischemic reperfusion injuries (Cerrato et al. 2019). It also has a PLLA coating for sirolimus delivery. In this case, the PLLA coating also solves the rapid degradation of the stent. A Magmaris implantation study in pigs showed a lower rate of acute thrombogenicity compared to the BVS Absorb, both with a similar thick structure (Waksman et al. 2017). As for clinical trials, Magmaris started in 2016 and will end in 2025. The partial results of the trial at post-surgery follow-up, 30 days after implantation and 6 months, are very promising and show no cardiac death, stent-induced thrombosis, or myocardial infarction. It is important to emphasize that these results are partial and in the long term the results may change. Additionally, this study was conducted with a small number of patients, simple lesions reducing the possibility of more serious complications, and without a control group using other stents already on the market. In the long-term follow-up, and results of Magmaris, the use of a control group is important to determine if its results are really promising (Rapetto and Leoncini 2017; Wlodarczak et al. 2019). According to Drelich and Goldman, Magmaris AMS competes very favorably with BVS with weaker mechanical properties, but it is unclear how quickly and by what mechanisms the footprint of calcified by-products could dissolve and leave the body. More importantly, there are no publicly available reports on the biological pathway of rare earth metals degradation and excretion (Drelich and Goldman 2022).

However, several other metals with soluble and nontoxic properties to the human body, such as Fe and Zn are considered. Lifetech Scientific's iron-based AMS IBIS is already in clinical trials. IBIS stent has a strut thickness of 53 μm, which is the smallest value for a fully bioresorbable stent created so far. This AMS is a drug-eluting whose coating is obtained by Zn electroplating plus a sirolimus-carrying amorphous PDLLA layer. Results obtained in preclinical tests in rabbit and pig models did not show corrosion and complete coverage of the stent with minimal neointima matrix and complete luminal endothelialization. However, the low resorption rate of this AMS in animal models appears to be a drawback. The first clinical trial of these AMS started in 2018 (Jinnouchi et al. 2019; Peng et al. 2020).

According to Drelich and Goldman, the development of fully bioresorbable metal stents has focused on improving mechanical properties to achieve a benchmark for tensile strength and ductility without compromising resorption performance. Recent reports on the beneficial bioactive properties of the metal ions leaching from biodegradable metal implants have opened a new path for biomaterial development, focusing on the effects of metal ions in the surrounding biological environment (Drelich and Goldman 2022).

14.4 Considerations Related to Biomaterials and Manufacturing Methods

Considering that coronary artery fully resorbable stents were initially designed to show functions like DES, but allow the recovery of vasomotor function, i.e., considering that such devices must work temporarily, the resorption rate of the biomaterial represents the basic quality factor of stent use. The process of degradation and resorption of these devices must be compatible with the healing of vascular lesions. The resorption of biomaterial must happen after it has fulfilled its purpose to support diseased vessel segments and provide effective support to the vessel wall for at least 6 months after implantation (Jezewski 2019).

In general, polymeric fully absorbable stents require a greater strut thickness than most metallic stents due to their low radial strength, which measures the radial resistance of the expanded stent. Thicker polymer stents cause greater arterial damage upon expansion, create greater surface area for thrombosis, and require larger catheters for delivery, making them difficult to use in adult and pediatric interventions. Therefore, new generations of BVS use polymers with improved mechanical properties, which allows for a reduction in the strut thickness. Table 14.1 shows the biomaterials and some structural features of the most important BVS and AMS that reached clinical trials throughout the 2000s. It can be observed that PLLA is the most used polymer among BVSs, while PDLLA is the most used in the outer layer as carrier for eluting antiproliferative drugs. The initial 6-month clinical results demonstrated that PLLA bioresorbable stents are feasible, safe, and effective. Later, other bioresorbable polymer materials were used, including ultra-high molecular weight PLLA, poly(lactide-co-glycolide) (PLGA), poly(ε-caprolactone) (PCL), poly (glycolic acid) (PGA), poly(D-lactide acid) (PDLA), were investigated for BVS application. And, since polymer stents also exhibit lack of radio-opacity, REVA Medical is developing a BVS named FANTON, which presents intrinsic radio-opacity due to the presence of iodinated tyrosine additive, allowing non-invasive radiological evaluation during the whole resorption period.

Table 14.1 Main characteristics and manufacturing processes of some BVS and AMS clinically tested so far

Type	Polymer base	Other components	Name and manufacturer	Characteristics	Year of CE marking approval	References
BVS	PLLA	–	Igaki-Tamai (Kyoto medical planning)	Thickness: 170 μm	–	Indolfi et al. (2016) and Jinnouchi et al. (2019)
	PLLA	PDLLA and Everolimus	Absorb last generation (Abbot vascular)	Thickness: 150 μm Resorption time: 3–4 years	2012	Jinnouchi et al. (2019), Peng et al. (2020), and Azzi and Shatila (2021)
	PLLA	PDLLA	ART-BRS (Arterial remodeling technologies)	Thickness: 170 μm Resorption time: 6 months	2015	Jinnouchi et al. (2019)
	PLLA	Novolimus	DESolve (Elixir medical)	Thickness: 120–150 μm Resorption time: 2 years	2014–2017	Peng et al. (2020) and Azzi and Shatila (2021)
	PTD-PC	Iodine and Sirolimus	Fanton (REVA medical)	Thickness: 125 μm Resorption time: 3 years	2017	Jinnouchi et al. (2019) and Peng et al. (2020)
	Ultra-high molecular weight PLLA	PDLLA and Sirolimus	Fortitude (Amaranth medical)	Thickness: 150 μm Resorption time: 1–2 years	–	Jinnouchi et al. (2019), Peng et al. (2020), and Azzi and Shatila (2021)
	PLLA	Cover of PDLLA and Sirolimus	MeRes 100 (Meril life sciences)	Thickness: 100 μm Resorption time: 2 years	–	Azzi and Shatila (2021)

	PLLA	PDLLA and Sirolimus	Mirage (Manli cardiology)	Thickness: 125–150 μm Resorption time: 14 months	–	Azzi and Shatila (2021)
	PLLA	PLDLLA and Sirolimus	Firesorb (Shanghai MicroPort medical)	Thickness: (100–125) μm Resorption time: 3 years	–	Jinnouchi et al. (2019), Peng et al. (2020), and Azzi and Shatila (2021)
	PLLA	PDLLA	NeoVas (Lepu medical technology)	Thickness: 170 μm Resorption time: 3 years	–	Jinnouchi et al. (2019) and Peng et al. (2020)
	PLLA	PDLLA and Sirolimus	Xinsorb (HuaAn biotechnology)	Thickness: 160 μm Resorption time: 2–3 years	–	Jinnouchi et al. (2019) and Peng et al. (2020)
AMS	Magnesium alloy	PLLA and Sirolimus	Magmaris (Biotronik)	Thickness: 150 μm Resorption time: 1 year	2015–2016	Jinnouchi et al. (2019) and Peng et al. (2020),
	Iron	Zn, PDLLA and Sirolimus	IBIS (Lifetech scientific)	Thickness: 53–70 μm Resorption time: 1 year	–	Jinnouchi et al. (2019) and Peng et al. (2020)

BVS Bioresorbable Vascular Stent, *AMS* Absorbable Metallic Stent, *PLLA* poly (L-acid lactic), *PTD-PC* desaminotyrosine-polycarbonate, *PDLLA* poly(DL-acid lactic)

In AMSs, corrodible metallic materials, such as Mg, Fe, and Zn alloys, are the biodegradable metals that have been used until today.

For both BVS and AMS, design-related features are important for good performance, so the manufacturing methods of these devices play a crucial role in their development. Due to the development of technologies used in the health field and the increasing demand for customized methods, appropriate production methods are essential to optimize the design of vascular stent. Currently, design optimization focuses on aspects of structural and surface characteristics, in addition to patient-specific structure. According to Canfield and Totary-Jain (2018), along with technological advances in stent design comes the need for personalized care that takes in account an individual's genetic lifestyle and environment adapted to healthcare decisions and individualizes treatment (Canfield and Totary-Jain 2018). In addition, it is important to emphasize that modeling and simulation tools using numerical methods are essential to optimize the geometry and biomechanical performance of stents.

In a recent paper, Jiang et al. discuss the technology of vascular stents in terms of biomaterials devices, fabrication methods, and post-processing methods. They present various vascular stent fabrication methods, future perspectives, and make suggestions for the development of stent manufacturing technology. According to these authors, the main methods used include braiding technology, micro-injection molding, laser slicing technique, and additive manufacturing. After the stent is processed, various post-treatment techniques must be used, including drug coating and surface modification, because the microstructure of the surface can significantly affect the biocompatibility of the stent. Therefore, the effectiveness and safety of vascular stents largely depend on the accuracy of the technology used in manufacturing and post-processing (Jiang et al. 2022; Ahadi et al. 2023).

14.5 Final Considerations

The introduction of fully bioresorbable stents is one of the most important innovations in the emerging field of PCI because it brings undeniable advantages, but so far no one has been able to solve the problems that arise. Expectations were high for BVS, but clinical trials were disappointing, so DESs remained as the first choice and BVS did not achieve this status.

First-generation BVS, which were mostly based on PLLA struct, presented performance limitations such as restenosis or inflammation and thrombosis at the stent site. As for the AMS models, there are few options on the market that have started clinical trials and basically comes down to the use of the Mg and Fe alloys. These stents are becoming attractive because of their potential to overcome the limitations of BVSs, with higher radial force and thinner struts. Partial results from clinical trials are quite encouraging; however, it is too early to say anything definitive.

However, although the future of fully absorbable stents remains uncertain, we believe that the motivation for innovation can drive the efforts dedicated to the development of the technology to overcome the current limitations, allowing the routine use of these devices in the clinical practice of coronary vascular restoration.

And finally, it is important to emphasize that fully bioresorbable stents are also an option for arterial revascularization in infants and children because the vessel segment containing the stent cannot adapt to the growth of the child, resulting in a relative constriction of the vessel. Especially in infants and children, when the bioresorbable stent disappears after a few months, the vessel can be repaired and grow naturally into adulthood, avoiding the need for repeated stent dilation surgeries.

Funding Declaration/Acknowledgments Federal University of ABC for Camila Cliquet fellowship.

References

Ahadi F, Azadi M, Biglari M, Bodaghi M, Khaleghian A (2023) Evaluation of coronary stents: A review of types, materials, processing techniques, design, and problems, Heliyon, Volume 9, Issue 2, 2023, e13575, https://doi.org/10.1016/j.heliyon

Azzi N, Shatila W (2021) Update on coronary artery bioresorbable vascular stents in percutaneous coronary revascularization. Rev. Cardiovasc. Med., 22(1), 137–145. https://doi.org/10.31083/j.rcm.2021.01.225

Canfield J, Totary-Jain H (2018) 40 Years of Percutaneous Coronary Intervention: History and Future Directions. J Pers Med. Oct 1; 8(4):33. https://doi.org/10.3390/jpm8040033

Cassar A, Holmes DR Jr, Rihal CS, Gersh BJ (2009) Chronic coronary artery disease: diagnosis and management. Mayo Clin Proc. Dec; 84(12):1130–46. https://doi.org/10.4065/mcp.2009.0391

Centemero M, Cano M, Maldonado G, Domingos de Almeida J. G. M. R., Sousa A. M. R., Sousa J (1993) Angioplastia coronária com uso de duplo balão em artéria de grande calibre. Arquivo Brasileiro de Cardiologia, 61(1), p. 37–39.

Cerrato E, Barbero U. A, Gil Romero J, Quadri G, Mejia-Renteria H, Tomassini F, Ferrari F, Varbella F, Gonzalo F, Escaned J (2019) Magmaris™ resorbable magnesium stent: state-of-art review. Future cardiology: Device Evaluation, 15(4). https://doi.org/10.2217/fca-2018-0081

Drelich JW, Goldman G (2022) Bioresorbable vascular metallic scaffolds: Current status and research trends, Current Opinion in Biomedical Engineering, Volume 24, 100411, https://doi.org/10.1016/j.cobme.2022.100411.

Hospital Israelita Albert Einstein (2020) Angioplastia coronária ou intervenção coronária percutânea. Disponível em: https://www.einstein.br/especialidades/cardiologia/exames-tratamento/angioplastia-coronaria-intervencao-coronaria-percutanea#:~:text=A%20Angioplastia%20Coron%C3%A1ria%20ou%20Interven%C3%A7%C3%A3o,de%20sangue%20para%20o%20cora%C3%A7%C3%A3o. Acesso em: 07 de fev. de 2022

Indolffi C, De Rosa S, Colombo A (2016) Bioresorbable vascular scaffolds – basic concepts and clinical outcome. Nature Reviews, 13, p. 719–729. https://doi.org/10.1038/nrcardio.2016.151

Jezewski M (2019) Bioresorbable Vascular Scaffold – Dead End or Still a Rough Diamond?, Journal of Clinical Medicine, 8, p. 2167–2188. https://doi.org/10.3390/jcm8122167

Jiang W, Zhao W, Zhou T, Wang L, Qiu T (2022) A Review on Manufacturing and Post-Processing Technology of Vascular Stents. Micromachines (Basel). Jan 16;13(1):140. https://doi.org/10.3390/mi13010140
Jinnouchi H, Torii S, Sakamoto A. D, Kolodgie F, Virmani R. V, Finn A (2019) Fully bioresorbable vascular scaffolds: lessons learned and future directions. Nature, 16, p. 286–304. PMID: 34781610
Kumar V, Kumar V, Yadav MS, Dhir S, Nayak P (2021) Efficacy and safety of Bioresorbable Vascular Scaffold (BVS) – Absorb in an Acute Myocardial Infarction – A 45 Month Follow up Study. Journal of the association of physicians of India, 69, p. 41–44.
Laçin NT and Utkan GG (2014) Role of biomaterials in prevention of in-stent restenosis, Journal of Biomedical Materials Research, 102(5): 1113–1120. https://doi.org/10.1002/jbm.b.33083
Oliveira GMM et al. (2022) Cardiovascular Statistics – Brazil 2021 Arq Bras Cardiol, 118(1):115–373. https://doi.org/10.36660/abc.20211012
Omar W and Kumbhani D (2019) The current literature on Bioresorbable Stents: a review. Current atherosclerosis Reports, 21(54), 1–9. https://doi.org/10.1007/s11883-019-0816-4
Pant S, Bressloff NW, Limberrt G and Curzen N. (2011) Multiobjective design optimization of coronary stents, Biomaterials, 32, p. 7755–73. https://doi.org/10.36660/abc.20211012
Peng X, Qu W, Jia Y, et al. (2020) Bioresorbable Scaffolds: Contemporary Status and Future Directions. Frontiers in Cardiovascular Medicine. 7:589571. https://doi.org/10.3389/fcvm.2020.589571
Rapetto C, Leoncini M (2017) Magmaris: a new generation metallic sirolimus-eluting fully bioresorbable scaffold: present status and future perspectives. J Thorac Dis, 9:S903. https://doi.org/10.21037/jtd.2017.06.34
Scafa Udriște A, Niculescu AG, Grumezescu AM, Bădilă E (2021) Cardiovascular Stents: A Review of Past, Current, and Emerging Devices. Materials (Basel). May 12;14(10):2498. https://doi.org/10.3390/ma14102498
Sigwart U, Puel J, Mirkovitch V, Jofre F and Kappenberger L. (1987) Intravascular Stents to prevent occlusion and restenosis after transluminal angioplasty, N Engl J Med, 316: 7016. https://doi.org/10.1056/NEJM198703193161201
Tamai H, Igaki K, Kyo E, Kosuga K, Kawashima A, Matsui S, Komori H, Tsuji T, Motohara S, Uehata H (2000) Initial and 6-Month Results of Biodegradable Poly-*l*-Lactic Acid Coronary Stents in Humans. Circulation, 102(4), p. 399–404. https://doi.org/10.1161/01.cir.102.4.399
Tomberli B, Mattesini A, Baldereschi GL, Di Mario C (2018) A Brief History of Coronary Artery Stents, Revista Española de Cardiología (English Edition), Volume 71, Issue 5, Pages 312–319. ISSN 1885-5857. https://doi.org/10.1016/j.rec.2017.11.022.
Toong D, Chen Koon Ng J, Huang Y, En Hou Wong P (2020) Bioresorbable metals in cardiovascular stents: Material insights and progress. Materialia, 12, p. 2589–2604. https://doi.org/10.1016/j.mtla.2020.100727.
Waksman R, Lipinski MJ, Acampado E, Cheng Q, Adams L, Torii S, Gai J, Torguson R, Hellinga DM, Westman PC, Joner M, Zumstein P, Kolodgie FD, Virmani R (2017) Comparison of Acute Thrombogenicity for Metallic and Polymeric Bioresorbable Scaffolds – Magmaris Versus Absorb in a Porcine Arteriovenous Shunt Model. Circulation: Cardiovascular Interventions, 15(10). https://doi.org/10.1161/CIRCINTERVENTIONS.116.004762
Wlodarczak A, Lanocha M, Jastrzebski A, Pecherzewski M, Szudrowicz M, Jastrzebski W, Nawrot J, Lesiak M (2019) Early outcome of magnesium bioresorbable scaffold implantation in acute coronary syndrome—the initial report from the Magmaris-ACS registry. Catheter Cardiovasc Interv. 93: E287–E292. https://doi.org/10.1002/ccd.28036
WHO. Cardiovascular diseases (CVDs). (2021). https://www.who.int/news-room/fact-sheets/detail/cardiovascular-diseases-(cvds). Accessed in: August, 6, 2023.
Yue R, Huang H, Ke G, Zhang H, Pei J, Xue G, Yuan G (2017) Microstructure, mechanical properties and in vitro degradation behavior of novel Zn-Cu-Fe alloys, Materials Characterization 134 p. 114–122. https://doi.org/10.1016/j.matchar.2017.10.015

Chapter 15
Organs-on-a-Chip: Principles and Applications

Christiane Bertachini Lombello, Laurent Rodrigues Rezende, Andressa Francine Martins, and João Lameu

Abstract Organs-on-a-Chip (OOC) corresponds to an innovative technology that mimics the complex systemic interactions of the human body, with different cells or microtissue constructs on microfluidic platforms. Chip design allows tissue integration simulating systemic interactions, and the microenvironment is controlled with sensors, pumpers, and heaters. Isolated cells, spheroids, microtissue constructs, or even human tissue samples are cultured on the chips, and the platform is designed to maintain specific tissue function and, simultaneously, integrate all the tissues on a chip. Common designs of microdevices applied to Organs-on-a-Chip include double or multichannel and modular systems, porous membrane, open channels, and liquid-wall devices. The chips can be used for toxicological or toxicokinetics evaluations of biomaterials, medical devices, and drugs, improving the cell culture systems usually applied, and diminishing, or even avoiding, the use of animal models. But the Organs-on-a-Chip applications are not limited to evaluation, it can be mentioned developing disease models, diagnosis, and basic physiology research. Some organs and tissues accessed by this technology include, for example, epithelia, heart, bone, liver, kidney, eye, and blood–brain barrier.

Keywords Cells · Devices · Microfluidic · Organs-on-a-Chip · Tissue

---

The original version of the chapter has been revised. A correction to this chapter can be found at https://doi.org/10.1007/978-3-031-38743-2_16.

Authors Christiane Bertachini Lombello and João Lameu have equally contributed to this chapter.

C. B. Lombello (✉) · L. R. Rezende · A. F. Martins
Graduate Program in Biomedical Engineering, Federal University of ABC – UFABC, São Bernardo do Campo, SP, Brazil
e-mail: christiane.lombello@ufabc.edu.br

J. Lameu
Undergraduate Course in Biomedical Engineering, Federal University of ABC, São Bernardo do Campo, SP, Brazil

C. B. Lombello, P. A. da Ana (eds.), *Current Trends in Biomedical Engineering*,
https://doi.org/10.1007/978-3-031-38743-2_15

15.1 Introduction

The health sector is one of the largest sectors in the world economy. In such an important and lucrative area, the development of health technologies, such as tissue engineering (TE) and regenerative medicines (TERM), drugs, and medical devices development, takes years from the project to the product commercialization.

Despite the applicability and efficiency, one of the main challenges for the area is the analysis of adverse local or systemic effects. Testing of medical devices used in animal models is required by regulatory agencies and follows straight directives, which comprise the preclinical assays that investigate the necessary systemic responses related to biotransformation, bioavailability, excretion, among others. In this direction, the 3R principles, which stand for refinement, reduction, and replacement of animal experiments, have been adopted and incentivized (Flecknell 2002; Tannenbaum and Bennett 2015; Petetta and Ciccocioppo 2021). However, in addition to the development of the ethical discussion about the use animal models, it is known that these models have limitations and do not reproduce entirely the human pathophysiology, presenting several discrepancies in the estimation of action and side effects when compared to tests in humans, which lead to an unsuccessful clinical stage (Gasiorowski et al. 2013; Stengelin et al. 2022).

In vitro testing is also preconized in the international biocompatible standards applied by the regulatory agencies, usually previous to the in vivo testing (ISO 10993-1 2018). Moreover, there is an actual tendency to improve the in vitro testing in order to strengthen the evidence provided by these assays, and so propose the substitution of the in vivo (Doke and Dhawale 2015; Stengelin et al. 2022). The advantages of the in vitro systems are reduced time, lower costs, reproducibility, controlled physicochemical conditions, avoids organisms heterogeneity and unwanted systemic interactions. The difficulties in maintaining and handling the animals must also be considered, including the professionals involved, and installations and standards to be followed (Kirkpatrick and Mittermayer 1990; Inayat-Hussain et al. 2009; Doke and Dhawale 2015).

To guarantee the safety application of medical devices, drugs, cosmetics, and others, it is necessary to consider, in an integrative way, aspects such as biocompatibility, systemic, immune, and allergic responses. It is known that in vitro 2D assays, using a cell monolayer, play an important role for screening, but they are not representative in predicting the effects and conditions within a 3D organism, with physiology and metabolisms that work in an integrated way. One example of an objective use of 2D systems is the cytotoxicity assays, one of the first mentioned testing in biological evaluation of medical devices (ISO 10993-5 2009; Nascimento et al. 2017). The cytotoxicity assays consist of exposing the test material, usually a biomaterial or a medical device sample, to direct or indirect contact with a cell culture. It can be used for the in vitro system testing cell lineages or primary culture, and, after appropriate exposure time and assay conditions, the cytotoxicity is evaluated by the existence and severity of morphological cellular response, such as cell lysis, vacuolation, inhibition of proliferation, among others (ISO 10993-5 2009; USP 2019; Müller 2008; Rach et al. 2013). Quantitative methods can also be used to evaluate cell cytotoxicity, such as 2,3-bis [2-methoxy-4-nitro-5-sulfopheny]-2H-tetrazolium-5-carboxyanilide (XTT), 3-(4,5-dimethylthiazol-2-yl)-2,5-diphenyltetrazolium (MTT), neutral red, and crystal violet (Reppetto et al. 2008; Huyck et al. 2012; Almutary and Sanderson 2016). The cytotoxicity can be applied as an in vitro early stage screening using cells on 2D substrates, avoiding more complex testing or even animal experimentation (Fig. 15.1).

Overmore, 3D constructs to achieve a spatial cell distribution onto scaffolds are used to mimic tissue and organ characteristics. These culture models are more representative of the organism architecture, cell distribution on extracellular matrix (ECM), cell interactions, cell physiology, nutrient and oxygen disponibility. But these 3D culture systems that are developed for tissue engineering (TE) and regenerative medicine (TERM) applications are not usually used as evaluation models; on the contrary, scaffolds are tested with the 2D protocols mentioned (ISO 10993-1 2018; Abdelaziz et al. 2023).

Nayak and Gupta (2015) developed a porous scaffold from a binary blend of keratin/agar for skin tissue engineering. The scaffolds presented interconnected porous with dimensions ranging from 50 to 300 μm, hydrophilic nature and adequate mechanical properties, the authors mentioned tensile strength of 00.154 ± 0.031 MPa, and percent of elongation at break with 16.33 ± 2.52%. The

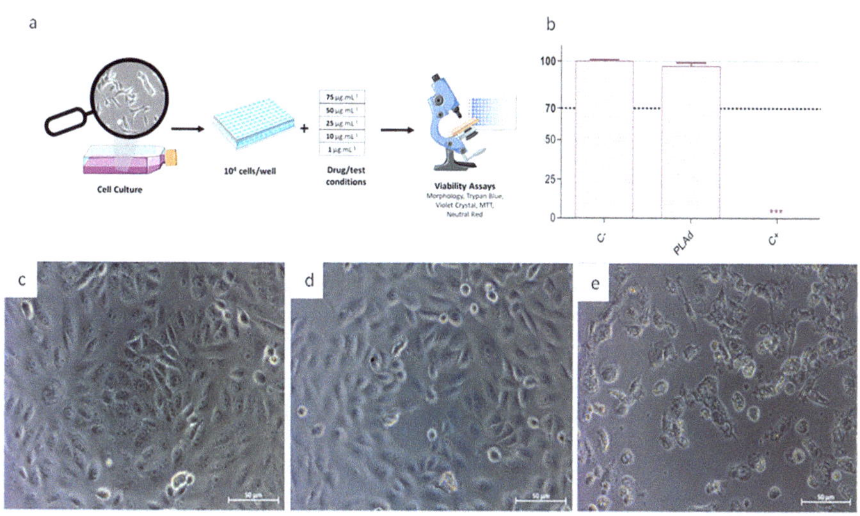

Fig. 15.1 Cytotoxicity testing on 2D cell culture systems. (**a**) Graphical abstract of cytotoxicity testing; (**b–e**) elution cytotoxicity testing of dense PLA (poly(lactic acid)) (PLAd) on Vero cell culture, positive control using standard culture conditions (C+), negative control using phenol 0,25% (C−). (**b**) Quantitative crystal violet assay (cell viability %, considering C− as 100% viability, read absorbance at 540 nm). (**c–e**) Morphological cytotoxicity evaluation: (**c**) C−; (**d**) PLAd; e) C−

negative cytotoxicity tested with mammalian myofibroblast cell line resulted in the conclusion that the scaffolds could be used for skin regeneration.

Similarly, the characterization of a scaffold of positively charged chitosan (CS) and methacrylated gelatin (GelMA) polymers, associated with the chondro-inductive kartogenin (KGN), was characterized viewing cartilage regeneration (Zhang et al. 2022). The study of Bahrami et al. (2021) presented a 3D scaffold made of collagen (Col) and reduced graphene oxide coated collagen (Col-rGO) for bone tissue repair presented data on surface topography, the mechanical and chemical properties adequate for the intended use. However, the 2D cytotoxicity testing was handled in both examples as for the biological scaffold evaluation (Bahrami et al. 2021; Zhang et al. 2022). 3D poly(lactic acid) scaffolds showed differential cytotoxicity using CD133+ cells, to 3D poly(lactic acid) scaffolds if compared to 2D systems (Biagini et al. 2021), reinforcing the development of technologies more reliable to access the human organism complexity. Microtissue in vitro systems have been developed in order to be used as testing systems for preclinical studies, simulating high complex organisms, improving the actual 2D in vitro tests, and viewing animal models substitution (Danku et al. 2022; Stengelin et al. 2022; Zarrintaj et al. 2022).

Thus, Organs-on-a-Chip (OOC) systems emerge as a technological alternative that would reduce the use of animal models, providing more adequate and reproducible data on behavior in human systems. The technology proposes the use of chips with circulating microfluidics that connect microtissue constructs, corresponding to natural human tissue or manufactured ones, through tissue engineering (TE) techniques (Ayuso et al. 2021; Stengelin et al. 2022; Zarrintaj et al. 2022).

15.2 Organs-on-a-Chip (OOC) Concept

The Organs-on-a-Chip (OOC) system is designed to integrate cellular microenvironments, or the corresponding microtissues, and maintain specific tissue functions in a coordinated action. Therefore, the design and microfluidic are fundamental. The idea is to integrate semipermeable membranes generating a continuous flow of fluids in microchannels and chambers with different cell types or microtissue constructs, in order to simulate a systemic integrated flow, similar to the functioning of an organ human, all inside a small device, or a chip (Ingber 2022; Stengelin et al. 2022) (Fig. 15.2).

With the application of techniques such as 3D printing, incorporated into the use of microdevices (sensors, stimulators, and heaters), Organs-on-a-Chip technology aims to enable the connection of these simulation platforms, generating an integrated system of different tissues and organs, such as lung, heart, arteries, liver, and kidneys, connected by a constant micro flow of fluid. This composition would allow the study of biological evaluation of medical devices, the reproduction of inflammatory, allergenic and tumorigenic systems, pharmacokinetic and pharmacodynamic relationships, in a similar pattern than in vivo specimens, into

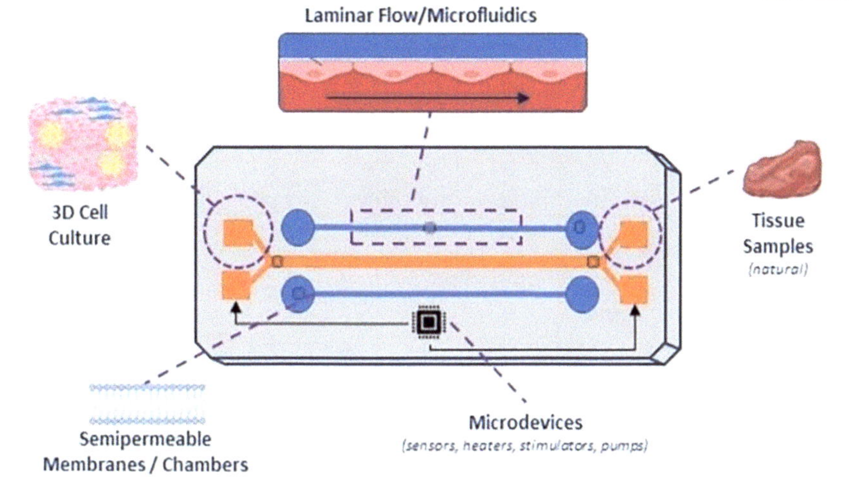

Fig. 15.2 Concept of Organs-on-a-Chip. This system uses microfluidics techniques in conjunction with cells, tissues, or microtissues engineering to mimic complex interactions

a homeostatic scenario, in normal or pathological and/or altered conditions in the models (Ayuso et al. 2021; Purtscher et al. 2021). This similarity would significantly increase the probability of success in the clinical phase, reducing or even eliminating animal model testing and diminishing the risk of adverse reactions in the clinical evaluation phase (Danku et al. 2022; Ingber 2022; Leung et al. 2022; Stengelin et al. 2022).

Another expressive advantage of the Organs-on-a-Chip technology is the possibility of recording and rigorously controlling the variables applied to the system, resulting in detailed data collection, in real time. Conditions such as concentrations of biomolecules, ions, temperature, pH, various exogenous stimulants, flow rate, and local forces in cell behavior can be controlled and monitored (Ayuso et al. 2021; Ingber 2022).

15.3 Organs-on-a-Chip (OOC) Applications

The scientific published data on Organs-on-a-Chip technique has increased in the last 10 years, as advances in 3D cell culture and tissue engineering have increased the possibilities and brought new applications for Organs-on-a-Chip systems (Fig. 15.3). Currently, there are systems described to test models for transdermal and intravenous absorption covering the main routes of drug administration (Wang et al. 2016; Dos Santos et al. 2023), oral/rectal absorption via intestine systems (Pocock et al. 2017), or model of absorption via inhalation from lung constructs (Park et al. 2018). In addition to toxicologic, toxicokinetics, and pharmacological applicability,

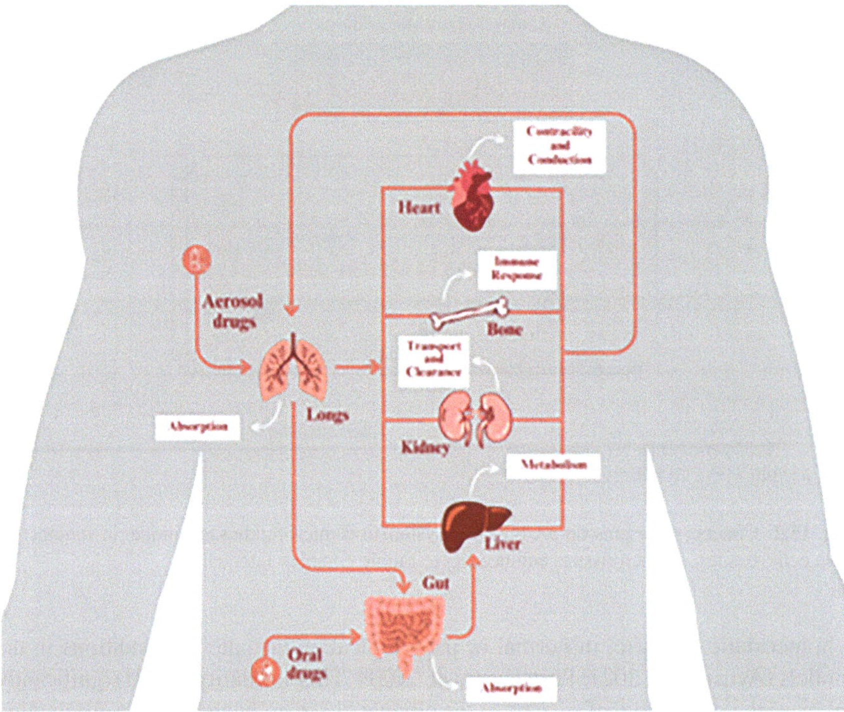

Fig. 15.3 Organs-on-a-Chip (OOC) applications

models for the study of pathologies, such as neurodegeneration with simulation of the blood–brain barrier, and tumor models are already being developed. Thus, this technique has been widely applied within studies in cell biology and human physiology, bringing more precision and speed, not only for the development of medical devices, drugs, and cosmetics, but also in basic research studies of human pathophysiology in general, understanding their origin and also the influence of external factors for their establishment in human being.

The Organs-on-a-Chip's wide potential for biomedical application, focusing on the diagnosis of diseases, their treatment, and, mainly, on precision regenerative medicine, is based on the purpose of biomimicry techniques to reproduce the anatomy and physiology of organs and tissues, so that safety and efficacy tests are possible, and also to improve understanding of the microenvironment where they are allocated, along with the cellular experiments to be carried out (Ayuso et al. 2021; Ingber 2022; Leung et al. 2022). Finally, it is possible to mention the use of Organs-on-a-Chip technologies of multi or specific organs, such as liver, kidney, lung, intestine, eye, bone, blood–brain barrier, and transdermal systems.

15.3.1 Multi-Organ-on-a-Chip

One of the possible applications of the Organs-on-a-Chip refers to the potentially toxic biomaterial, medical devices, drugs or cosmetics, and the chronic toxicological studies associated. The systemic physiological responses may be modeled in vitro using this technology, assembling multi-organ through microfluidic flow in human body-on-chips systems. However, not all the organism responses are represented, so the results must be carefully analyzed. Still, the protocols may be more directed to a certain metabolic or physiologic aspect.

These models can integrate the response of 2–10 different organs on a chip (Ingber 2022). Coculture cell systems can be seen as Organs-on-a-Chip precursors. An example is the system developed simultaneously to the multi-12-transwell-like constructs. The different tissue interactions are based on coculturing different human cell types in each side of the system. It can be mentioned, for example, hepatocytes and liver sinusoidal endothelial cells (Domansky et al. 2010). Edington et al. (2018) developed microphysiological systems (MPSs) as in vitro models for specialized culture microenvironments, including 3D scaffolds and microperfusion flow. The construct could be performed in different configurations, allowing specific responses as for liver protein distribution observed with phenotypic markers. The system used and approach the authors called coculture multiple different microphysiological systems (MPSs) and "Physiome-on-a-Chip," with the main objective to study and direct drug discovery.

While the first devices used separated cells in a coculture system, from different sources, as differentiated tissue or stem cells, other devices were based on spheroids. These microtissue are formed of high-density cell deposities, with or without micro and nano structured scaffolds (Kim et al. 2020). To achieve a spheroid construct it is maximized cell–cell cohesion by using centrifugal forces, f.ex., and interaction with the substrate is not enhanced. This system aims to direct cell differentiation mainly of stem cells into specific types. These models are widely utilized in Organs-on-a-Chip systems, and the use of bioprinting technology enhances its structure (Bovard et al. 2017; Stengelin et al. 2022).

Usually the Multi-Organ-on-a-Chip devices usually connect a liver system to other tissue, viewing the characterization of toxicological responses. TissUse is based in a chip that connects liver systems, in the form of spheroids of HepaRG and hepatic stellate cells, with human skin samples, obtained through biopsy, through microfluidic channels connecting the compartments (Wagner et al. 2013; Bovard et al. 2017). In this example, the original tissue is applied into the device, avoiding the need to develop 3D constructs or spheroids.

Another example, the emulate (Varone et al. 2021) is a two parallel channel construct made with soft lithography onto polydimethylsiloxane (PDMS) clear matrices. Each channel can settle different cell types, separated by the central extracellular matrix (ECM)-coated membrane. The interface of cell–cell interaction is settled on micrometer-sized pores allowing transmigration and mechanical forces can be applied, such as comprehensive ones, viewing primary human skin, and alveolar epithelial cells.

15.3.2 Bone-on-a-Chip

The development of Bone-on-a-Chip may have different directions. Schoon et al. (2020) developed a system of human decellularized bone to evaluate the release of metals derived from arthroplasty implants, such as titanium (Ti), chromium (Cr), and cobalt (Co). The system used human mesenchymal stem cells, osteoblasts, and bone marrow cells. The system showed similarity to in vivo observations in peri-implant areas, with the Co and Cr integration into peri-implant bone trabeculae, a result not seen for Ti. The results stand for the application of the system to toxicokinetics studies, in a reliable way.

On the contrary, other studies guide bone remodeling and models for bone diseases (Mansoorifar et al. 2021; Zhang et al. 2023). Zhang et al. used biosensor integration in organ chips to bone regeneration, mentioning next-generation technologies for bone systems, once the biosensor allows the integration to monitor all experimental or use conditions, with possible translational research. Galvan-Chacon et al. (2022) also developed a biomimetic Bone-on-a-Chip system to be applied in tissue formation and remodeling. The platform made of two-photon polymerization (2PP) from a 3D nano-computed tomography scan of trabecular bone, coated with bone mineral-like calcium phosphate, to achieve the biomimetics was associated with human mesenchymal stromal cells, cultured for 21 days in this system, with promising results, maintaining cellular viability, and secretion of extracellular matrix elements, characteristic of bone tissue.

15.3.3 Eye-on-a-Chip

Organs-on-a-Chip to applied to eye tissues are among recent developments. Tissue engineering of eye has some particularity, and eye models can be applied to different situations, from ophthalmic drug screening to ocular microenvironment biomimetism (Haderspeck et al. 2019; Peng et al. 2020; Aghmiuni and Keshel 2023; Ingber 2022). Promising results were obtained for understanding and in vitro simulating eye physiology and anatomy. The animal model substitutions are again one of the goals for these developments. Achberger et al. (2019) developed a retina chip using induced pluripotent stem cells (iPS) cells to obtain retinal cells. The system uses a vascular perfusion and mimics the mature photoreceptor segments and retinal pigmented epithelium interactions, and also can be applied to drug toxicity evaluation protocols. In the same direction, cornea models are also developed, reproducing its multilayered avascular structure, and based on human corneal epithelial cells (Bennet et al. 2018).

15.3.4 Heart-on-a-Chip

The platform of Heart-on-a-Chip has the singularity of contractile function and electrical activity coupled to the entire system. The development of cardiomyocytes obtained from iPSC is a fundamental cell type in these systems (Zhang et al. 2016) and can be associated with fibroblasts (Zhao et al. 2019). The distinction of atrial and ventricular tissues, based on the electrophysiology, was obtained with specific chambers design and can be applied not only to study drug responses but also differential gene expression (Zhao et al. 2019). Heart-on-a-Chip can also reproduce the interface of cardiomyocytes and endothelial cells to be applied to understand drug action and toxicity on heart myocardium (Zhang et al. 2016).

15.3.5 Brain-on-a-Chip and Blood–Brain Barrier

Brain models developed in a two-channel human chip were developed by Pediaditakis et al. (2021). The system included iPS-cell-derived dopaminergic neurons, astrocytes, microglia, and pericytes, and also iPS cell-derived microvascular brain endothelium, mimicking the interface between cell types. Such a complex study reproduced in a controlled device stands for drug screening, toxicokinetics, and also disease modeling, mainly those derived from genetic sources or degenerative processes, such as Parkinson disease (Ingber 2022).

In the same direction, Organs-on-a-Chip principles were applied to develop human blood–brain barrier chip, representative of the neurovascular unit. These systems are applied to understand the events and possible treatments to neurological disease that affects this barrier. The chips, two-channel based, may use derived iPS cells or even primary cells (Pediaditakis et al. 2021; Ingber 2022). The permeable barrier reconstitution can be used in simulated protocols of injury and neuroinflammation, contributing to basic research data and clinical situation. The difficulty in accessing this information from in vivo models, as brain homeostasis and barrier protection, states the importance of this development (Leung et al. 2022).

15.3.6 Liver-on-a-Chip

Organs-on-a-Chip technology used here can help detect the possible damage caused by specific drugs in liver cells, demonstrating its importance in hepatotoxicity studies (Danku et al. 2022). The liver is a parenchymal organ that presents a certain complexity, due to the biological functions essential to human beings. Various types of cells, such as hepatocytes, hepatic stellate cells, Kupffer cells, and endothelial cells, ensure the functionality of this tissue, forming its metabolic environment (Deng et al. 2019). The Liver-on-a-Chip is a microphysiological system that aims to recreate the conditions of this organ microscopically, and ideally, it is a system that can mimic the conditions of hepatocytes and also their physical–chemical environment (Hassan et al. 2020).

Another important application, in addition to the toxicity, deals with the development of microfluidic chips for the establishment of models of liver diseases, a system that allows the review of the stages of the life cycle of these diseases, including their replication (Deng et al. 2019).

15.3.7 Kidney-on-a-Chip

The Kidney-on-a-Chip has several applications that can aid in biomedical developments, such as the evaluation of nephrotoxic drugs from kidney injury models (Kim and Takayama 2015), and the use of a multilayer microfluidic system that can help simulate a fluid environment for the study of renal filtration by maintaining cell polarity (Wu et al. 2020; Chambers et al. 2023).

Petrosyan et al. (2019) developed a Glomerulus-on-a-Chip as a renal disease model and also for drug testing, with a system highly reproducible and possible to validate to be used clinically. Different types of human podocytes were used to mimic glomerular epithelium in a three-channel device creating a functional in vitro tissue.

15.3.8 Lung-on-a-Chip

Considered the most important organ of the human respiratory system, the lung is essentially responsible for gas exchange, which has the alveoli as its regulator (Huang et al. 2021; Wu et al. 2020). In addition, this structure does not have a high regenerative capacity, which can make the damage caused permanent (Danku et al. 2022). Thus, due to its main function, it is an organ that has contact with external agents and can be affected by a range of related pathologies, in addition, one of the most commonly seen diseases in this tissue is lung cancer (Huang et al. 2021; Danku et al. 2022). This set of diseases can lead to pulmonary insufficiency, one of the main causes of death in the world, which increases the search for reliable physiological models that represent their anatomy for the development of drugs and the study of pathophysiology (Huang et al. 2021).

Through this scenario, the Lung-on-a-Chip can be developed through a microfluidic system coupled to lung cells by simulating the behavior of the lungs by applying pressure under the device, which facilitates the evaluation of the reactions of this tissue in relation to cells tumors, external agents, and also exposure to smoke in order to end the side effects caused (Wu et al. 2020). The use of the Lung-on-a-Chip can also mimic the lung parenchyma for the simulation of breathing and also the simulation of fluid exchange for understanding the epithelial barrier of this tissue (Park et al. 2018; Leung et al. 2022).

Park et al. (2018) used a decellularized lung extracellular matrix and hydrogel to a vascularized two-channel construct of Cancer-on-a-Chip. As cell source it used spheroids of A549 cells, HUVECs, and human lung and the main applications related to the action of Doxorubicin, as in vitro drug screening tool.

15.3.9 Gut-on-a-Chip

Responsible for promoting the digestion and absorption of nutrients and also for having a regulating role in the function of the other organs of the human body, the intestine has as its main function to establish itself as a barrier that limits the locomotion of microorganisms (such as microbes), within of the digestive system (Ashammakhi et al. 2020). This structure also participates in the immune and endocrine systems for the maintenance of human health (Ashammakhi et al. 2020). Regarding its structure, the intestine is composed of villi and secretory mucus that is composed of differentiated epithelial cells that protect the intestinal mucosa (Shin and Kim 2022) and also of enterocytes that are present in the intestinal epithelium (Beaurivage et al. 2020).

The Intestine-on-a-Chip comes as an alternative to understand the mechanism of diseases that affect the gastrointestinal tract, such as Crohn's disease, colorectal cancer, and ulcerative colitis, whose pathogenesis is not entirely known (Beaurivage et al. 2020). Thus, a multichannel perfusion platform was developed with the cells of interest and the extracellular matrix, for example, with adenocarcinoma cells (Beaurivage et al. 2020). The development of a microfluidic system lined by intestinal epithelial and vascular endothelial cells could also facilitate the modeling of radiation injuries and evaluate drugs for their treatment. Finally, the use of stem cells from patients of interest in Organs-on-a-Chip technology also promotes the advancement of personalized medicine (Ashammakhi et al. 2020).

15.4 Microfluidics and Organs-on-a-Chip (OOC)

Microfluidics is the field that studies and handles fluids at microscale. The transport phenomena are enhanced at this length scale (Virumbrales-Muñoz and Ayuso 2022; Folch 2016), mostly due to the elevated surface area to volume ratio. Fluid flow at microscale is marked by the predominance of capillary and viscous forces over inertial ones (e.g., gravitational field force) (Virumbrales-Muñoz and Ayuso 2022; Folch 2016), resulting in an improved control of fluid flow, a key factor to precisely study biological systems. The fluid flow regime is characterized by the nondimensional number of Reynolds, given by the ratio of momentum transport by convection (inertial contribution) to diffusion (viscous molecular contribution). For microchannel diameter below 1 millimeter the laminar flow regime prevails (Bhatia and Ingber 2014).

The well-controlled fluid flow regime allows the generation of chemical and physical gradients mimicking the physiological environment, as required in Organs-on-a-Chip (OOC) (Fig. 15.4). The cell environment is sensible to fluid shear stresses, inherent from the viscous interaction between the liquid medium and the microchannel walls, where the cell surfaces are exposed. Also, the mass transport regime of soluble factors is coupled to the fluid flow regime. Changes in flow rate, or flow velocity, alter the shear stresses and molecular signaling. Besides, the channel dimensions and geometry design can be used to alter such characteristics (Carraro et al. 2008). These advantages allow to evaluate noninvasively biological processes, by creating a precisely controlled cell microenvironment similar to in vivo physiology. This environment fundamentally includes the interaction of cells with soluble factors, extracellular matrices (ECM), and 2D or 3D structures with specific mechanical and biological properties. Microfluidics application in biomedical area includes cardiac tissue formation (Radisic et al. 2005), cell migration (Li Jeon et al. 2002), among others. Also, cell separation can be performed in the microfluidic devices by using micro-engineered restrictions for cell passage (Lee et al. 2007) or nanostructured porous membranes (Carraro et al. 2008). Fluid dynamic and mechanical modeling and simulation tools can be used to optimize the microdevice design and intensify molecular transport (nutrient delivery and waste removal), thus enhancing cell survival and function (Carraro et al. 2008).

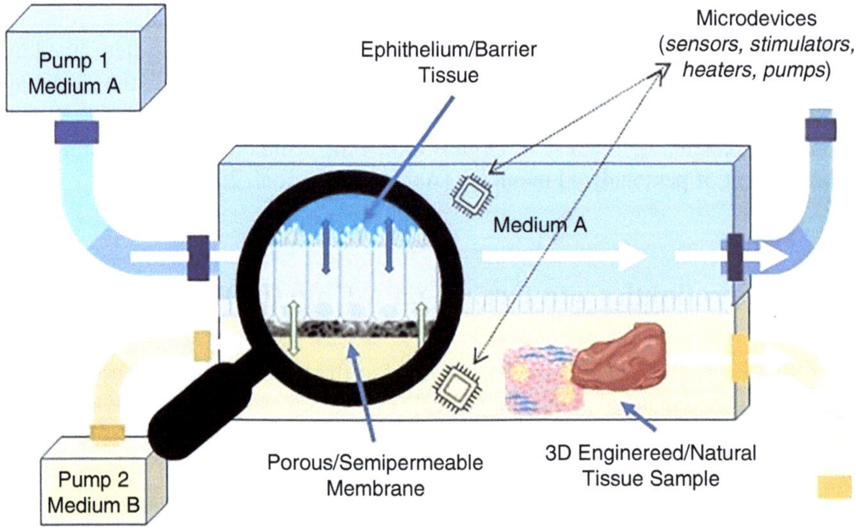

Fig. 15.4 Controlled fluid flow, Organs-on-a-Chip (OOC) system, mimics the physiological environment. Perfusion condition, semipermeable membranes, cell/tissue samples, and controlling microdevices, such as sensors, simulators, heaters, and pumps, can be used in different OOC constructs

The media perfusion is a key factor for Organs-on-a-Chip. This feature must mimic the circulatory system in order to provide the concentration gradients for nutrient transport and waste removal. Due to restricted low volume of fluid in the Organ-on-a-Chip, the fraction and frequency of sampling (volume removal) for offline analysis is also an important factor, especially to keep the blood/interstitial fluid to body mass ratio physiologically realistic (Bhatia and Ingber 2014). Partial fluid replacement, about only 25–50 μl per day, is recommended to emulate more realistic conditions and to sustain the Organs-on-a-Chip operating for several days, as often need to achieve the physio-pathological condition.

Due to the reduced microscale, the transport phenomena are enhanced, and the conventional characteristics of macroscale cell culture, including nutrient consumption and bubble formation, can drastically change critical aspects of the cell environment, for example, nutrient availability and pH (Leung et al. 2022). Accordingly, it is fundamental to understand the role of physical forces acting in the system, including fluid flow and its relationship with molecular transport. In addition to the nondimensional Reynolds number, we can highlight two additional nondimensional numbers: the Peclet number, Pe, and the Damköhler number, Da. The Peclet number clarifies the molecular transport mechanism, given by the ratio of convective transport to the diffusive transport. The Damköhler number is generally defined as the ratio between (bio)chemical timescale to the mass transport timescale. For continuous systems, defined by the predominant mass transfer mechanism, being convection timescale is defined as the predominant mechanism of mass transfers, while molecular diffusion timescale is defined for systems involving mass transfer through interfaces. For $Da \gg 1$, mass transport delimitates the process. For the design of Organs-on-a-Chip or microfluidic devices for biology applications, all three nondimensional numbers (Re, Pe, and Da) must be analyzed as a starting point, providing insights about the relationship between the transport phenomena and the (bio)chemical behavior (Leung et al. 2022; Bhatia and Ingber 2014).

Several pumping strategies have been proposed and developed to achieve the required media perfusion. These include peristaltic, syringe, and membrane pumps, microvalve-driven actuators, and also pump-free systems, based in hydrostatic pressure (gravity-driven flows) (Kurth et al. 2020; Yu et al. 2020). Gravity-driven flow presents some restrictions, including that the flow rate is not constant over time, once the hydrostatic pressure inherent from the liquid height continuously change in time. Strategies to overcome this issue include the use of a tilting plate where the microfluidic platform are included. The amplitude and frequency of plate motion are adjusted to achieve the desired throughput. Peristaltic pumps deliver pulsatile flow and are the simplest solution to recirculate the media. Syringe pumps provide a very stable and controlled flow rate, however, limited by the syringe volume. Membrane pumps presents a cavity, a flexible membrane (diaphragm), inlet and outlet valves, and an actuator. Different force principles can be used to control the diaphragm, including pneumatic, electromagnetic, thermal, and piezoelectric forces, the most common. The flow rate is controlled by the cavity volume, affected by the flexible membrane, delivering a pulsatile flow (Kurth et al. 2020). The ideal pump system depends on the specific Organs-on-a-Chip media, including the necessity of one pass or recirculation passes of perfusion medium and cell environment, including the proper shear stress level. For instance, one pass perfusion can be achieved by using syringe pump or gravity driven flow. One single pass allows stable supply of nutrients, however, restricted to only one direction for tissue communication (Leung et al. 2022; Kurth et al. 2020). For Organs-on-a-Chip requiring molecular signaling, a key factor for multiorgan devices, also known as physical coupling of Organs-on-a-Chip (Leung et al. 2022), flow recirculation is mandatory. For such systems, peristaltic pumps can be used. An important issue of such systems is that the media recirculation leads to the accumulation of waste products, not allowing continuous supply of nutrients. Another strategy to develop more complex cell environment in Organs-on-a-Chip is the functional coupling, by transferring a conditioned medium from one Organs-on-a-Chip to another in a sequential step (Novak et al. 2020), eliminating the technical issues of physical coupling (Leung et al. 2022).

15.4.1 Microfluidic Designs and Devices

In this section, the most common designs of microdevices applied to Organs-on-a-Chip are described, including multichannel and modular system, porous membrane, open channels, and liquid-wall devices (LWD) (Fig. 15.5).

Multichannel Designs This configuration employs two or more parallel channels. Usually, cells are embedded in 3D hydrogel structures, acting as porous membrane, surrounded by liquid media. The interface created by the 3D hydrogel and the perfusion media from the parallel channels provide biochemical gradients, mimicking the biological environment. Several configurations of multichannel design have been proposed to investigate other biological systems, including brain (Spijkers et al. 2021) and vascular biology, e.g., vasculogenesis and angiogenesis (Baker et al. 2013; Nguyen et al. 2013). One of the restrictions of this configuration is the integration of microvascular networks with other tissues (Virumbrales-Muñoz and Ayuso 2022). The multichannel design also supports the development of modular configuration, or multiculture systems. These advanced designs can group in a single-device different cell culture. Multiple cell population are selectively isolated in spatial regions and the interaction among them is controlled in space and time by the media perfusion and modular channel design (Yu et al. 2021). The modular system can be arranged in vertical or horizontal channels. Major limitations of modular systems are the low volume of each layer and cell migration and biochemical gradients in vertical direction, hindering the microscopy analysis (vertical modules) (Yu et al. 2021) and difficulties to reassemble the horizontal platform to continue the experiment (Ayuso et al. 2021).

Porous Membrane Designs This Organs-on-a-Chip configuration integrates two parallel microchannels separated by porous substrates. The porous membrane allows analysis of tissue barrier functions, absorption, secretion, and transcellular transport. This configuration allows the culture of different cell types on the opposite

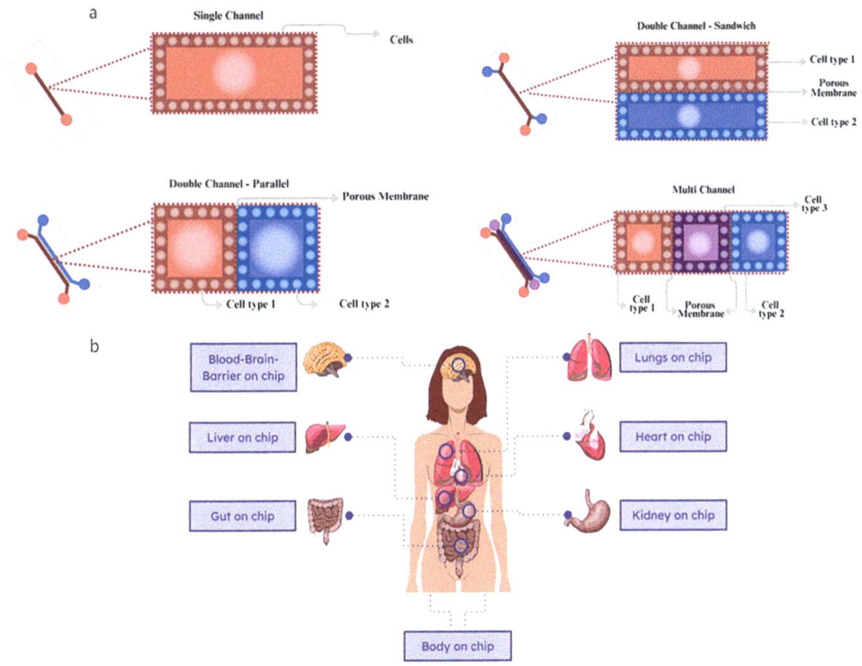

Fig. 15.5 Common designs of Organs-on-a-Chip devices (**a**), single channel, double channel (parallel or sandwich), and multichannel; and examples of applications (**b**)

substrate sites, then creating tissue–tissue interfaces that mimics the realistic in vivo cell environment of an organ (Shin et al. 2021). Porous membrane microdevices has been employed in 2D monolayers or 3D hydrogel structures (Leung et al. 2022). The complex mechanical microenvironment can be reproduced by these designs, including cyclic mechanical strain and stresses by using deformable membranes along air chambers, that promote rhythmic stretching and relaxing of the attached membrane wall and the adjacent microchannel zone. These features are a fundamental key to reproduce reliably cell environment of cardiovascular and breathing cycling and peristalsis, among others (Shin et al. 2021; Sung et al. 2019). In addition to the mechanical stimuli, electrical fields can be used to promote contractile cells to reproduce cardiac cycle and also to stimulate wound healing (Leung et al. 2022; Sung et al. 2019).

Open Channels (OCs) This design present one or more interphases directly contacting the environment, resulting in air–liquid interphase. OCs stand out to mimic biological structures naturally exposed to air, including skin or lung epithelium (Sutterby et al. 2020; Ayuso et al. 2021). Main limitations of OCs application to biological system include the high rate of evaporation, once the cell is commonly cultured for several days or weeks. Design optimization is still required to improve the OCs reliability for Organs-on-a-Chip application (Virumbrales-Muñoz and Ayuso 2022).

Liquid-Wall Devices (LWD) This design implies cell culture in aqueous solution (e.g., collagen hydrogel or cell media) with the geometry controlled by using an immiscible liquid (e.g., oil). The device design is then generated by pipetting the hydrogel droplet on the substrate made by polymers or glass. Several microchannels and accessories can be developed using this method, including mixing structures, reservoirs, and gradient generators (Virumbrales-Muñoz and Ayuso 2022). The oil-based structure provides lower evaporation rate of LWD concerning OCs, representing an interesting and simple alternative to control nutrient and drug concentration within the system. Also, LWD presents advantages over other microdevices, including easy accessibility due to the wall-less configuration and flexibility for system reconfiguration. Such characteristics have been employed to explore different mechanobiological phenomena, mostly related to shear stress and biochemical gradient responses (Virumbrales-Muñoz and Ayuso 2022; Li et al. 2020).

15.5 Conclusion

Organs-on-a-Chip appears as interesting alternatives to conventional cell culture and animal models, with promising results for different applications in recent years. Cell culture systems, spheroids, and microtissue constructs have improved to achieve more reliable results. However, several challenges remain. Some limitations are related to the time period for pathology development (about less than 1 month) and spatial cell distribution within the specific microarchitecture of the organ or tissue. Also, the next generation of Organs-on-a-Chip must integrate in-line sensors for critical and functional parameters (e.g., temperature, pH, and fluid flow) and include microscopic imaging tools allowing real-time monitoring. The perspective of models to evaluate pathologic outcomes and to screen drugs for personalized therapies, using patient-specific primary cell culture or with gene editing technologies, is still ongoing and, combined with design and microfluidic improvement, represents future innovations for the Organs-on-a-Chip.

References

Abdelaziz AG, Nageh H, Abdo SM et al (2023) A Review of 3D Polymeric Scaffolds for Bone Tissue Engineering: Principles, Fabrication Techniques, Immunomodulatory Roles, and Challenges. Bioengineering 10(2):204. https://doi.org/10.3390/bioengineering10020204

Achberger K, Probst C, Haderspeck J et al. (2019) Merging organoid and organ-on-a-chip technology to generate complex multi-layer tissue models in a human retina-on-a-chip platform. Elife. 8:e46188. https://doi.org/10.7554/eLife.46188.

Aghmiuni AI, Keshel SH (2023) Chapter 10 – Eye-on-a-chip. In Woodhead Publishing Series in Biomaterials, Principles of Human Organs-on-Chips, Woodhead Publishing, p.315–369, https://doi.org/10.1016/B978-0-12-823536-2.00001-8

Almutary A, Sanderson BJS (2016) The MTT and Crystal Violet Assays: Potential Confounders in Nanoparticle. Toxicity Testing International Journal of Toxicology, 35(4) 454–462. https://doi.org/10.1177/1091581816648906

Ashammakhi N, Nasiri R, Roberto NB, et al. (2020) Gut-on-a-chip: Current progress and future opportunities,. Biomaterials 255:120196: https://doi.org/10.1016/j.biomaterials.2020.120196

Ayuso JM, Sadangi S, Lares M, et al. (2021) Microfluidic model with air-walls reveals fibroblasts and keratinocytes modulate melanoma cell phenotype, migration, and metabolism. Lab Chip 21(6):1139–1149. https://doi.org/10.1039/d0lc00988a

Bhatia SN, Ingber DE (2014) Microfluidic organs-on-chips. Nat Biotechnol 32(8):760–72. https://doi.org/10.1038/nbt.2989

Baker BM, Trappmann B, Stapleton SC, et al. (2013) Microfluidics embedded within extracellular matrix to define vascular architectures and pattern diffusive gradients. Lab Chip 13(16):3246–52. https://doi.org/10.1039/c3lc50493j

Bahrami S, Baheiraei N, Shahrezaee M. (2021, August 18) Biomimetic reduced graphene oxide coated collagen scaffold for in situ bone regeneration. Sci Rep. 11(1):16783

Beaurivage C, Kanapeckaite A, Loomans C et al. (2020) Development of a human primary gut-on-a-chip to model inflammatory processes. Sci Rep 10:21475. https://doi.org/10.1038/s41598-020-78359-2

Bennet D, Estlack Z, Reid T, et al. (2018) A microengineered human corneal epithelium-on-a-chip for eye drops mass transport evaluation. Lab Chip. 18(11):1539-1551. https://doi.org/10.1039/c8lc00158h

Biagini G, Senegaglia AC, Pereira T et al (2021) 3D Poly(Lactic Acid) Scaffolds Promote Different Behaviors on Endothelial Progenitors and Adipose-Derived Stromal Cells in Comparison With Standard 2D Cultures. Front Bioeng Biotechnol 9:700862. https://doi.org/10.3389/fbioe.2021.700862

Bovard D, Iskandar A, Luettich K, et al. (2017) Organs-on-a-chip. Toxicol Res Appl 1:239784731772635. https://doi.org/10.1177/2397847317726351

Carraro A, Hsu WM, Kulig KM et al. (2008) In vitro analysis of a hepatic device with intrinsic microvascularbased channels. Biomed Microdevices 10:795–805. https://doi.org/10.1007/s10544-008-9194-3

Chambers BE, Weaver NE, Wingert RA (2023) The "3Ds" of Growing Kidney Organoids: Advances in Nephron Development, Disease Modeling, and Drug Screening. Cells 12(4):549. https://doi.org/10.3390/cells12040549

Danku AE, Dulf EH, Braicu C et al. (2022) Organ-On-A-Chip: A Survey of Technical Results and Problems. Front Bioeng Biotechnol 10:840674. https://doi.org/10.3389/fbioe.2022.840674

Deng J, Wei W, Chen Z et al. (2019) Engineered Liver-on-a-Chip Platform to Mimic Liver Functions and Its Biomedical Applications: A Review. Micromachines (Basel) 10(10):676. https://doi.org/10.3390/mi10100676

Domansky K, Inman W, Serdy J et al. (2010) Perfused multiwell plate for 3D liver tissue engineering. Lab Chip 10(1):51–8. https://doi.org/10.1039/b913221j

Doke SK, Dhawale SC (2015) Alternatives to animal testing: A review. Saudi Pharm J 23(3):223–9. https://doi.org/10.1016/j.jsps.2013.11.002

Dos Santos JF, Freitas-Marchi BL, Reigado GR et al. (2023) Mesenchymal stem cells express epidermal markers in an in vitro reconstructed human skin model. Front Cell Dev Biol 10:1012637. https://doi.org/10.3389/fcell.2022.1012637

Edington CD, Chen WLK, Geishecker E et al. (2018) Interconnected Microphysiological Systems for Quantitative Biology and Pharmacology Studies. Sci Rep 8(1):4530. https://doi.org/10.1038/s41598-018-22749-0

Flecknell P (2002) Replacement, reduction and refinement. ALTEX 19(2):73-8. Available at: https://www.altex.org/index.php/altex/article/view/1106. Accessed: May 2023

Folch A (2016) Introduction to bioMEMS. CRC Press, Boca Raton, FL, USA. 528p. https://doi.org/10.1201/b12263

Galvan-Chacon VP, Zampouca A, Hesse B et al. (2022) Bone-on-a-Chip: A Microscale 3D Biomimetic Model to Study Bone Regeneration. Advanc Eng Mater, 24: 2101467. https://doi.org/10.1002/adem.202101467

Gasiorowski JZ, Murphy CJ, Nealey PF (2013) Biophysical cues and cell behavior: the big impact of little things. Annu Rev Biomed Eng 15:155–76. https://doi.org/10.1146/annurev-bioeng-071811-150021

Haderspeck JC, Chuchuy J, Kustermann S et al. (2019) Organ-on-a-chip technologies that can transform ophthalmic drug discovery and disease modeling. Expert Opin Drug Discov 4(1):47–57. https://doi.org/10.1080/17460441.2019.1551873.

Hassan S, Sebastian S, Maharjan S, et al. (2020) Liver-on-a-Chip Models of Fatty Liver Disease. Hepatology. 71(2):733–740. https://doi.org/10.1002/hep.31106

Huang D, Liu T, Liao J, et al. (2021) Reversed-engineered human alveolar lung-on-a-chip model. Proc Natl Acad Sci 118(19):e2016146118. https://doi.org/10.1073/pnas.2016146118

Huyck L, Ampe C, Van Troys M. The XTT cell proliferation assay applied to cell layers embedded in three-dimensional matrix. Assay Drug Dev Technol. 2012 Aug;10(4):382–92.

Inayat-Hussain S, Rajab N, Siew, EL (2009). In vitro testing of biomaterials toxicity and biocompatibility. Cellul Resp Biomat, 2009:508–537. https://doi.org/10.1533/9781845695477.3.508.

Ingber DE (2022) Human organs-on-chips for disease modelling, drug development and personalized medicine. Nat Rev Genet 23:467–491. https://doi.org/10.1038/s41576-022-00466-9

ISO – International Organization for Standardization 10993-1: 2018. Biological evaluation of medical devices — Part 1: Evaluation and testing within a risk management process p.41

ISO – International Organization for Standardization 10993-5: 2009. Biological Evaluation of Medical Devices. Part 5: Tests for in Vitro Cytotoxicity p.34

Kim SJ, Kim EM, Yamamoto M et al. (2020) Engineering Multi-Cellular Spheroids for Tissue Engineering and Regenerative Medicine. Adv Healthc Mater 30:e2000608. https://doi.org/10.1002/adhm.202000608

Kim S, Takayama S. (2015, September) Organ-on-a-chip and the kidney. Kidney Res Clin Pract. 34(3):165–9.

Kirkpatrick CJ, Mittermayer C (1990) Theoretical and practical aspects of testing potential biomaterials *in vitro*. J Mater Sci: Mater Med 1:9–13. https://doi,org/10.1007/BF00705347

Kurth F, Györvary E, Heub S et al. (2020) Organs-on-a-chip engineering, In: Hoeng J. et al. Organ-on-a-chip: Engineered Microenvironments for Safety and Efficacy Testing, pp. 47–130, Academic Press. https://doi.org/10.1007/s00204-022-03234-0

Lee PJ, Hung PJ, Lee LP (2007) An artificial liver sinusoid with a microfluidic endothelial-like barrier for primary hepatocyte culture. Biotechnol Bioeng 97(5):1340-1346. https://doi.org/10.1002/bit.21360

Leung CM, de Haan P, Ronaldson-Bouchard K et al. (2022) A guide to the organ-on-a-chip. Nat Rev Methods Primers 2:33. https://doi.org/10.1038/s43586-022-00118-6

Li C, Hite Z, Warrick JW, et al. (2020) Under oil open-channel microfluidics empowered by exclusive liquid repellency. Sci Adv 6(16):eaay9919. https://doi.org/10.1126/sciadv.aay9919

Li Jeon N, Baskaran H, Dertinger S. et al. (2002) Neutrophil chemotaxis in linear and complex gradients of interleukin-8 formed in a microfabricated device. Nat Biotechnol 20:826–830. https://doi.org/10.1038/nbt712

Mansoorifar A, Gordon R, Bergan R et al., (2021). Bone-on-a-chip: microfluidic technologies and microphysiologic models of bone tissue. Adv Funct Mater 31(6):2006796. https://doi.org/10.1002/adfm.202006796

Müller U (2008) In vitro biocompatibility testing of biomaterials and medical devices. Med Device Technol 19(2):30, 32–4

Nascimento MHM, Ferreira M, Malmonge SM, Lombello CB (2017) Evaluation of cell interaction with polymeric biomaterials based on hyaluronic acid and chitosan. J Mater Sci: Mater Med 28(5):68. https://doi.org/10.1007/s10856-017-5875-x

Nayak KK, Gupta P (2015) In vitro biocompatibility study of keratin/agar scaffold for tissue engineering. Int J Biol Macromol 81:1–10. https://doi.org/10.1016/j.ijbiomac.2015.07.025

Nguyen DH, Stapleton SC, Yang MT et al. (2013) Biomimetic model to reconstitute angiogenic sprouting morphogenesis in vitro. Proc Natl Acad Sci USA 110:6712–6717. https://doi.org/10.1073/pnas.1221526110

Novak R, Ingram M, Marquez S et al. (2020) Robotic fluidic coupling and interrogation of multiple vascularized organ chips. Nat Biomed Eng 4:407–420. https://doi.org/10.1038/s41551-019-0497-x

Park JY, Ryu H, Lee B et al. (2018) Development of a functional airway-on-a-chip by 3D cell printing. Biofabrication 11(1):015002. https://doi.org/10.1088/1758-5090/aae545

Pediaditakis I, Kodella KR, Manatakis DV et al. (2021) Modeling alpha-synuclein pathology in a human brain-chip to assess blood-brain barrier disruption. Nat Commun 12(1):5907. https://doi.org/10.1038/s41467-021-26066-5

Peng Z, Zhou L, Wong JKW et al. (2020) Eye-on-a-chip (EOC) models and their role in the future of ophthalmic drug discovery. Expert Rev Ophthalm 15:5259–261. https://doi.org/10.1080/17469899.2020.1788388

Petetta F, Ciccocioppo R (2021) Public perception of laboratory animal testing: Historical, philosophical, and ethical view. Addict Biol 26(6):e12991. https://doi.org/10.1111/adb.12991

Petrosyan A, Cravedi P, Villani V, et al. (2019) A glomerulus-on-a-chip to recapitulate the human glomerular filtration barrier. Nat Commun 10(1):3656. https://doi.org/10.1038/s41467-019-11577-z

Pocock K, Delon L, Bala V, et al. (2017). Intestine-on-a-chip microfluidic model for efficient in vitro screening of oral chemotherapeutic uptake. ACS Biomat Scienc Eng 3(6): 951–959. https://doi.org/10.1021/acsbiomaterials.7b00023

Purtscher M, Rothbauer M, Kratz SRA et al. (2021) A microfluidic impedance-based extended infectivity assay: combining retroviral amplification and cytopathic effect monitoring on a single lab-on-a-chip platform. Lab Chip 21(7):1364–1372. https://doi.org/10.1039/d0lc01056a

Rach J, Halter B, Aufderheide M (2013) Importance of material evaluation prior to the construction of devices for in vitro techniques. Experim Toxic Pathol 65(7–8):973–978. https://doi.org/10.1016/j.etp.2013.01.011

Radisic M, Deen W, Langer R, Vunjak-Novakovic G (2005) Mathematical model of oxygen distribution in engineered cardiac tissue with parallel channel array perfused with culture medium containing oxygen carriers. Am J Physiol Heart Circ Physiol 2005 Mar;288(3):H1278–89. https://doi.org/10.1152/ajpheart.00787.2004

Reppetto G, Peso AD, Zurita JL (2008) Neutral red uptake assay for the estimation of cell viability/cytotoxicity. Nature Protocols 3:7. https://doi.org/10.1038/nprot.2008.75

Schoon J, Hesse B, Rakow A et al. (2020) Metal-Specific Biomaterial Accumulation in Human Peri-Implant Bone and Bone Marrow. Adv Sci (Weinh). 7(20):2000412. https://doi.org/10.1002/advs.202000412

Shin J, Ko J, Jeong S et al. (2021) Monolithic digital patterning of polydimethylsiloxane with successive laser pyrolysis. Nat Mater 20:100–107. https://doi.org/10.1038/s41563-020-0769-6

Shin W, Kim HJ (2022) 3D in vitro morphogenesis of human intestinal epithelium in a gut-on-a-chip or a hybrid chip with a cell culture insert. Nat Protoc 17:910–939. https://doi.org/10.1038/s41596-021-00674-3

Spijkers XM, Pasteuning-Vuhman S, Dorleijn JC et al. (2021) A directional 3D neurite outgrowth model for studying motor axon biology and disease. Sci Rep 11:2080. https://doi.org/10.1038/s41598-021-81335-z

Sung JH, Wang YI, Narasimhan Sriram N, et al. (2019) Recent Advances in Body-on-a-Chip Systems. Anal Chem 91(1):330–351. https://doi.org/10.1021/acs.analchem.8b05293

Sutterby E, Thurgood P, Baratchi S, et al. (2020) Microfluidic Skin-on-a-Chip Models: Toward Biomimetic Artificial Skin, Small 16(39):e2002515. https://doi.org/10.1002/smll.202002515

Stengelin E, Thiele J, Seiffert S (2022) Multiparametric Material Functionality of Microtissue-Based In Vitro Models as Alternatives to Animal Testing. Adv Sci (Weinh) 9(10):e2105319. https://doi.org/10.1002/advs.202105319.

Tannenbaum J, Bennett BT (2015) Russell and Burch's 3Rs then and now: the need for clarity in definition and purpose. J Am Assoc Lab Anim Sci 54(2):120–32.
USP – UNITED STATES PHARMACOPEIA. USP 43. General Chapter 87: Biological Reactivity Tests, in vitro. Rockville 2019.
Varone, A, Nguyen JK, Leng L (2021) A novel organ-chip system emulates three-dimensional architecture of the human epithelia and the mechanical forces acting on it. Biomaterials 275:120957. https://doi.org/10.1016/j.biomaterials.2021.120957
Virumbrales-Muñoz M, Ayuso JM (2022) From microfluidics to microphysiological systems: Past, present, and future. Organs-on-a-Chip 4:100015.https://doi.org/10.1016/j.OoC.2022.100015
Wagner I, Materne E-M, Brincker S, et al. A dynamic multiorgan-chip for long-term cultivation and substance testing proven by 3D human liver and skin tissue co-culture (2013) Lab Chip 13: 3538–3547
Wang X, Phan DTT, Sobrino A et al. (2016) Engineering Anastomosis between Living Capillary Networks and Endothelial Cell-Lined Microfluidic Channels. Lab Chip 16(2):282–90. https://doi.org/10.1039/c5lc01050k
Wu Q, Liu J, Wang X et al. (2020) Organ-on-a-chip: recent breakthroughs and future prospects. BioMed Eng OnLine 19:9. https://doi.org/10.1186/s12938-020-0752-0
Yu F, Kumar NDS, Foo LC, et al. (2020) A pump-free tricellular blood-brain barrier on-a-chip model to understand barrier property and evaluate drug response. Biotechnol Bioeng 117(4):1127–1136. https://doi.org/10.1002/bit.27260
Yu J, Piazza A, Sparks S et al. (2021) A reconfigurable microscale assay enables insights into cancer-associated fibroblast modulation of immune cell recruitment. Integr Biol (Camb). 13(4):87–97. https://doi.org/10.1093/intbio/zyab004
Zarrintaj P, Saeb MR, Stadler FJ et al. (2022) Human Organs-on-Chips: A Review of the State-of-the-Art, Current Prospects, and Future Challenges. Adv Biol (Weinh) 6(1):e2000526. https://doi.org/10.1002/adbi.202000526
Zhang W, Chen R, Xu X et al. (2022) Construction of Biocompatible Hydrogel Scaffolds With a Long-Term Drug Release for Facilitating Cartilage Repair. Front Pharmacol 13:922032. https://doi.org/10.3389/fphar.2022.922032
Zhang Y, Yu T, Ding J et al. (2023) Bone-on-a-chip platforms and integrated biosensors: Towards advanced in vitro bone models with real-time biosensing. Biosensors Bioelectron 219:114798. https://doi.org/10.1016/j.bios.2022.114798
Zhang YS, Arneri A, Bersini S et al. (2016) Bioprinting 3D microfibrous scaffolds for engineering endothelialized myocardium and heart-on-a-chip. Biomaterials 110:45–59. https://doi.org/10.1016/j.biomaterials.2016.09.003
Zhao Y, Rafatian N, Feric NT et al. (2019) A Platform for Generation of Chamber-Specific Cardiac Tissues and Disease Modeling. Cell 176(4):913–927.e18. https://doi.org/10.1016/j.cell.2018.11.042

Correction to: Current Trends in Biomedical Engineering

Christiane Bertachini Lombello and Patricia Aparecida da Ana

Correction to:
C. B. Lombello, P. A. da Ana (eds.), *Current Trends in Biomedical Engineering*, https://doi.org/10.1007/978-3-031-38743-2

The book was inadvertently published with an incorrect spelling of the author's name in Chapter 9 as Kaline Ferreira Nascimento whereas it should be Kaline do Nascimento Ferreira.

In addition to this, the name of the author in Chapter 15 should be Andressa Francine Martins instead of Andressa Fracnine Martins.

The updated versions of the chapters can be found at
https://doi.org/10.1007/978-3-031-38743-2_9
https://doi.org/10.1007/978-3-031-38743-2_15

C. B. Lombello, P. A. da Ana (eds.), *Current Trends in Biomedical Engineering*,
https://doi.org/10.1007/978-3-031-38743-2_16

Index

C. B. Lombello, P. A. da Ana (eds.), *Current Trends in Biomedical Engineering*,
https://doi.org/10.1007/978-3-031-38743-2

MIX
Papier aus verantwortungsvollen Quellen
Paper from responsible sources
FSC® C105338

If you have any concerns about our products,
you can contact us on
ProductSafety@springernature.com

In case Publisher is established outside the EU,
the EU authorized representative is:
Springer Nature Customer Service Center GmbH
Europaplatz 3, 69115 Heidelberg, Germany

Printed by Libri Plureos GmbH
in Hamburg, Germany